COMPUTATIONAL MODELING
From Chemistry to Materials to Biology

Proceedings of the 25th Solvay Conference on Chemistry

COMPUTATIONAL MODELING
From Chemistry to Materials to Biology

Proceedings of the 25th Solvay Conference on Chemistry

Brussels, Belgium   16 – 19 October 2019

Scientific Editors

Kurt Wüthrich
The Scripps Research Institute, California, USA
ETH Zürich, Switzerland

Bert Weckhuysen
Utrecht University, The Netherlands

International Solvay Institutes Editors

Laurence Rongy
International Solvay Institutes, Belgium

Anne De Wit
International Solvay Institutes, Belgium

 **World Scientific**

NEW JERSEY · LONDON · SINGAPORE · BEIJING · SHANGHAI · HONG KONG · TAIPEI · CHENNAI · TOKYO

Published by

World Scientific Publishing Co. Pte. Ltd.

5 Toh Tuck Link, Singapore 596224

USA office: 27 Warren Street, Suite 401-402, Hackensack, NJ 07601

UK office: 57 Shelton Street, Covent Garden, London WC2H 9HE

British Library Cataloguing-in-Publication Data
A catalogue record for this book is available from the British Library.

COMPUTATIONAL MODELING: FROM CHEMISTRY TO MATERIALS TO BIOLOGY
Proceedings of the 25th Solvay Conference on Chemistry

ISBN 978-981-122-820-9 (hardcover)
ISBN 978-981-123-369-2 (paperback)
ISBN 978-981-122-821-6 (ebook for institutions)
ISBN 978-981-122-822-3 (ebook for individuals)

For any available supplementary material, please visit
https://www.worldscientific.com/worldscibooks/10.1142/12039#t=suppl

Desk Editor: Nur Syarfeena Binte Mohd Fauzi

Typeset by Stallion Press
Email: enquiries@stallionpress.com

Printed in Singapore

CONTENTS

I. The International Solvay Institutes

Board of Directors

Members

Jean-Marie Solvay
President

Paul Geerlings
Vice-President & Treasurer - Emeritus Professor VUB

Gino Baron
Secretary - Emeritus Professor VUB

Nicolas Boel
Chairman of the Board of Directors of Solvay Group

Eric Boyer de la Giroday
Chairman of the Board of Directors ING Belgium

Eric De Keuleneer
Former Chairman of the Board of Directors of the ULB

Pierre Gurdjian
Chairman of the Board of Directors of the ULB

Daniel Janssen
Former Chairman of the Board of Directors of Solvay S.A.

Eddy Van Gelder
Chairman of the Board of Directors of the VUB

Honorary Members

Franz Bingen
Emeritus Professor VUB
Former Vice president and Treasurer of the Solvay Institutes

Philippe Busquin
Minister of State

Jean-Louis Vanherweghem
Former Chairman of the Board of Directors of the ULB

Irina Veretennicoff
Emeritus Professor VUB

Lode Wyns
Emeritus Professor VUB, Former Deputy Director for Chemistry of the
Solvay Institutes

Guest Members

Gert Desmet
Professor VUB, Deputy Director for Chemistry

Anne De Wit
Professor ULB and Scientific Secretary of the International Committee
for Chemistry

Freddy Dumortier
Secretary of the Royal Flemish Academy for Science and the Arts of Belgium

Franklin Lambert
Emeritus Professor VUB

Alexander Sevrin
Professor VUB and Deputy Director for Physics and Scientific Secretary
of the International Committee for Physics

Marina Solvay

Didier Viviers
Secretary of the Royal Academy for Science and the Arts of Belgium

Director

Professor Marc Henneaux
Professor ULB

II. Solvay Scientific Committee for Chemistry

Professor Kurt Wüthrich (Chair)
Scripps Research Institute, La Jolla, USA and ETH Zurich, Switzerland

Professor Joanna Aizenberg, Harvard University, Cambridge (USA)

Professor Thomas Cech, University of Colorado Boulder (USA),
Nobel Laureate 1989

Professor Gerhard Ertl, Fritz-Haber-Institut der Max-Planck-Gesellschaft
Berlin (Germany), Nobel Laureate 2007

Professor Ben Feringa, University of Groningen (The Netherlands),
Nobel Laureate 2016

Professor Robert H. Grubbs, California Institute of Technology,
Pasadena (USA), Nobel Laureate 2005

Professor Stefan Hell, Max Planck Institute (Germany), Nobel Laureate 2014

Professor JoAnne Stubbe, Massachusetts Institute of Technology Cambridge (USA)

Professor Bert Weckhuysen, Utrecht University (The Netherlands)

Professor George M. Whitesides, Harvard University, Cambridge (USA)

Professor Anne De Wit, Université Libre de Bruxelles, Belgium, Scientific Secretary

III. Acknowledgements

The organization of the 25th Solvay conference has been made possible thanks to the generous support of the Solvay Family, the Solvay Group, the Université libre de Bruxelles, the Vrije Universiteit Brussel, the Belgian National Lottery, the Brussels-Capital Region, the Fédération Wallonie-Bruxelles, the Vlaamse Regering, the City of Brussels and the Hôtel Métropole.

IV. Participants

Joanna	Aizenberg	Harvard University, Cambridge, USA
Alán	Aspuru-Guzik	University of Toronto, Canada
Dean	Astumian	University of Maine, Orono, USA
Mischa	Bonn	MPI for Polymer Research, Mainz, Germany
Thomas	Cech	University of Colorado, Boulder, USA
Markus	Covert	Stanford University, Palo Alto, USA
Leroy	Cronin	University of Glasgow, UK
Xavier	Darzacq	University of California, Berkeley, USA
Winfried	Denk	MPI of Neurobiology, Martinsried, Germany
Mark	Ellisman	University of California, San Diego, USA
Jörg	Enderlein	Göttingen University, Germany
Ben	Feringa	University of Groningen, The Netherlands
Laura	Gagliardi	University of Minnesota, Minneapolis, USA
Bartosz	Grzybowski	UNIST, Ulsan, Korea
Bernd	Hartke	University of Kiel, Germany
Stefan	Hell	MPI for Biophysical Chemistry, Göttingen, Germany
Thomas	Hermans	University of Strasbourg, France
Wilhelm	Huck	Radboud University, Nijmegen, The Netherlands
Jan	Huisken	Morgridge I. for Research, Madison, USA
Sinan	Keten	Northwestern University, Evanston, USA
Henk	Lekkerkerker	Utrecht University, The Netherlands
Matthew	Lew	Washington University, St. Louis, USA
Todd	Martínez	Stanford University, Palo Alto, USA
Angelos	Michaelides	University College London, UK
Eva	Nogales	University of California, Berkeley, USA
Raimund	Ober	Texas A&M University, College Station, USA
David	Quéré	ESPCI, Paris, France
Bernd	Rieger	Delft University of Technology, The Netherlands
Eugene	Shakhnovich	Harvard University, Cambridge, USA
Yoav	Shechtman	Technion University, Haifa, Israel
Berend	Smit	EPFL Lausanne, Switzerland
Miquel	Solà	University of Girona, Spain

Katarina	Stanciakova	Utrecht University, The Netherlands
Annette	Taylor	University of Sheffield, UK
Veronique	Van Speybroeck	University of Gent, Belgium
Andreas	Walther	University of Freiburg, Germany
Bert	Weckhuysen	Utrecht University, The Netherlands
Kurt	Wüthrich	Scripps Research, La Jolla, CA, USA & ETH Zürich Switzerland
Julia	Yeomans	Oxford University, UK

V. Auditors

Emilie	Cauët	Université libre de Bruxelles, Belgium
Gino	Baron	Vrije Universiteit Brussel, Belgium
Benoît	Champagne	U. Namur, Belgium
Yannick	De Decker	Université libre de Bruxelles, Belgium
Anne	De Wit	Université libre de Bruxelles and Solvay Institutes, Belgium
Geneviève	Dupont	Université libre de Bruxelles, Belgium
Frank	De Proft	Vrije Universiteit Brussel, Belgium
Gert	Desmet	Vrije Universiteit Brussel, Belgium
Rouslan	Efremov	Vrije Universiteit Brussel, Belgium
Paul	Geerlings	Vrije Universiteit Brussel, Belgium
Jeremy	Harvey	KU Leuven, Belgium
Alain	Jonas	Université catholique de Louvain, Belgium
Martine	Prévost	Université libre de Bruxelles, Belgium
Roberto	Lazzaroni	U. Mons, Belgium
Bortolo	Mognetti	Université libre de Bruxelles, Belgium
Yoann	Olivier	U. Namur, Belgium
Han	Remaut	Vrije Universiteit Brussel, Belgium
Laurence	Rongy	Université libre de Bruxelles, Belgium

Opening Address by Professor Marc Henneaux
Director of the International Solvay Institutes

MÉTROPOLE — 16 OCT 2019

Dear Marina, Dear Jean-Marie,
Dear Colleagues, Dear Friends,

It is my great honour and pleasure to open the Solvay Conference on Chemistry that starts today. Its theme is "Computational Modeling: From Chemistry to Materials to Biology".

This is the 25th Solvay Conference in Chemistry, the 25th in a very prestigious list:

1. 1922   "Cinq Questions d'Actualité"
2. 1925   "Structure et Activité Chimique"
3. 1928   "Questions d'Actualité"
4. 1931   "Constitution et Configuration des Molécules Organiques"
5. 1934   "L'Oxygène, ses réactions chimiques et biologiques"
6. 1937   "Les Vitamines et les Hormones"
7. 1947   "Les Isotopes"
8. 1950   "Le Mécanisme de l'Oxydation"
9. 1953   "Les Protéines"
10. 1956   "Quelques Problèmes de Chimie Minérale"
11. 1959   "Les Nucléoprotéines"
12. 1962   "Transfert d'Energie dans les Gaz"
13. 1965   "Reactivity of the Photoexited Organic Molecule"
14. 1969   "Phase Transitions"
15. 1970   "Electrostatic Interactions and Structure of Water"
16. 1976   "Molecular Movements and Chemical Reactivity as conditioned by Membranes, Enzymes and other Molecules"
17. 1980   "Aspects of Chemical Evolution"
18. 1983   "Design and Synthesis of Organic Molecules Based on Molecular Recognition"
19. 1987   "Surface Science"
20. 1995   "Chemical Reactions and their Control on the Femtosecond Time Scale"

This list shows great diversity in the themes, ranging from physical chemistry to biochemistry.

The normal rhythm is one conference every 3 years, but this rhythm has been interrupted by the war, and perhaps by a lack of energy after the 1980's.

The 3-year periodicity was resumed in 2007 and has been strictly adhered to since then. The next conference will take place in 2022.

The theme of the 25th Conference is broad, original and quite unusual. It covers all of chemistry but with a particular angle, that of computational modelling and artificial intelligence methods. We have therefore participants with different backgrounds, the common thread being that of the methods used. We hope that many interesting scientific developments will come from this peculiar mix.

Let me recall the format of the Solvay Conferences. These are conferences by invitation-only, with a limited number of participants. There is a small number of presentations but a lot of discussions. People come to the Solvay Conferences mainly for the scientific interactions, which are indeed privileged, not for giving a talk.

For the discussions to be fruitful, at least two conditions must be met: (1) First, participants should attend the full conference. (2) Second, an extremely careful preparation is clearly needed. We all know that spontaneous discussions only work if there is preparation!

This careful preparation was done by the chair and co-chair of the conference, Professors Kurt Wüthrich (who is also the chair of the International Solvay Scientific Committee for Chemistry) and Bert Weckhuysen, as well as by the chairs of the various scientific sessions. They did a lot of work and put up a splendid program. I would like to express the deepest gratitude of the International Solvay Institutes to all of them!

Before giving the floor to Kurt Wüthrich, let me make one announcement concerning the proceedings. We have an editorial committee that already worked remarkably well since many of the contributions have been received before the meeting. This has to be particularly praised and I would like to thank our colleagues Anne De Wit and Laurence Rongy for the splendid job that they have done. Furthermore, since the discussions are important, they will be included in the proceedings. This is again a distinctive feature of the Solvay Conferences. To facilitate the work of our scientific secretariat, please give your name each time you intervene in the discussions.

Thank you very much for your attention.

Preface by Professor Kurt Wüthrich
Chair of the 25th Solvay Conference on Chemistry

During the last two decades, the use of ever more powerful computers with more and more sophisticated software had a dominating impact on research in the natural sciences. With the title "Computational Modeling: From Chemistry to Materials to Biology" this Conference brought together scientists working in widely different specialized areas of chemistry, materials research, biology and biomedical sciences, who are united by common interests in applications of informatics tools. To approach this broad integrative subject we were able to assemble an illustrious group of scientists specialized in different areas of information technology, who were joined by colleagues who use these techniques to validate and interpret their experimental data. The combination of highly different research areas united by common interests in front-line computational techniques resulted in a most exciting week of scientific exchange.

The 25th Conference followed the format of the Conferences in 2013 and 2016, which ensured that all invited participants had formal assignments in the scientific program. For a start, in advance of the Conference, all participants contributed short articles presenting their current views on the theme of their session. Most of these papers arrived in good time, so that they could be sent to the participants during the week before the Conference. Of six scientific sessions, each one was then opened with a general introduction by the chairperson and short presentations by five to seven panelists. This was followed by a discussion among the panelists, and then the floor was opened to all invited participants. The discussions, which covered two thirds of the time allotted to the scientific program, were recorded and transcribed by "Auditors" recruited from Belgiun Universities. In this volume, the transcripts have been edited by these Auditors, who attended the Conference, and then further checked by the discussion contributors for correct scientific content.

The Proceedings are organized in the order of the six Sessions, where the transcripts of the discussions are preceded by the reports prepared by the session chair and the panelists prior to the Conference. Session 1 on "Artificial Intelligence/Machine Learning in Chemistry", chaired by Prof. Kurt Wüthrich and introduced by Prof. Alán Aspuru-Guzik, provided an introduction to the ongoing transfer of current use of machine learning for image recognition and vehicle automation to applications in chemistry, for example, the planning of multi-step organic syntheses. Session 2 on "Modeling of Functional Materials", chaired by

Prof. Bert Weckhuysen, presented an impressive illustration of the principles introduced in Session 1, with examples from rational materials design, heterogeneous catalysis and chemical process development. Session 3 on "Models and Experimental Data on Water Dynamic Complexity of Solid/Liquid Interfaces", chaired by Prof. Joanna Aizenberg, entertained exciting views of processes at such interfaces, the role of turbulence in such processes, and the impact of mechanical stress on the shape of glasses and higher-order chemical and biological structures. Session 4 on "Computational Modeling in High-resolution Imaging", chaired by Prof. Stefan Hell, introduced the important role of modern computational techniques in support of super-high-resolution microscopy. Various aspects of transfers from conventional machine learning to deep learning were illustrated with problem solving in single particle averaging and in automated pattern recognition. Session 5 on "Modeling of Non-equilibrium Systems and Simulation of Molecular Machines", chaired by Prof. Ben Feringa, introduced the treatment of dynamic functional systems that are out of equilibrium. This included, for example, treatments of catalysis by molecular machines and the importance of feedback loops, especially with reference to bioinspired behavior of molecular machines. Session 6 on "Computers in Interactive Structural Biology Leading to Modelling of an Intact Biological Cell", chaired by Prof. Thomas Cech, extended the scope of the Conference beyond "core chemistry" to front-line biological research. A large part of this session provided direct links to Session 4, with extensive discussion of data validation in cryo-electron microscopy of macromolecular structures and higher-order biological entities. The combination of the six Sessions resulted in innovative insights into relations between widely different specialized fields of scientific research which were here united by common interests in the development and applications of emerging bioinformatics tools. The meeting thus presented a unique opportunity for interdisciplinary exchanges on anticipated developments during the coming years and in the more distant future.

It was a special privilege for me to chair this Solvay Conference on Chemistry, and I want to acknowledge the co-Chair of the Conference, Prof. Bert Weckhuysen, and the chairpersons of the individual Sessions for their support in generating the scientific program. To do proper justice to the widely different aspects of computational modeling to be covered by the Conference, the panelists in the individual Sessions were selected by the respective chairpersons, so that I personally got the benefits of in-depth education in widely different fields and could focus primarily on solving organizational issues arising with the nominations received.

We are grateful to the Solvay family for their continued support, which enabled the organization of the present Conference. Mr. Jean-Marie Solvay, who is the President of the Solvay Institutes, and Mrs. Marina Solvay showed their broad interests by attending a large part of the scientific program. Mrs. Marie-Claude Solvay de la Hulpe joined us for the social parts of the meeting, and we will keep fond memories of the warm welcome for dinner at her beautiful home. We greatly appreciate the generous support by the Solvay Institutes represented by Prof. Marc Henneaux.

Special thanks go to Prof. Anne De Wit and Prof. Laurence Rongy, who supervised the recording and transcription of the discussions, and to the scientists who served as "Auditors" during the Conference and contributed to the discussion transcriptions, which are now a major part of these Proceedings. Dominique Bogaerts and Isabelle Van Geet from the Solvay Institutes made our lives easy and enjoyable with their commitment and dedicated assistance; we owe them our heartfelt thanks.

Kurt Wüthrich
Chair of the Conference

Session 1

Artificial Intelligence/Machine Learning in Chemistry

ARTIFICIAL INTELLIGENCE AND CHEMISTRY

ALÁN ASPURU-GUZIK

Chemical Physics Theory Group, Department of Chemistry and Department of Computer Science, University of Toronto, Toronto, ON, M5S 3H4, Canada
Vector Institute for Artificial Intelligence, Toronto, ON, M5S 3H4, Canada
Canadian Institute for Advanced Research, Toronto, ON, M5S 3H4, Canada

What is Artificial Intelligence?

Defining intelligence, let alone its artificial sibling, has been a hard task over the years. Several definitions have been proposed [1], of which I will adopt the succinct definition of D. Poole *et al.* [2] for the purposes of this article: "[An intelligent agent does what] is appropriate for its circumstances and its goal, it is flexible to changing environments and changing goals, it learns from experience, and it makes appropriate choices given perceptual limitations and finite computation." As a broad field with such a comprehensive goal, the development of artificial systems with intelligence is poised to help chemistry, a vast science itself.

The field of artificial intelligence (AI) has gained a resurgence of interest, and even potentially hype from the community at large due to the convergence of the availability of large amounts of data together with the access to large computational resources, perhaps first instantiated to the ubiquitous access to general purpose graphics processing units (GPGPU).

It is worth discussing the scope and definition of some of the subfields of AI that impact chemistry before we delve into a few of the applications of the field. Figure 1 provides such a brief map. Needless to say, in this short contribution, I am omitting the work of many researchers and mainly using examples of our own research. In addition, I am omitting many branches of the tree of AI that could be or have been relevant to chemistry. The several reviews that I cite provide a broader set of this rapidly growing literature.

General AI, a machine that is as powerful at general cognitive tasks as a human, is perhaps decades away. Shallow AI is the development of AI for a narrow application. For certain tasks, shallow AI has already shown superhuman advantage, for example, the famous example of playing complex strategy board games such as Go [3].

Machine learning (ML), arguably, is the subfield of AI that has gained the most attention in this decade due to the huge advances in areas such as image recognition and natural language processing. ML is built on tools from statistical inference that

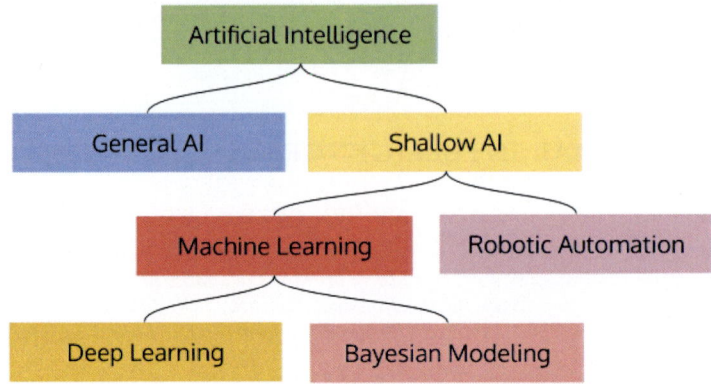

Fig. 1.   A simple hierarchy of the subfields of AI that are relevant to chemistry and covered in this brief paper.

allow computer programs to make decisions based on data and not specific to a rigid procedural code.

One of the methods in the field of machine learning is the field of *deep learning*, which is called that because it involves calculations that flow data through several (deep) layers of artificial neurons connected to each other. These artificial neurons are parameterizable mathematical functions that have a nonlinear response to their input(s). I refer the reader to reviews and textbooks on deep learning for chemistry to learn more [4–7].

An important point to make is that machine learning has a long history in chemistry represented by the field of cheminformatics. The difference between the more traditional quantitative structure–activity relationships (QSARs) [8] and deep learning models is that the latter do not employ handcrafted features for the molecules but rather infer or learn them from the data. The downside of this approach is that more data may be needed to train them. A scheme for the difference between these approaches is shown in Fig. 2.

In the next sections, as an example, I will briefly describe some of the contributions from our group and others to chemistry.

Molecular Representations

Molecular information can be represented at different scales. At smaller scales, a full quantum description of the many-body state is usually employed. At the largest scales, perhaps the molecule is represented as a sequence of fragments, for example, the simplest protein sequence representation. Therefore, the field of representation is still an open and virgin one, on the one hand, learning from the wave function itself to, on the other hand, learning from molecular fingerprints.

For the purposes of molecular design and for rationalizing organic chemistry mechanisms, humans have employed "2D" molecular graph representations that may or may not contain 3D information for the position of the atoms.

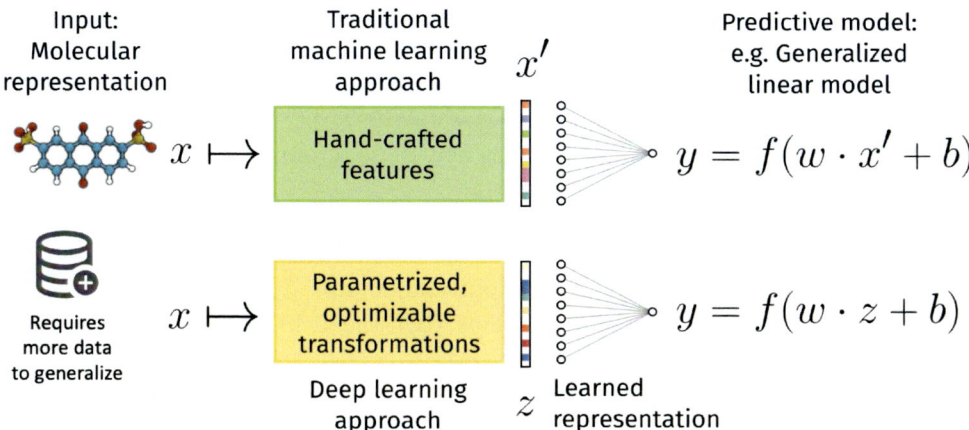

Fig. 2. The difference between traditional machine learning approaches in chemistry such as those used in QSARs and deep learning involves the use of handcrafted features. (top) Usually arranged as a vector x from which a generalized linear model is used to infer the molecular property y. In the field of deep learning (bottom), usually larger amounts of data are employed to build a representation vector z, which then can be employed to predict y by means of a generalized linear model. Figure credit: Benjamín Sanchez-Lengeling.

To carry out deep learning applications, as described in the previous sections, one way to proceed would be to obtain a fixed-length vector z from an arbitrarily sized molecular graph. For this purpose, we developed a graph convolutional neural network GCN to represent molecules [9]. Since that original work, the GCN has been employed as one of the commonplace deep learning representations for chemistry [10]. The reader should refer back to Fig. 2. The graph convolutional network learns the representation vector z rather than an algorithmic fingerprint algorithm that is not optimized for the dataset at hand.

Generative Models and Inverse Design for Chemistry

Arguably, one of the most desired applications in the field of chemistry is the *molecular design* of a molecule with a practical application. This can be translated into a desired structure–property relationship. The *inverse design* problem [6] entails the search of a very large chemical space of a desired molecule with a given property. If the inverse problem is solvable efficiently, a significantly smaller number of chemical experiments or trials may be conducted to test the computational predictions.

The family of *generative models* [11] is a very active field of machine learning. These models aim to reproduce the joint (z, y) distribution. In particular, tools such as *autoencoders* [12], *generative adversarial networks* [13, 14], and *evolutionary strategies* [15, 16] have been employed in the field to aim to accelerate the design of molecules.

We introduced a molecular autoencoder that showed that inverse design for molecules was possible and we employed it using a computational example related to drug-likeness and synthesizability [12]. Later, in collaboration with inSilico AI, we employed a similar model to suggest a DDR1 kinase inhibitor in 46 days [17].

Generative models may be important in the *ideation* and *design* phases of research, but they need to be integrated with the larger experimental endeavor, and this is where self-driving laboratories may play an important role.

Self-driving Chemical Laboratories

Arguably, the promise of high-throughput combinatorial chemistry (HTCC) fell relatively short with regard to the claims that the field made at a time when automation was introduced to chemistry. How is accelerating the raw *rate* of experimentation considered a downside? A first potential reason is that the cost to set up, maintain, debug, and analyze high-throughput systems is large. Second, substantial data science work has to be done in terms of compound library selection and process optimization. Finally, a multidisciplinary training in statistics, computation, and experiment may be required for successful high-quality personnel in this area. Given these three reasons, we posit that HTCC is not as prevalent as it could be, and that the human-driven serial approach to science continues to perhaps be the main *modus operandi* in fields such as organic or materials chemistry.

Laboratories that actively choose experiments "on the fly" using data science and AI have early notable examples in the work of King and co-workers [18] and Maruyama and co-workers [19]. These self-driving laboratories aim to improve upon many of the deficiencies of HTCC as they aim to maximize the information gain per unit experiment and make quick decisions on what next to explore in an automated fashion.

With the recent resurgence and attention to machine learning methods, the possibility of accelerating the discovery process at the computational, screening, optimization, device manufacturing, and scale-up levels promises a reduction in times to discovery [20]. In particular, automated reaction optimization approaches [21] can be carried out using Bayesian approaches [22] that respect multi-objective optimization [23] for continuous variables and as well as discrete variables [24].

Recent examples of applications of self-driving laboratories of my research group include the optimization of thin films for hole transport layers of organic materials [25] as well as the optimization of quaternary blends for organic photovoltaics [26].

Outlook

To make further progress in the field of AI for chemistry and materials science, one should ask oneself the question of what higher forms of understanding can computers take us to. Together with Lindh and Reiher [27], we posed a few challenges for theoretical chemistry in the 21st century that include the *interpretability* of the models we produce. This will help the irreducible pair of self-driving laboratory/human to accelerate scientific progress and hence help solve several of the problems that humanity currently faces.

Acknowledgments

I thank Anders G. Frøseth for his generous support. I also acknowledge the generous support of Natural Resources Canada and the Canada 150 Research Chairs program. Tata Steel and The Office of Naval Research have also generously supported our work on self-driving laboratories.

References

[1] M.H. Shane Legg, *Advances in Artificial General Intelligence: Concepts, Architectures and Algorithms*, Volume 157 of Frontiers in Artificial Intelligence and Applications (2007).

[2] D. Poole, A. Mackworth, and R. Goebel, *Computational Intelligence: A Logical Approach*, Oxford University Press (1998).

[3] D. Silver, T. Hubert, J. Schrittwieser, I. Antonoglou, M. Lai, A. Guez, M. Lanctot, L. Sifre, D. Kumaran, T. Graepel, T. Lillicrap, K. Simonyan, and D. Hassabis, *Science* **362** 1140–1144 (2018).

[4] I. Goodfellow, Y. Bengio, and A. Courville, *Deep Learning*, Cambridge: MIT Press (2016).

[5] D.P. Tabor, L.M. Roch, S.K. Saikin, C. Kreisbeck, D. Sheberla, J.H. Montoya, S. Dwaraknath, M. Aykol, C. Ortiz, H. Tribukait, C. Amador-Bedolla, C.J. Brabec, B. Maruyama, K.A. Persson, and A. Aspuru-Guzik, *Nat. Rev. Mater.* **3**, 5–20 (2018).

[6] B. Sanchez-Lengeling and A. Aspuru-Guzik, *Science* **361**, 360–365 (2018).

[7] G. Carleo, I. Cirac, K. Cranmer, L. Daudet, M. Schuld, N. Tishby, L. Vogt-Maranto, and L. Zdeborová, *Rev. Mod. Phys.* **91**, 045002 (2019).

[8] E.N. Muratov, J. Bajorath, R.P. Sheridan, I.V. Tetko, D. Filimonov, V. Poroikov, T.I. Oprea, I.I. Baskin, A. Varnek, A. Roitberg, O. Isayev, S. Curtalolo, D. Fourches, Y. Cohen, A. Aspuru-Guzik, D.A. Winkler, D. Agrafiotis, A. Cherkasov, and A. Tropsha, QSAR without borders, *Chem. Soc. Rev.*, **49**, 3525–3564 (2020).

[9] D.K. Duvenaud, D. Maclaurin, J. Iparraguirre, R. Bombarell, T. Hirzel, A. Aspuru-Guzik, R.P. Adams, in: C. Cortes, N.D. Lawrence, D.D. Lee, M. Sugiyama, and R. Garnett (Eds.), *Advances in Neural Information Processing Systems*, Vol. 28 (Curran Associates, Inc. 2015), pp. 2224–2232.

[10] D.C. Elton, Z. Boukouvalas, M.D. Fuge, and P.W. Chung, *Mol. Syst. Des. Eng.* **4**, 828–849 (2019).

[11] D. Polykovskiy, A. Zhebrak, B. Sanchez-Lengeling, S. Golovanov, O. Tatanov, S. Belyaev, R. Kurbanov, A. Artamonov, V. Aladinskiy, M. Veselov, A. Kadurin, S. Nikolenko, A. Aspuru-Guzik, and A. Zhavoronkov, arXiv [cs.LG] (2018).

[12] R. Gómez-Bombarelli, J.N. Wei, D. Duvenaud, J.M. Hernández-Lobato, B. Sánchez-Lengeling, D. Sheberla, J. Aguilera-Iparraguirre, T.D. Hirzel, R.P. Adams, and A. Aspuru-Guzik, *ACS Cent Sci.* **4**, 268–276 (2018).

[13] I.J. Goodfellow, J. Pouget-Abadie, M. Mirza, B. Xu, D. Warde-Farley, S. Ozair, A. Courville, and Y. Bengio, *Proceedings of the 27th International Conference on Neural Information Processing Systems — Volume 2* (MIT Press, Cambridge, MA, USA, 2014), pp. 2672–2680.

[14] G.L. Guimaraes, B. Sanchez-Lengeling, C. Outeiral, P.L.C. Farias, and A. Aspuru-Guzik, arXiv [stat.ML] (2017).

[15] A. Nigam, P. Friedrich, M. Krenn, and A. Aspuru-Guzik, *International Conference in Learning Representations*, Addis Ababa, Ethiopia, April 26–30, 2020.

[16] A. Slowik and H. Kwasnicka, *Neural Comput. Appl.* **32**, 12363–12379 (2020).

[17] A. Zhavoronkov, Y.A. Ivanenkov, A. Aliper, M.S. Veselov, V.A. Aladinskiy, A.V. Aladinskaya, V.A. Terentiev, D.A. Polykovskiy, M.D. Kuznetsov, A. Asadulaev, Y. Volkov, A. Zholus, R.R. Shayakhmetov, A. Zhebrak, L.I. Minaeva, B.A. Zagribelnyy, L.H. Lee, R. Soll, D. Madge, L. Xing, T. Guo, and A. Aspuru-Guzik, *Nat. Biotechnol.* **37**, 1038–1040 (2019).

[18] R.D. King, J. Rowland, S.G. Oliver, M. Young, W. Aubrey, E. Byrne, M. Liakata, M. Markham, P. Pir, L.N. Soldatova, A. Sparkes, K.E. Whelan, and A. Clare, *Science* **324**, 85–89 (2009).

[19] P. Nikolaev, D. Hooper, F. Webber, R. Rao, K. Decker, M. Krein, J. Poleski, R. Barto, and B. Maruyama, *NPJ Comput. Mater.* **2**, 16031 (2016).

[20] F. Häse, L.M. Roch, and A. Aspuru-Guzik, *TRECHEM* **1**, 282–291 (2019).

[21] A.D. Clayton, J.A. Manson, C.J. Taylor, T.W. Chamberlain, B.A. Taylor, G. Clemens, and R.A. Bourne, *React. Chem. Eng.* **4**, 1545–1554 (2019).

[22] F. Häse, L.M. Roch, C. Kreisbeck, and A. Aspuru-Guzik, *ACS Cent Sci.* **4**, 1134–1145 (2018).

[23] F. Häse, L.M. Roch, and A. Aspuru-Guzik, *Chem. Sci.* **9**, 7642–7655 (2018).

[24] F. Häse, L.M. Roch, and A. Aspuru-Guzik, arXiv [stat.ML] (2020).

[25] B.P. MacLeod, F.G.L. Parlane, T.D. Morrissey, F. Häse, L.M. Roch, K.E. Dettelbach, R. Moreira, L.P.E. Yunker, M.B. Rooney, J.R. Deeth, V. Lai, G.J. Ng, H. Situ, R.H. Zhang, M.S. Elliott, T.H. Haley, D.J. Dvorak, A. Aspuru-Guzik, J.E. Hein, and C.P. Berlinguette, *Sci. Adv.* **6**, eaaz8867 (2020).

[26] S. Langner, F. Häse, J.D. Perea, T. Stubhan, J. Hauch, L.M. Roch, T. Heumueller, A. Aspuru-Guzik, and C.J. Brabec, *Adv. Mater.* **32**, e1907801 (2020).

[27] A. Aspuru-Guzik, R. Lindh, and M. Reiher, *ACS Cent Sci.* **4**, 144–152 (2018).

USING MACHINE LEARNING TO LEARN CHEMISTRY

TODD J. MARTINEZ

Department of Chemistry and the PULSE Institute, Stanford University,
Stanford, CA 94305, USA
SLAC National Accelerator Laboratory,
Menlo Park, CA 94025, USA

My View of the Present State of Research on Artificial Intelligence and Machine Learning in Chemistry

The recent successes of machine learning (ML) in image recognition [1], vehicle automation, and intelligent game play [2] have ignited a broad interest in the field. A critical question is how these ideas can spur progress in chemistry. If one views ML as a set of tools for data-driven nonlinear regression, the natural approach is to simply identify useful functional relationships and then "learn" these with the tools of ML. Examples might include predictions of the solubility of different molecules in different solvents or binding affinity of drug molecules to protein targets. This is clearly useful, although one must recognize that much of the recent success in ML hinges on the availability of a massive amount of data. Most experimental chemical data sources struggle to come anywhere close to the terabytes/day (or even petabytes/day) collected by modern social media companies. In contrast, simulation data can be produced at a large scale, and thus are a natural source for processing by ML. Indeed, an alternative perspective on ML views it as a collection of tools aimed at exposing and exploiting sparsity in data. If one can understand these sparsity patterns, they can be exploited directly to make simulations more efficient. This enables a productive feedback loop where ML is used to uncover sparsity that makes simulations more efficient and these more efficient simulations are fed into ML models that uncover further sparsity.

My Recent Research Contributions to Artificial Intelligence and Machine Learning in Chemistry

Recommendation systems, such as those used by Netflix and Amazon, attempt to predict what movies/products users will like on the basis of knowledge about their preferences for other products and knowledge about the preferences of many other customers. This is a "matrix completion" problem, where the matrix in question has rows labeled by customer and columns labeled by movie/product. Where user preference for a product is known, it is entered in the corresponding matrix element.

The vast majority of the entries in the matrix are unknown, as most users have not expressed any preference for most products. One can summarize that the number of total entries in the matrix is CP, where C is the number of consumers and P is the number of products. Perhaps 1% of the entries in this matrix are known and the rest are unknown. The task of the recommendation system (often called collaborative filtering) is to determine the unknown values (which are perhaps 99% of the matrix elements). In general, this task is obviously impossible, since there is no source for the information needed. Under what conditions would this problem be soluble? Only if the amount of information in the matrix is completely encapsulated by the entries given! One can quantify the amount of information in the matrix by the number N_{nz} of non-zero singular values, as $N_{nz}(C + P)$, because each non-zero singular value comes with a left and right singular vector. The ML strategy here is to *assume* that the problem is soluble [3]. Thus, the number N_{known} of elements which are known (products where consumers have exhibited a preference) defines N_{nz} as $N_{nz} = N_{\text{known}}/(C + P)$. One then solves for the singular vectors corresponding to the non-zero singular values such that the known entries are reproduced as closely as possible. The important thing to note is that these singular vectors and singular values will reconstruct a matrix that has non-zero elements throughout, i.e., it makes *predictions* for *all CP* elements of the matrix. One can surmise that the reason this works is that there is a compressed form of the preference matrix, listing only movie genres and personality types. If many movies are similar to each other (e.g., within a genre) and many people are similar to each other (few personality types, where each personality type enjoys a certain movie genre), then it would be reasonable that the full movie/person matrix is massively redundant and a compressed form should exist. The reason for the success of collaborative filtering is simply that the full set of product preferences for all persons is redundant/ non-informative — the true information content is far less than it appears.

We asked whether this same situation applies in the context of electronic structure theory, where one is solving the electronic Schrodinger equation that gives the potential energy surface (and its derivative, i.e., the forces on atoms) for any arrangement of atoms (in the Born–Oppenheimer approximation). The key components needed to solve this "quantum chemistry" problem are wavefunctions and operators. Applying the collaborative filtering ideas to the electronic wavefunction shows that there is massive redundancy in our usual representations — as little as one part in 10^{-5} of the wavefunction coefficients are needed to reproduce the solution to chemical accuracy [4, 5]. It is of course not enough to know that there exists a compressed representation — one must also be able to compute within the compressed representation without ever accessing the full representation (and this is possible for the cases we have treated so far). If the electronic wavefunction is largely redundant/ non-informative, one can then ask whether the operators are also similarly redundant. In some sense, the answer here is (at least in hindsight) more obvious. The Coulomb interaction is universal in the sense that it has the same form for any pair of electrons. Thus, the representation of many Coulomb interactions between many pairs of electrons should repeat much of the same information. Indeed, collaborative filtering ideas lead to the realization that the fourth-order tensor representing the Coulomb operator (the two-electron repulsion integrals) can be written in a compressed form involving only second-order tensors:

$$(ij|kl) = \sum_{PQ} X_{iP} X_{jP} Z_{PQ} X_{kQ} X_{lQ},$$

where i, j, k, l are atomic orbitals and the indices P and Q are auxiliary indices describing some abstract "types" of charge distributions (similar to the movie genres discussed above). The range of the P, Q indices is typically about 3x the range of the i, j, k, l indices. This tensor hypercontraction (THC) form leads directly to decreases in the scaling behavior (computational cost as a function of molecular size) of the computational methods [6–8].

Given the use of ML techniques to compress the electronic Schrodinger equation and accelerate its solution, we then introduced the "*ab initio* nanoreactor" that discovers chemical reactions by direct atomistic simulation based on *ab initio* molecular dynamics. This is a viable approach because new algorithms coupled with graphical processing units (GPUs) enable simulations of hundreds of atoms for time scales approaching a nanosecond on a desktop computer [9, 10]. The molecules that seed the nanoreactor are placed in a confining sphere (typically of $\approx$ 1-nm radius) and the molecular dynamics is followed in a canonical ensemble. Surprisingly, this is sufficient to see chemical reactions within 10 s of picoseconds for many systems, especially when the temperature is chosen to be near 1000 K. In order to accelerate reactions even further, one can introduce other biases such as periodic compression/decompression [11] or artificial forces such as stirring (analogous to a nanometer-sized stir bar). Because of the artificial forces introduced to accelerate reactions (such as elevated temperature, time-varying pressure, stirring), one cannot use the frequency of determining a reaction to infer its rate. Instead, we take the discovered reactions (which are determined automatically using graph representations of molecules and hidden Markov models) and determine minimum energy paths corresponding to them [12, 13] (seeded with initial guesses from the dynamics). These paths give the reaction barriers and energetics needed to construct transition state theory estimates of the corresponding reaction rate. The list of reactions and associated rates is then used to construct a microkinetic model which gives the concentration profile as a function of time. This can be used to step forward in time even further, by sampling from the concentration profile and running nanoreactors seeded with the concentrations of molecules appropriate to a given point in time. A schematic of the whole procedure is shown in Fig. 1.

1. Discovery

Use hyper-real forces to accelerate reactions

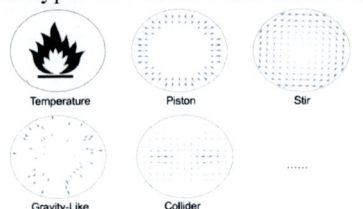

3. Refine

Reaction path optimization to remove bias

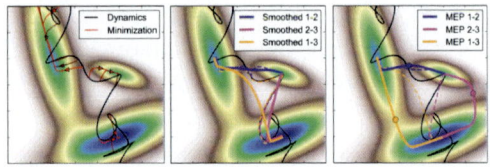

2. Event Identification/Extraction

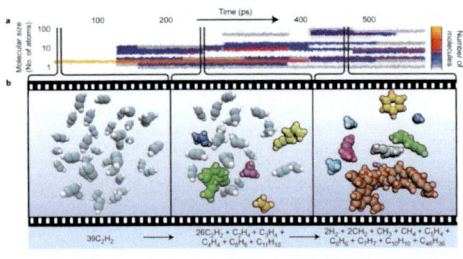

4. Kinetic Models

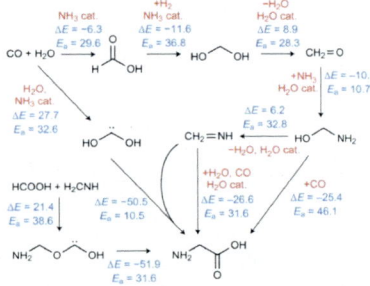

Fig. 1.   Schematic of the nanoreactor workflow. Reaction events are harvested by accelerated *ab initio* molecular dynamics. These are then refined to obtain minimal energy paths which give the information needed for transition state theory estimates of the reaction rates. From the list of discovered reactions and computed rates, one constructs a microkinetic model. Iterating over these steps to self-consistency allows one to discover reactions with many intermediates.

Outlook on Future Developments of Research on Artificial Intelligence and Machine Learning in Chemistry

The *ab initio* nanoreactor is now able to generate lists of thousands (or even tens of thousands) of reactions (where each reaction is an elementary step). The pressing problem now is to classify these reactions in order to compactly present the results and identify which reactions are most interesting. This kind of information could also be used to "steer" the nanoreactor to discover unknown chemistry. Algorithms used to characterize social networks are currently being investigated for this purpose [14]. It is clear that an exciting direction would connect the nanoreactor with "self-driving laboratories" to enable reinforcement learning [15]. The nanoreactor would discover chemical reactions, direct the robotic laboratory to carry them out, characterize the results, and then adjust the accelerating forces and/or accuracy of the underlying *ab initio* methods in order to effectively span the space of chemical reactions and improve its own accuracy. With full feedback, this system could teach itself chemistry. Fulfilling this goal would require developments in robotic laboratories, rapid and accurate characterization of reaction outcomes, and flexible ways to improve the *ab initio* methods used for discovery and/or refinement in the nanoreactor. It is a challenging goal on many fronts, but one which is sure to improve our understanding of chemistry.

Acknowledgments

This work was supported by the Office of Naval Research.

References

[1] A. Krizhevsky, I. Sutskever and G.E. Hinton, *Comm. ACM* **60**, 84 (2017).
[2] D. Silver, J. Schrittwieser, K. Simonyan, I. Antonoglou, A. Huang, A. Guez, T. Hubert, L. Baker, M. Lai, A. Bolton, Y. Chen, T. Lilicrap, F. Hui, L. Sifre, G. van den Driessche, T. Graepel and D. Hassabis, *Nature* **550**, 354 (2017).
[3] E.J. Candes and B. Recht, *Found. Comp. Math.* **9**, 717 (2009).
[4] B.S. Fales, S. Seritan, N.F. Settje, B.G. Levine, H. Koch and T.J. Martinez, *J. Chem. Theo. Comp.* **14**, 4139 (2018).
[5] R.M. Parrish, Y. Zhao, E.G. Hohenstein and T.J. Martinez, *J. Chem. Phys.* **150**, 164118 (2019).
[6] E.G. Hohenstein, R.M. Parrish and T.J. Martinez, *J. Chem. Phys.* **137**, 044103 (2012).
[7] C. Song and T.J. Martinez, *J. Chem. Phys.* **149**, 044108 (2018).
[8] C. Song and T.J. Martinez, *J. Chem. Phys.* **146**, 034104 (2017).
[9] A.V. Titov, I.S. Ufimtsev, N. Luehr and T.J. Martinez, *J. Chem. Theo. Comp.* **9**, 213 (2013).
[10] I.S. Ufimtsev and T.J. Martinez, *J. Chem. Theo. Comp.* **5**, 2619 (2009).
[11] L.-P. Wang, A. Titov, R. McGibbon, F. Liu, V.S. Pande and T.J. Martinez, *Nat. Chem.* **6**, 1044 (2014).
[12] L.-P. Wang, R.T. McGibbon, V.S. Pande and T.J. Martinez, *J. Chem. Theo. Comp.* **12**, 638 (2016).
[13] X. Zhu, K.C. Thompson and T.J. Martinez, *J. Chem. Phys.* **150**, 164103 (2019).
[14] W.L. Hamilton, R. Ying and J. Leskovec, I arXiv, 1706.02216 (2017).
[15] F. Hase, L.M. Roch and A. Aspuru-Guzik, *Trends Chem.* **1**, 282 (2019).

SYNTHESIS PLANNING, REACTION DISCOVERY, AND DESIGN OF CHEMICAL SYSTEMS USING COMPUTERS

BARTOSZ A. GRZYBOWSKI

Institute for Basic Science and Ulsan National Institute of Science and Technology, 50 UNIST-gill, Ulju-gun, Ulsan 44919, Ulsan South Korea and Polish Academy of Sciences, Institute of Organic Chemistry, ul. Kasprzaka 44/52, Warsaw 01-224, Poland

My View of the Present State of Research on Artificial Intelligence/Machine Learning in Chemistry

The first attempts to teach computers the planning of multi-step organic syntheses date to the 1960s and were undertaken by some of the most prominent chemists of our era: E.J. Corey [1], K. Djerassi [2], and Ivar Ugi [3]. In many ways, these pioneering efforts were ahead of their time as neither the computing power nor the algorithms suitable to address the problem were available back then. The problem itself is an extremely hard one — the machine must be taught not only the myriad of rules governing individual reaction types but also how to navigate the enormous networks of synthetic possibilities, and how to stitch together individual steps into synthetic sequences. Ideally, the computer code combining these components should be able to design "elegant" routes to arbitrary targets including complex ones whose synthesis is challenging to human experts. Of course, such routes should be free of any apparent chemical inconsistencies and, even better, should be executable in the laboratory.

Unfortunately, the abovementioned early efforts have largely failed to deliver on their promise. Even Corey's LHASA — arguably, the most advanced of these approaches — allowed for only step-by-step planning with a human operator controlling the choices at each synthetic step. LHASA was a useful "synthetic assistant" but not a platform for truly automated synthetic design. After the termination of the LHASA effort, the research in this area virtually came to a halt for almost two decades and it has only been in the past five years that the widespread interest in the problem was rekindled. What gave this renewed effort its impetus is (i) the development of methods to analyze and query large chemical networks [4, 5] and (ii) recent improvements in neural networks and other classification methods [6]. A combination of these methods with comprehensive databases of reaction rules (either expert coded [7–11] or machine learned [6]) and with related synthetic-logic heuristics [7, 8] has led to the emergence of first software platforms capable of autonomous synthetic planning — notably of our group's Chematica which is now an experimentally validated and commercially available software distributed as Synthia[TM] by the Sigma-Aldrich/Merck conglomerate. These exciting recent developments herald the dawn of a new era for synthetic chemistry, whereby computers will take increasingly important roles in synthetic design, ultimately reducing the number of failed experiments and thus the cost — both monetary and environmental — of organic–synthetic experimentation and production.

My Recent Research Contributions to Chemical AI

Our group has been pursuing the problem of computer-assisted synthetic design since the early 2000s. Much of the groundwork for our recent achievements has been laid earlier [4, 5] when we studied the topology of large synthetic networks and the algorithms that can be used to query them. Such networks are relevant to synthetic planning since each retrosynthetic step entails on the order of 100 viable options (Fig. 1(a)) giving, for n-step syntheses, networks of enormous sizes, $\sim 100^n$. Another all-important problem has been to teach the machine the underlying chemical rules.

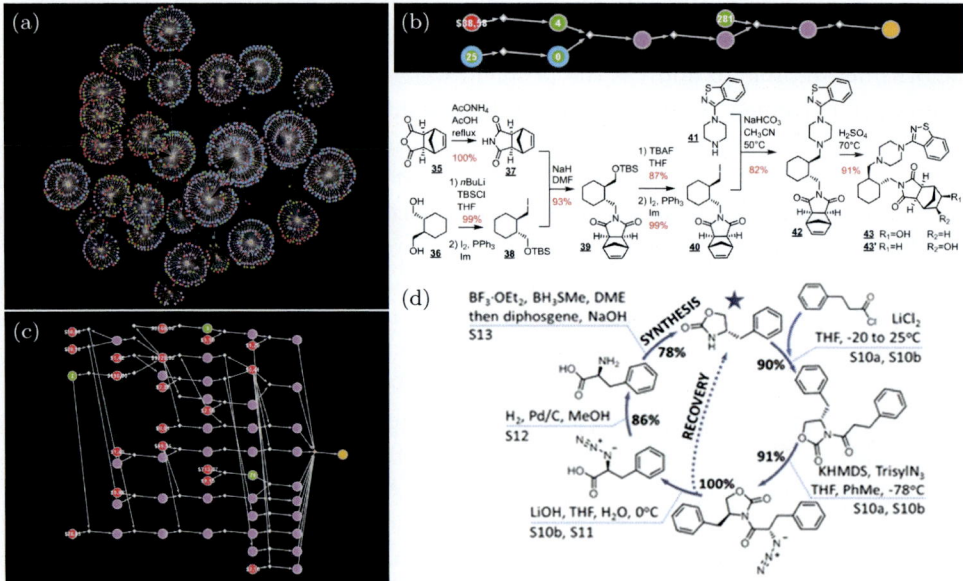

Fig. 1.   (a) A still small network of synthetic possibilities at the very beginning of Chematica's planning task.   (b) Complete synthetic pathway to 5β/6β-hydroxylurasidone designed autonomously by Chematica and successfully executed in the laboratory (reproduced from Ref. [8] with permission from Cell Press).   (c) The synthetic plan for a library of $^{13}C_6$ labelled anticoagulants. Notice many intermediates shared between different sub-pathways.   (d) Example of an autocatalytic cycle multiplying Evans' auxiliary used commonly in asymmetric synthesis (reproduced from Ref. [13] with permission from Wiley).

Here, one choice is to extract reaction "cores" from the published reaction precedents [6]. Although rapid, this method is not very accurate as the "rules" thus extracted come without the knowledge of reactivity conflicts and/or of stereoelectronic effects dictating reaction outcomes. A significantly more laborious but also much more reliable [11] way is to code reaction rules based on the underlying reaction mechanisms, covering more than existing literature precedents and capturing all the nuances of chemical reactivity — this is the approach we have been pursuing over the years and, by now, have largely captured the mechanistic basis of organic–chemical transformations (some 80,000+ of them).

Still, the ability to search the trees of retrosynthetic options by applying reaction rules one at a time, without any broader view of synthetic strategy, does not translate into efficient synthetic planning. At least three classes of improvements [7–11] are needed.

First, the search algorithm must be able to revert from dead ends (i.e., from unpromising excursions into the tree's depth). In Chematica, we achieve this by assigning penalties to synthetic attempts that keep repeating the same combinations of synthetic rules but do not yield viable synthetic plans. As such penalties accrue, the algorithm is ultimately forced to backtrack and seek chemically diverse alternatives. The algorithm must also be able to pursue multiple search strategies — akin to a human chemist, who might be trying to perfect one synthetic approach but has multiple other back-up plans in the back of his/her head. This we achieve by implementing multiple priority queues, some directing searches "deep" into the retrosynthetic trees and some "wide". These algorithms continuously communicate their results with one another and thus act synergistically.

Second, we teach the machine how to strategize over multiple steps and overcome local maxima of molecular complexity — this is especially important for planning syntheses of complex targets for which simply using structure-simplifying steps is unlikely to produce a viable synthetic plan. We address this problem by implementing functional group interconversions, FGIs, and also by using sequences of steps (so-called Tactical Combinations, TCs) in which the first (in the retrosynthetic direction) complexifies the structure, but the second simplifies it greatly, opening new valleys of downstream synthetic possibilities. Remarkably, while only ~500 TCs have been cataloged to date, our recent large-scale computer analyses have identified some 5 million of such sequences, extending by orders of magnitude the available strategic-synthetic toolbox. In one recent demonstration, we used one of the newly discovered TCs to design an efficient route to a small natural product — this route was approximately 50% shorter than all previously published syntheses of this molecule.

Third, for the machine to be able to deal with complex synthetic tasks, the reaction rules and the scoring functions navigating the retrosynthetic trees must be supplemented by AI, quantum mechanical, and even molecular-mechanics routines. For instance, when a reaction involves closure of a macrocycle, it is not enough to determine the "local" correctness of the reaction template (i.e., only in terms of atoms changing bonding patterns and their immediate neighborhoods), but one must also ensure that the ring closure does not entail excessive strain, which can be only modelled by QM/MM techniques. In a related genre, when the algorithm decides which of the multiple options to take, it can benefit from the knowledge of "synthetic habits" machine-learned from prior literature examples. However, one has to be very careful that such machine learning does not simply repeat known synthetic approaches and is not biased toward only very popular reactions while omitting the ones having few precedents in the literature. In Chematica, we address this problem by training our search-guiding scoring functions on literature examples *matched* onto mechanistic expert-coded rules. In this way, the functions are able to correct the frequencies of literature precedents for their objective synthetic "power" — in doing so, they offer accurate synthetic suggestions even for reaction types sparse in the literature but synthetically useful.

By implementing all of the above (and some other) solutions, we have been gradually improving Chematica's synthetic skill. In 2018, the program was challenged with designing syntheses of eight medicinally relevant targets that had previously proven difficult to make in good yields (or at all) by Sigma-Aldrich's chemists. In all cases, Chematica's solutions proved superior to prior approaches in terms of overall yields and/or costs [8] (Fig. 1(b)). More recently, Chematica has been designing syntheses of complex natural products. Two such syntheses have just been finalized and were part of my presentation at the Solvay Conference. The list is continuously growing as Chematica is warming up to compete with the top-level human experts. I project that within 1–2 years, the program will achieve a grand master level, outcompeting human contenders. This, in fact, is bound to happen as the machine continuously accumulates new knowledge, never forgets anything it has learned, and is perfecting its strategizing skill embodied in increasingly sophisticated search routines.

The Computerized Synthetic Chemistry of the Future

Having already included in the previous section some of the forward-looking statements about Chematica's future, I hasten to add that retrosynthetic design of individual pathways and retrosynthesis in general are not the only areas of organic chemistry in which computers will bring about revolutionary changes and perform tasks that are largely beyond the capabilities of a human chemist(s). Within five years at most, the following tasks — some already prototyped in our laboratory — will become mainstream synthesis-oriented uses of computers:

(1) Machines will simultaneously design syntheses of multiple targets and of entire target libraries [11]. In doing so, they will optimize global synthetic schemes making use of intermediates common to different sub-pathways (Fig. 1(c)) in effect lowering the cost of the global synthetic plan. Every

major chemical company will use such algorithms to minimize its cost and the use of materials.

(2) Computers will combine synthetic design with the knowledge of "green chemistry" to design synthetic plans that navigate around toxic and/or listed substances and will use only recyclable solvents.

(3) Computers will progress to design fundamentally new reaction types based on new reaction mechanisms. Some examples of how this can be done were shown in my Solvay talk.

(4) Working in the "forward" rather than the "retro" directions, machines will generate new and synthesizable chemical spaces populated with molecules possessing desired, user-specified sets of characteristics. Algorithms will generate myriads of synthesizable candidate molecules and, using AI routines, will evaluate their physico-chemical, material, and medicinal properties. Based on the advances under point (3), new types of scaffolds will become available, and this unchartered structural space will translate into new properties.

(5) Computers will then set out to construct entire systems of reactions — sequences of reaction that can be performed in one pot, reaction cascades amplifying weak chemical signals [12], and, above all, autocatalytic cycles multiplying valuable chemicals [13] (Fig. 1(d)). Organic chemistry will become a science of controlling *systems* rather than of tweaking individual reactions.

The next 5 years will be an exciting and transformative time for synthetic chemistry which will finally live up to Kant's criteria of an "exact science" grounded in algorithm rather than in trial-and-error experimentation. Forty years ago, few thought that computer repositories such as Reaxys or SciFinder will effectively displace departmental libraries and that meticulous literature studies could all be done by few clicks of a mouse. In much less than forty years from now, few will risk spending time on paper-and-pencil synthetic design when such design will have been available just a few clicks away on sophisticated computer platforms.

References

[1] E.J. Corey and W.T. Wipke, *Science* **166**, 178 (1969).
[2] T.H. Varkony, D.H. Smith, and C. Djerassi, *Tetrahedron* **34**, 841 (1978).
[3] I. Ugi, J. Bauer, R. Baumgartner, E. Fontain, D. Forstmeyer, and S. Lohberger, *Pure Appl. Chem.* **60**, 1573 (1988).
[4] M. Fialkowski, K.J.M. Bishop, V.A. Chubukov, C.J. Campbell, and B.A. Grzybowski, *Angew. Chem. Int. Ed.* **44**, 7263 (2005).
[5] M. Kowalik, C.M. Gothard, A.M. Drews, N.A. Gothard, A. Wieckiewicz *et al.*, *Angew. Chem. Int. Ed.* **51**, 7928 (2012).
[6] M.H.S. Segler, M. Preuss, and M.P. Waller, *Nature* **555**, 604 (2018).
[7] S. Szymkuć, E.P. Gajewska, T. Klucznik, K. Molga, P. Dittwald *et al.*, *Angew. Chem. Int. Ed.* **55**, 5904 (2016).

[8] T. Klucznik, B. Mikulak-Klucznik, M. P. McCormack, H. Lima, S. Szymkuć *et al.*, *Chem* **4**, 522 (2018).

[9] K. Molga, P. Dittwald, and B.A. Grzybowski, *Chem* **5**, 460 (2019).

[10] K. Molga, P. Dittwald, and B.A. Grzybowski, *Chem. Sci.* DOI: 10.1039/C9SC02678A (2019).

[11] K. Molga, E.P. Gajewska, S. Szymkuć, and B.A. Grzybowski, *React. Chem. Eng.* **4**, 1506 (2019).

[12] R. Roszak, M.D. Bajczyk, E.P. Gajewska, R. Holyst, and B.A. Grzybowski, *Angew. Chem. Int. Ed.* **58**, 4520 (2019).

[13] M.D. Bajczyk, P. Dittwald, A. Wolos, S. Szymkuć, and B.A. Grzybowski, *Angew. Chem. Int. Ed.* **57**, 2367 (2018).

EMERGING BIOPHYSICAL MECHANISM AND EVOLUTION: SYNERGISTIC APPROACHES TO PREDICT EVOLUTIONARY DYNAMICS TO FIGHT DRUG RESISTANCE

EUGENE I. SHAKHNOVICH

Department of Chemistry and Chemical Biology, Harvard University, Cambridge, MA 02138, USA

Predicting Evolution: Status and Challenges

Predicting the dynamics and outcomes of evolutionary selection is a key unsolved problem of modern biology. Being of fundamental importance to our understanding of evolution, it is also vital to the development of novel therapeutics and vaccines. However, two major gaps in our understanding of physical–chemical biology and evolutionary dynamics challenge this prediction. *First,* the genotype–phenotype relationship (GPR), often cast as the fitness landscape, is poorly understood even for simple organisms like viruses and bacteria. Existing approaches attempt to derive fitness landscapes of already observed strains from epidemiological data [2, 3] with the aim of predicting which of the existing variants will prove resistant to current drugs or dominate new viral epidemics. However, such empirical tools are not able to predict new pathogenic variants arising due to de novo mutations. *Second,* the structure of a population has a crucial effect on how it evolves in the fitness landscape. Theoretical population genetics predicts that population size controls the balance between the forces of selection and random genetic drift; this eventually determines the direction of evolution in the fitness landscape. Earlier studies appreciated the role of population size, deriving an "effective population size" to fit the genetics or evolutionary data to existing population genetics models, resulting in a somewhat circular reasoning [4]. Nevertheless, direct experimental evidence for this fundamental concept is very scarce due to the inherent difficulty of simultaneously controlling population size and obtaining tractable molecular readouts of evolutionary processes in a configurable and reproducible environment of a laboratory experiment or in multi-scale organism-based simulations.

Both these challenges call for a multi-scale and multi-tool research effort that encompasses both theory and experiment. It has been a long-standing dogma among evolutionary biologists that fitness landscapes cannot be reduced to few simple molecular traits so only phenomenological approaches that postulate certain distribution of fitness effects of mutations are possible. While most previous effort

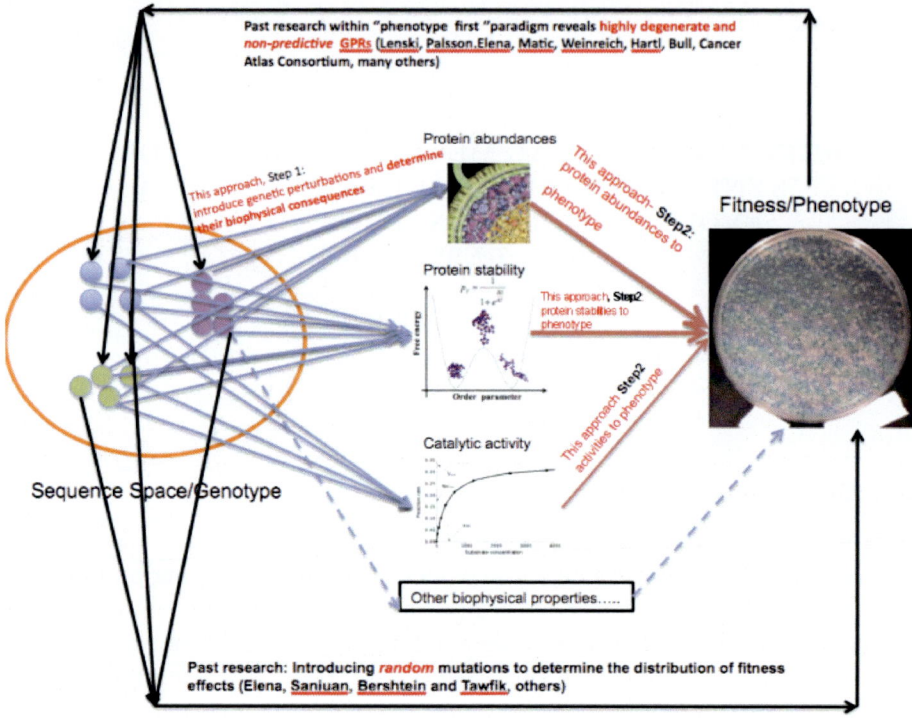

Fig. 1. Biophysical properties of proteins serve as a "hidden layer" in GPR. We hypothesize that the relation between intermediate phenotype — molecular properties of proteins — and phenotype is "smooth". Black arrowed lines illustrate past efforts to study GPR "in one jump" — from sequences to phenotype — while the conceptual foundation of our approach is illustrated in the middle of the figure.

treated molecular biophysics and evolutionary theory as separate or even unrelated disciplines, studies conducted by us and others demonstrated the power of unifying multiscale approaches that provide evolutionary rationale and guidance to the phenomena observed at a molecular level. Recent work from our lab and others' and future research offer a direct challenge to this dogma.

Specifically, we postulated that complexity of the GPR mostly stems from the highly complex effect of mutations on *protein* molecular phenotype (e.g., stability, activity) while mapping between changes in protein phenotype and ensuing fitness effects might be "smooth" and predictable from biophysical principles (Fig. 1). The new concept of Biophysical Fitness Landscape (BFL) is a map of protein/nucleic acid molecular properties to fitness. [5]. We demonstrated the conceptual validity of BFL by discovering a simple and accurate quantitative relationship between fitness of *E. coli* and molecular properties of important core metabolic enzymes.

These successes notwithstanding, several limitations of this approach should be appreciated: First, previous studies (see below) showed that *changes* of fitness upon genetic variation in specific enzymes can be predicted from biophysics and biochemistry, while *ab initio* prediction of fitness itself from genomic sequences remains beyond reach. Second, while simple relationships between molecular and fitness variation have been shown for monomeric enzymes, they are not known for protein complexes or other functional groups of proteins.

Here, we briefly discuss recent studies grounded in the proven conceptual foundation of BFL to explore exciting problems on the interface of biophysics and evolutionary biology and outline perspectives for future research. These works, besides their fundamental contribution to mechanistic, molecular-level understanding of key biological problems in molecular evolution such as epistasis and tradeoffs in evolutionary dynamics, will address important public health challenges concerning the evolution of drug resistance and the approaches to effectively block it, as well as evolutionary adaptation of microorganisms to environmental stress.

Recent Progress in Development of Predictive Biophysical Models of Evolution: Kinetic Flux Theory Closes the Genotype–Phenotype Gap for Essential Enzymes in *E. coli*

A predictive GPR is a prerequisite for a truly mechanistic multi-scale understanding of evolution. Our experimental approach to establish the relationship between molecular and fitness effects is bottom-up (see Fig. 1 of (6)) whereby we do the following: (1) introduce a controlled genetic variation into a locus of interest, by editing the *E. coli* genome using the bacterial genome editing technology based on a variant of lambda red homologous recombination that we developed in the lab [6] and, more recently, using CRISPR technology; (2) determine *in vitro* molecular properties of protein variants (stability, catalytic activity, cellular abundance, molten globule content, etc.); (3) determine effect of the mutations on organismal fitness (lag times, growth rates of *E. coli* in log phase, or the outcome of competition with wild-type strains); (4) analyze the relationship between the variation of molecular properties and fitness effects. Initially, we used point mutations to explore the fitness effect of changes in stability of a central core metabolic enzyme DHFR [6]. However, later, we realized that this approach is limited to a relatively narrow range of variation of other key molecular properties such as catalytic activities. To broaden the range, we resorted to a novel alternative strategy whereby we replaced, on the *E. coli* chromosome, the endogenous *E. coli* enzyme DHFR by DHFRs from a diverse set of 35 mesophilic bacteria [7]. In total, 43 orthologous DHFRs were purified and biophysically characterized (we succeeded in chromosomal incorporation of 35 out of 43 orthologous DHFRs). The activity range spans almost 2 orders of magnitude from roughly 0.1 of the *E. coli* value to 10 times more active enzymes. In addition, we determined fitness of *E. coli* where wild-type or trimethoprim-resistant strains of DHFR were expressed from pFLAG plasmids.

We found (Fig. 2, details in (1)) that metabolic flux through the enzyme [8] accurately predicts fitness in terms of molecular properties of DHFR in free form

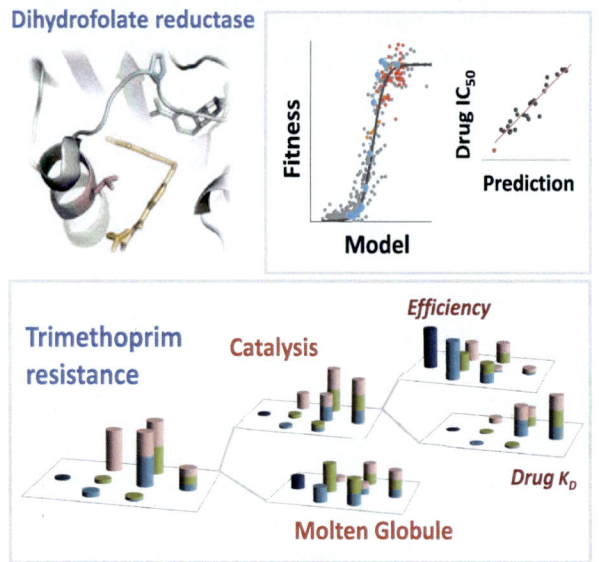

Fig. 2.   Fitness landscape and phenotype- trimethoprim resistance (IC50) can be predicted accurately (top panel) from the set of molecular properties of DHFR (bottom panel) using Eq. (1). Furtherdetails are presented in (1).

and in the presence of inhibitor trimethoprim:

$$fitness \sim flux = \frac{V_{DHFR}}{(B + V_{DHFR})} \quad \text{where}$$

$$V_{DHFR} = \frac{k_{cat}^{mut}}{K_M^{mut}} \cdot [DHFR]^{mut} \cdot \frac{1}{\left(1 + \frac{\alpha \cdot [TMP]_{medium}}{K_i^{mut}}\right)} \tag{1}$$

Here, V_{DHFR} is the velocity of the reaction catalyzed by DHFR, $\frac{k_{cat}^{mut}}{K_M^{mut}}$ is the mutant enzyme catalytic efficiency, $[DHFR]^{mut}$ is the intracellular abundance of a mutant variant, $[TMP]_{medium}$ is the concentration of inhibitor in the medium, K_i is the enzyme inhibition constant for that inhibitor, and α is the ratio of intracellular to extracellular drug concentration, which was independently determined to be 0.1. This model maps the contribution of individual protein properties onto fitness (Fig. (2) bottom) and thus provides a molecular rationale for elusive phenomena such as epistasis and pleiotropy. However, Eq. (1) links both molecular and cellular — enzyme concentration [DHFR] — properties to phenotype. In our earlier work [9], we developed a kinetic model of enzymatic turnover in the cell in the presence of Protein Quality Control that allowed accurate prediction of intracellular DHFR concentration by relating in vitro molecular property of the protein — ANS binding which characterizes the equilibrium population of the molten globule [10] — to cellular abundance of DHFR. With this advance, we were able to predict phenotypic trait IC50 of TMP entirely from variation of molecular properties of the target enzyme. This fundamental discovery forms a basis for rational design of "evolution drugs" — compounds that prevent or dramatically delay the emergence of resistance in pathogens as will be described below.

To explore generality of these results beyond DHFR, we studied the relationship between molecular and fitness effects for another protein, Adenylate Kinase (AdK). In contrast to DHFR, AdK is a high copy number enzyme [11]. We engineered 21 *E. coli* strains that carry mutations in the *adk* locus. Next, we biophysically characterized the mutant forms of AdK and measured fitness of the corresponding strains [12]. To extend the dynamic range of conditions, in a separate experiment, we increased the abundance of AdK by expressing it, in addition to the chromosomal copy, from pBAD plasmid. As in the case of DHFR, we found that fitness at various temperatures is well described by flux dynamics Eqs. (1) and (12) sans the TMP-dependent term, which is not relevant to AdK. Further, we established that not only growth rate but also lag time is an important component of fitness under evolutionary selection. In a subsequent study [13], we made a surprising discovery that stabilized AdK variants show strong substrate inhibition causing noticeable loss of fitness in chromosomally incorporated *E. coli* variants These findings point to an interesting interplay between protein conformational flexibility and function. It also adds another example of evolutionary forces constraining stability of some proteins.

Variation of DHFR Activity Causes Broad Proteomic and Metabolomics Effects

To explore another level of GPR, we carried out a high-throughput proteomics LC/MS study to determine how the abundance of all proteins in the *E. coli* proteome responds to mutations, (14) orthologous transfers (7), or TMP inhibition [14]. We found, surprisingly, that partial loss of DHFR activity as well as gain of extra activity leads to thousands of proteins, mostly unrelated to the folate pathway, changing their abundance. Furthermore, comparing proteomics effects to changes in the transcriptome under the same conditions, we concluded that the broad effect is mostly at the proteome level: Variations of numerous protein abundance are mostly due to changes in their folding-degradation turnover balance rather than expression changes [14]. An addition to the medium of the "folA mix" — combination of small molecule compounds that rescue the loss of DHFR function [15] — results in further dramatic changes in the *E. coli* proteome. Furthermore, variation of DHFR function causes profound metabolomics "domino"-like effects that extend beyond the immediate chemical vicinity of DHFR products, and widespread proteomic and metabolomics changes appear to be linked.

Predicting Evolutionary Dynamics on Biophysical Fitness Landscapes and using AI to Develop Optimal Anti-resistance Regimens

Earlier, we developed quantitative agent-based simulation tools that reliably model evolutionary dynamics at constant and variable population sizes without

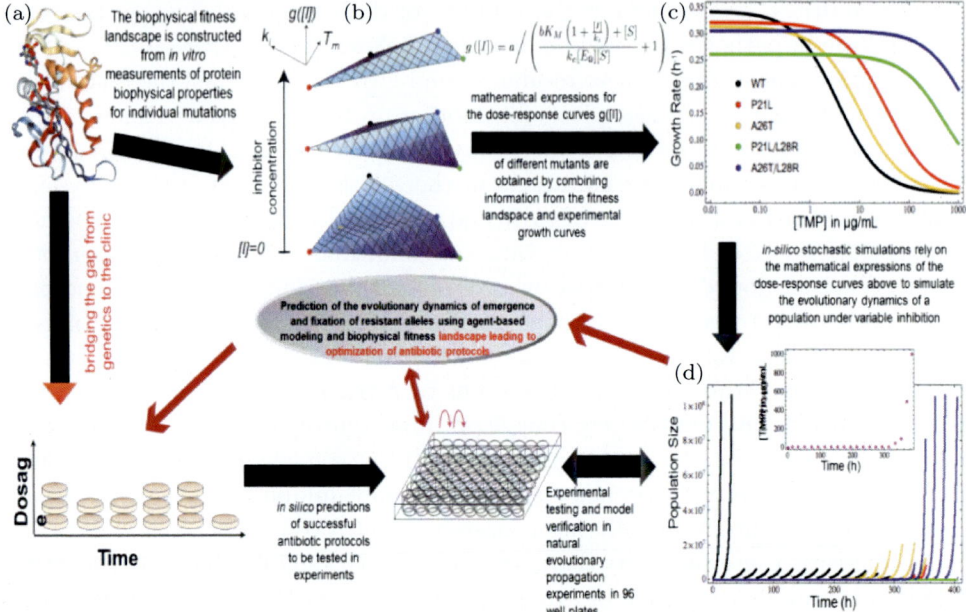

Fig. 3.   Schematics of our synthetic approach to study evolutionary dynamics on BFL. Starting from experimental biophysical characterization of the effects of individual mutations on protein target and fitness (a) and ending with biologically relevant and specific predictions about the expected success of input antibiotic protocols. Intermediate steps involve (a) use of biophysical modeling of the fitness landscape (b) resulting in highly accurate dosage response curves for each DHFR TMP escape variant (c), computational stochastic modeling of system-level population dynamics (d). Comparison to experimental data will inform modeling efforts at multiple steps. Shown in d are the preliminary results of a run of a simulation algorithm for a serial passaging experiment, where the inhibition is increased if the growth rate exceeds a certain percentage of original growth rate (here set to 60%); due to stochasticity, the dilution times and the inhibitor level increases may differ between runs. The inset panel shows the particular antibiotic protocol of this simulation. Different colors show rise and fall of different resistant variants.

assuming monoclonality of populations (i.e., beyond the weak mutation, strong selection regime) [16–18]. Next, we merged advanced evolutionary simulations with BFL-based GPR to carry out full multi-scale predictive analysis of evolutionary dynamics. Our main model is evolution of resistance under antibiotic stress (Fig. 3). Since predicting molecular effects of mutations remains challenging, as a first step, we used DHFR variants whose molecular properties have been fully evaluated by us earlier (1) as an interconverting set of possible resistant strains. Computational predictions are concurrently tested in experimental evolution using a new robotic algorithm and technology as described in our recent publication [19] (see Fig. 3).

When new mutations appear in resistant strains in our pipeline, we experimentally evaluate their full BFL and add to the ensemble of resistant variants. In parallel, we will use an extensive computational toolbox from our lab to predict molecular effects of mutations in term of stability, appearance of long-living molten globule folding intermediates, mRNA stability and translation initiation, catalytic efficiency, and inhibitor binding [20–27]. We use these predictions in a twofold manner. First, we predict BFL (Eq. (1)) computationally for de novo mutations. Second, in a related effort, we evaluate molecular biophysics of some of these de novo variants experimentally and build an extensive database to use for training of the models. Next, we developed a new AI approach, which uses thousands of short classical dynamics and DFT calculations to train the model in the space of 64 atomic features and atomic coordinates presented as Gaussians centered on a specific atom. The thermodynamics effects from multiple simulations are fed into an AI neural network to predict thermodynamic effects of mutations [20, 27–29]. This approach provides a fast pre-trained tool to evaluate molecular effects of mutations on the fly in evolutionary simulations which otherwise can be obtained only from much longer trajectories, which is not practical in multi-scale evolutionary models.

In a related effort, we are developing reinforcement learning AI protocol (see Fig. 4) to determine the most optimal regiments under the constraint that minimizes the total antibiotic regiment (Dalit Engelhardt, ES, work in progress). The aim of this approach is to develop an optimal protocol that maximizes the probability of successful administration of drug (i.e., extinction of bacterial population) while minimizing the total drug dose. An optimal protocol should increase dosage only when new resistant mutation arrives. However the challenge is to "predict" when a mutation would arise. In our approach, we developed a policy through feedback learning of several experimental protocols and developing an agent-based training

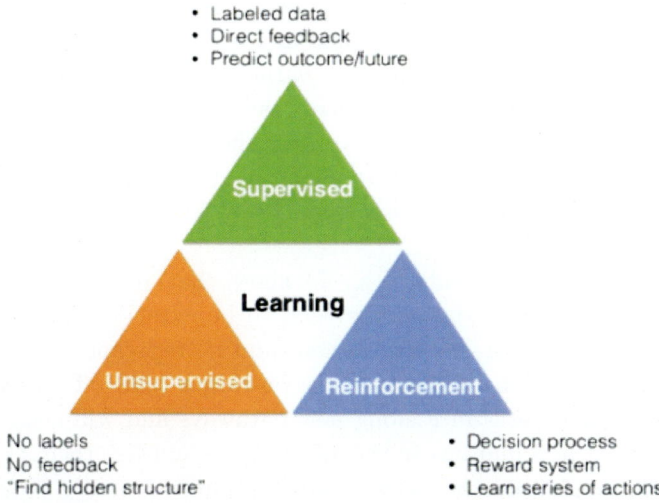

Fig. 4. The schematics of reinforcement learning whereby the decision process is based on a reward system that learns a series of actions — in this case, administration of antibiotics after resistance mutations arrive.

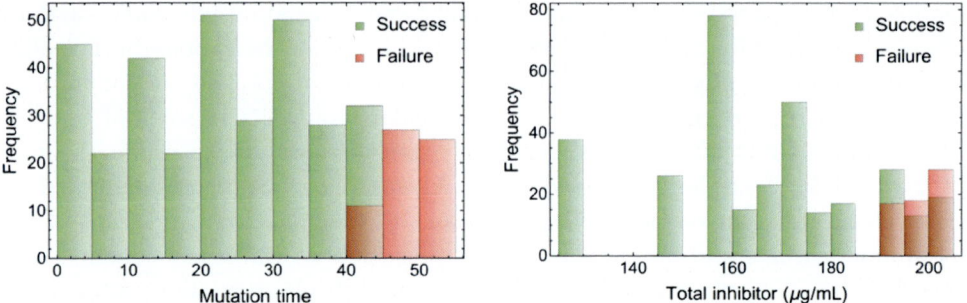

Fig. 5.   Results of agent-based modeling of drug optimization for 500 simulated drug protocols where trimethoprim-resistant variants with known BFL (see Fig. 3) arose stochastically. In the vast majority of realizations, the algorithm successfully reached clearance reaching target dosage regiment.

policy protocol with feed forward neural net. Omitting details, we show in Fig. 5 that the algorithm is capable of administering optimal drug protocols.

Using AI Algorithms to Design "Evolution Drugs"

Knowledge of potential evolutionary paths of escape from antibiotic pressure suggests a new approach to design multi-functional compounds or combinations that block all known escape paths. As a first step, we used our molecular fragment-based design tool OpenGrowth (OG) [30]. One of the key features of OG is its ability to design concurrently against several targets (i.e., wild-type and resistant mutant variants of the same target protein). This is crucial for the current application aimed at developing antibacterial agents against multiple resistance strains. In our first application of this technology, we designed and tested a library of compounds that are active against multiple known escape variants of DHFR. We tested these and other compounds extensively both in vitro (as inhibitors of wt DHFR and escape variants) and in vivo and found that several of them are active in both sets of tests (Fig. 6). Next, we plan to carry out an evolutionary propagation experiment using TMP and designed compounds to ascertain the evolutionary dynamics of emerging resistance.

Our preliminary experiments (Fig. 7) show that resistant strains have not emerged so far, but these experiments need to be extended under the control of the computational protocol (Fig. 5). When (and if) resistance arises, we will fully analyze resistant strains to determine new on-target (and off-target if necessary) mutations that confer resistance along new pathways and will design a new set of compounds against new mutants in the affected loci and test them as before in vivo and in vitro. Through iterative cycles of propagation with analysis outlined in Fig. 3 and molecular design, we will be able to explore a full set of evolutionary pathways and assess their mechanisms to maximally delay the emergence of resistance.

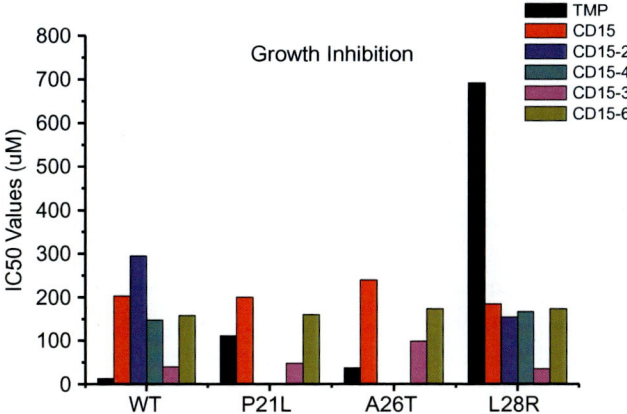

Fig. 6. Biological activity of a series of designed compounds (internal notation CDx) against wt and TMP-resistant mutants (*x*-axis notation indicates mutations in DHFR conferring TMP resistance). Remarkably, our compounds are very active against L28R mutation, which stabilizes DHFR conferring especially strong and intractable (for *E. coli*) resistance.

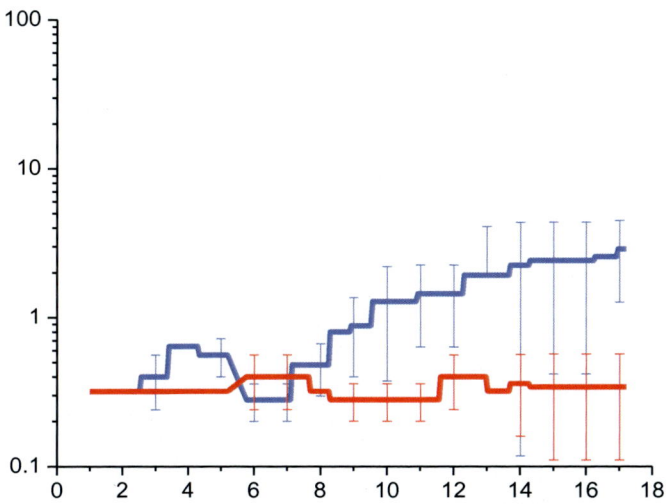

Fig. 7. Emergence of resistance against TMP (blue) when the IC50 (*y*-axis) increases almost 2 orders of magnitude over 2 weeks (*x*-axis) of evolution. In contrast, the new compound (red curve) appears to suppress resistance on the same time scale

Outlook

"Nothing in Biology makes sense except in the light of evolution" (see Ref. [31]). The new approaches that integrate molecular biophysics of protein folding and function, system-level analysis of metabolomics and proteomics with agent-based learning and other AI approaches to predict dynamic trajectories provide a unique opportunity to gain deep insight into the fundamental biophysical underpinnings of evolutionary processes. Despite encouraging initial results, the field is still in its infancy and many interesting new discoveries and surprises lie ahead. In fact, the BFLs presented here are very simple as they assume the mutational effects in response to stress limited to the target locus. In reality, evolutionary dynamics could be affected by epistasis, dead-end scenarios [19], and additional stochasticity due to fluctuating environments and variation in protocols. Addressing these issue would call for models of ever increasing complexity where a significant part of the metabolome beyond the immediate vicinity of affected enzymes should be included in considerations. As biophysical models of evolution become more complex, the need in methods of AI and machine learning to extrapolate long time trajectories becomes more pressing. As noted here, these methods can and should be applied on multiple scales — from predicting molecular effects of mutations to predicting the approximate timing of rising and fixation of stochastic mutations. Another area ripe for AI approaches is metabolomics analysis that combines flux balance approaches [32–34] with approaches based on kinetic modeling of intracellular processes [35]. The high dimensionality of corresponding objective functions renders reductionist AI tools extremely valuable in these future applications.

References

[1] J.V. Rodrigues *et al.*, *Proc. Natl. Acad. Sci. USA* **113**(11), E1470 (2016).

[2] M. Luksza and M. Lassig, *Nature* **507**(7490), 57 (2014).

[3] R.A. Neher, C.A. Russell, and B.I. Shraiman, *eLife* **3** (2014).

[4] M. Lynch and J.S. Conery, *Science* **302**(5649),1401 (2003).

[5] S. Bershtein, A.W. Serohijos, and E.I. Shakhnovich, *Curr. Opin. Struct. Biol.* **42**, 31 (2017).

[6] S. Bershtein, W. Mu, and E.I. Shakhnovich, *Proc. Natl. Acad. Sci. USA* **109**(13), 4857 (2012).

[7] S. Bershtein *et al.*, *PLoS Genetics* **11**(10), e1005612 (2015).

[8] H. Kacser and J.A. Burns, *Genetics* **97**(3-4), 639 (1981).

[9] S. Bershtein, W. Mu, A.W. Serohijos, J. Zhou, and E.I. Shakhnovich, *Molecular Cell* **49**(1), 133 (2013).

[10] O.B. Ptitsyn, V.E. Bychkova, and V.N. Uversky, *Philos. Trans. Roy. Soc. Lond. B Biol. Sci.* **348**(1323), 35 (1995).

[11] Y. Taniguchi *et al.*, *Science* **329**(5991), 533 (2010).

[12] B.V. Adkar *et al.*, *Nat. Ecol. Evol.* **1**(6), 149 (2017).

[13] B.V. Adkar, S. Bhattacharyya, A.I. Gilson, W. Zhang, and E.I. Shakhnovich, *Proc. Natl. Acad. Sci. USA* **116**(23), 11265 (2019).

[14] S. Bershtein, J.M. Choi, S. Bhattacharyya, B. Budnik, and E.I. Shakhnovich, *Cell Reports* **11**(4), 645 (2015).

[15] S. Singer, R. Ferone, L. Walton, and L. Elwell, *J. Bacteriol.* **164**(1), 470 (1985).

[16] A. Rotem *et al.*, *Mol. Biol. Evol.* **35**(10), 2390 (2018).

[17] N. Cheron, A.W.R. Serohijos, J.M. Choi, and E.I. Shakhnovich, *Protein Sci.* **25**(7), 1332 (2016).

[18] A.W. Serohijos and E.I. Shakhnovich, *Mol. Biol. Evol.* **31**(1), 165 (2014).

[19] J.V. Rodrigues and E.I. Shakhnovich, *eLife* **8** (2019).

[20] J. Tian, J.C. Woodard, A. Whitney, and E.I. Shakhnovich, *PLoS Comput. Biol.* **11**(4), e1004207 (2015).
[21] A. Bitran, W.M. Jacobs, X. Zhai, and E.I. Shakhnovich, *Proc. Natl. Acad. Sci.* in press.
[22] S. Bhattacharyya *et al., Molecular Cell* **70**(5), 894-905 e895 (2018).
[23] M.H. Olsson, W.W. Parson, and A. Warshel, *Chem. Rev.* **106**(5), 1737 (2006).
[24] A.J. Adamczyk, J. Cao, S.C. Kamerlin, and A. Warshel, *Proc. Natl. Acad. Sci. USA* **108**(34), 14115 (2011).
[25] S. Vicatos, M. Roca, and A. Warshel, *Proteins* **77**(3), 670 (2009).
[26] M. Roca, H. Liu, B. Messer, and A. Warshel, *Biochemistry* **46**(51), 15076 (2007).
[27] S. Yin, F. Ding, and N.V. Dokholyan, *Nat. Meth.* **4**(6), 466 (2007).
[28] V. Potapov, M. Cohen, and G. Schreiber, *Protein Eng. Des. Sel.* **22**(9), 553 (2009).
[29] G. Gilis and M. Rooman, *Protein Eng.* **13**(12), 849 (2000).
[30] N. Cheron, N. Jasty, and E.I. Shakhnovich, *J. Med. Chem.* **59**(9), 4171 (2016).
[31] T. Dobzhansky, *Zygon* **8**(3-4), 261 (1973).
[32] N.E. Lewi, H. Nagarajan, and B.O. Palsson, *Nat. Rev. Microbiol.* **10**(4), 291 (2012).
[33] J.S. Edwards, R.U. Ibarra, and B.O. Palsson, *Nat. Biotechnol.* **19**(2),125 (2001).
[34] A.M. Feist and B.O. Palsson, *Curr. Opin. Microbiol.* **13**(3), 344 (2010).
[35] E.T. Powers, D.L. Powers, and L.M. Gierasch, *Cell Reports* **1**(3), 265 (2012).

SESSION 1: ARTIFICIAL INTELLIGENCE/MACHINE LEARNING IN CHEMISTRY

CHAIR: K. WÜTHRICH
AUDITORS: E. CAUËT[1], B.M. MOGNETTI[2]

[1] *Spectroscopy, Quantum Chemistry and Atmospheric Remote Sensing (SQUARES),
CP 160/09, Université Libre de Bruxelles (ULB), Av. F. Roosevelt 50, 1050 Brussels, Belgium*
[2] *Center for Nonlinear Phenomena and Complex Systems, CP 231,
Université Libre de Bruxelles (ULB), Boulevard du Triomphe, 1050 Brussels, Belgium*

Discussion among the panel members

Kurt Wüthrich: There are some things that may be interesting to clarify. I have heard that you make use of big data, but you also weigh the data or rank the data. It is not clear to me to what extent this can lead to innovation. Is the critical point to weigh or rank of the data that you use as input? That the protein databank (PDB), which is the largest data set that I use in my own research, is not big data was a revelation for me. Maybe we should look into this again considering how much information there is in each individual entry in the PDB. Maybe I can also add that it is no surprise to me that artificial intelligence can help to speed up known processes. I am not surprised that there is a 34-fold increase in efficiency. In contrast, it is still not clear to me how breakthroughs and real novelty can come from the discussed approaches.

Todd Martinez: I would try to point out what I was talking about first. If these kinds of techniques lead you to understand better what the fundamental structure is behind e.g. protein structure, the electronic wavefunction or whatever that leads to a new paradigm, that can itself lead to understanding of new things and novelties. So, in other words, the construction that Alán (Aspuru-Guzik) was talking about, taking a large amount of data and then compactifying it to a simple model, is exactly what we have been doing in Science for the last 200 years or more. It is coming up with some simple model that then gives you ideas of things, of what to do next. And that's where I think the novelty is. The generation of these simplified models from the data is something that machine learning tools give you, a way that you did not have before. Where you did it before by sheer creativity or luck, now you have a tool to help you do it.

Alán Aspuru-Guzik: So Kurt, I published a paper in American Chemical Society with Roland Lindh and Markus Reiher called "(R)evolution" with six directions of theoretical chemistry that we think that will go on in the 21th century. One of the six challenges we post is this idea of interpretable AI for new discovery. We want AI to give us new directions in Science, new insights and of course maybe accelerate the production of molar prices, for example can we come up with something that took Suzuki forever, but can we come up with a reaction faster. So, I made the mistake (last challenge) citing Woodward–Hoffmann rules. Is it possible to rediscover something like the Woodward–Hoffmann rules? Roland Hoffmann wrote back to me and said that he was not so happy about this, that "I do not think your computer can discover something that needs our great insight". So, I took the challenge, and soon will publish a paper, attacking the "(R)evolution" paper, where we will come back, and actually show him, collecting places where AI already has come up with new insights in Science. One of the examples is quantum optics. One of the post docs in my lab Mario Krenn, has discovered new quantum optical setups, completely novel to generating entanglement by using AI, discovering new optical setups in quantum optics. So perhaps in physics or in biology things will come first and later chemistry. But definitely there are new concepts coming out of AI already in Science and we are collecting them to respond to that article of Roland to show that it is possible to actually come up with new things and they are amazingly worthy, not yet, but they will come. So, I predict that new reactions will come out of this as Bartosz (Grzybowski) already found and new things will be discovered in chemistry but right now we are learning the technology. So, give it some time, it is a new tool.

Eugene Shakhnovich: I think this scientific discussion is important. However, I have been somewhat sceptical throughout my scientific carrier about this application of machine learning, although due respect for its efficiency, but the question is how will it facilitate real discoveries? To me probably I am a little snobbish on that, because for me a discovery of a new reaction is not a scientific discovery, but it is a kind of a huge cultural or a huge contribution to our wellbeing in that sense. Because this reaction can be important for many things, but it is still somewhat empirical. It is something a physicist will not appreciate. It is not an *ab initio* thing which you can understand from first principle. It is still state of the art. To draw an analogy with having Rembrandt painting a picture and have a computer doing a similar job but still it is state of the art, it is art. What is still an open question, where I agree with Todd (Martinez) that generating simple intuitive models and generalizing data to build our intuiting in a greater power then our brains can do is probably a very important thing. Let me give an example which is close to my heart and probably also Kurt can sympathies with that. For a certain sense the protein folding problem has been solved ans recent AI approaches have contributed a lot to this. Because, now there are complex neural networks that can certainly out do any human algorithms in terms of how to predict the structure from the sequence. So, in this sense the problem was solved, or partly solved. For sure not for every protein. Did it aid in our understanding of how proteins folds, what makes them foldable, what kind of effects there are on evolutions of proteins sequences, etc.? Many things which I think are really exciting in protein sciences. The answer is yes and no. In several aspects of, for example, sequence signatures, they are quite important but there are quite a bit of generalizations of things of which people have already known about hydrophobic effect and things like this. We will talk about it in another section. On the other hand, I think the real kind of disconnect between this technological advance of predicting complex structures from say sequences and intellectual advances in understanding of the mechanistic aspects of the process, still exists. Though I think that we are all optimistic that this disconnect will be closed somehow by making more examples, etc. But what I really want to say, to conclude my contribution to this discussion, is that it is not a trivial problem. It is not because we can do something better, faster, etc. that we understand it from the greater depths. I think the challenges in the future is that these computation algorithms will be so far ahead of our mental capacity as a scientist to build models to facilitate our understanding that we have to address it in a certain sense. I think that if we can discuss that, it will be very interesting at least to me.

Bartosz Grzybowski: To me it is a little bit depressing because I was raised in a culture where making a Nature product was called human-readable art. It was the pinnacle of chemical creativity, this 40-step synthesis of Nature products, you know. At least for myself, I am discovering this is completely amenable to algorithms, these things are a succession of algorithmic steps. So, when we talk about creativity, I do not want to take a stand here because on the one hand many of Nobel prizes were given for total syntheses and this was considered very creative. On the other hand, we have machines beating Nobel prize winner level of sciences. Is this creative or maybe is it just technology? I will let you be the judge. But in terms of creativity I will tell you one thing: this progression of retrosynthesis to discovery of new reactions and systems of reactions, cycles and all that. Wilhelm (Huck) is here and probably discusses this later. How many autocatalytic cycles has mankind discovered over the last 50 years? Probably 5, 10 so as machines can do this, and they have already proven they can. This an example were computers are going beyond what humans can do. And whether we call it creative or not, it is just combinatorics. This could mean chemistry is a simple science. It is just a succession of synthesis steps arranged into reactions that are then arranged into systems. It might be so. So, I do not know if the question of creativity is relevant here but it is still very useful for organic chemistry to find something to do else then total synthesis.

Kurt Wüthrich: I would like to mention that it struck me that in all of the presentations we heard about combining artificial intelligence approaches based on big data with, on the one hand, expert knowledge and, on the other hand, quantum mechanics or quantum chemistry. Do you have some additional comments on this point?

Alán Aspuru-Guzik: The mixture of quantum chemistry and AI? There is a convergence. There is something Todd Martinez mentioned. Quantum chemistry is formally unsolvable, but AI is making it practical. Eventually we will need quantum computers and many of us are working on that as well. So, hopefully one day on a Solvay conference we will be talking about quantum computers for chemistry. However, the problem is that direct simulations will always be complemented with statistical aspects. So, machine learning is a statistical science and direct simulation is a forward solution of the Schrödinger equation, they both complement each other. I do not know that this is the question. That is where I think the two fields stand. They have an interface but the challenges for the right simulation and statistical models are going to be complementary.

Todd Martinez: There is another thing to say here which is that one of the roles of machine learning should be that we do not spend a lot of time to make machines learn the things we already know. One of the reasons why quantum chemistry is naturally paired with artificial intelligence in the context of any chemical application is that quantum chemistry gives you some encapsulation of the physics you already do know about chemistry. There are actually two aspects. The first is that you should take advantage of that. The second aspect is to try to leverage that in some way. You could imagine what Alán (Aspuru-Guzik) talked about: you try to go from molecular coordinates and atom types in rare elements into energies directly. That is just a function, a huge class of functions that go from the 3N-coordinates of the atoms into a single scalar. Or you could imagine an approach from those coordinates into some representation of a Hamiltonian and from that into some energy that can be corrected in different ways. This last approach is what most people will now call "physics-based machine learning". It is really quite clearly to me the right way to go forward. Not only does it prevent the machine from learning a lot of things you already know, it also stages in the learning. In this way, a deep network for example can learn a little bit at a time rather than having to learn everything at once. Therefore, you need much less data in order to actually train it. This concept of terabits per day comes mostly out of the traditional Silicon Valley I am immersed in. The Silicon Valley approach to machine learning violates the rule I said at the beginning. They do not care if the machine learns things they already know. They actually view it as a badge of honor that the machine learns everything they already knew. Somehow, they think that is better. So, if you put any information in, about the simple fact that negative numbers are less then positive numbers, they say that it is extra data, the machine should learn that. There is a purist machine learning approach which actually is in contrast to what I have said.

Eugene Shakhnovich: I think that all this many excesses of machine learning are intrinsically related to the fact that the system is being learned intrinsically simple. There is a certain way to project this complicity into a limited number of excess for example a Hamilton that should be a continued function and should be such and such, so that you can from raw space of data points learn the Hamilton. Imagine you have a mapping which is completely random. Then the machine learning algorithm will be predictive. Then you have some intrinsic simplicity. The machine can of course learn that all numbers should be positive or that all functions should be differentiable, etc. In this sense machine learning is a way to discover. Or the discovery of intrinsic kinds of degeneracies and simplicities in your system make machine learning possible. Again, if it is an NP-complete problem it is impossible to learn it because it will not converge to any generalization of that. I think we can probably use machine learning to learn about intrinsic simplicity and redundancies of the models that we are working with and also similar things about biological evolution. However, we do not have the time to go over these kinds of examples. But indeed, there are certain examples where evolution is rationalizable backwards. There are intrinsic simplicities that are intrinsically rational into a complete stochastic set of mutations which arise from some pressure. How reactive as it happens in reactions, as reactive it is in biological systems, there are proactive ways, some kind of learning the simple rules and then teaching the algorithm to reproduce them.

Bartosz Grzybowski: On the quantum chemistry front I actually do not know. AI could even be better than quantum mechanics. Let me mention two examples. Someone mentioned the Woodward–Hoffmann rules. These are heuristic, these were guessed. They do not work in most complex cases, like complex molecules. There is a solid piece of evidence that for pericyclic reactions like Diels-Alder, the best quantum mechanical methods can predict regioselectivity in about 80% of the cases. AI can predict regioselectivity in about 95% of the cases. So, I actually do not know. This might be the question Dirac asked: that quantum chemistry, quantum mechanics tells us everything, but we do not understand its consequences and we cannot interpret the results. So, I do not know, it is heuristic in many ways. We are doing the Wittgenstein experiment. We are observing a game. We are observing examples and learning from it. But the alpha goal this tells you, this is as good as any low-level quantum rule that we at some point have stopped understanding. In chemistry proper, I am still waiting to see the first example that quantum chemistry that tells us everything produces the first example of a rationally discovered reaction. Have you ever seen this?

Todd Martinez: That has already been done Bartosz. The question should be to go do experiments to prove those reactions actually exist. But there are already predictions.

Bartosz Grzybowski: Predictions like every time we read an organic chemical paper. There is a new reaction somewhere, there is a Figure 5 that justify the HOMO-LUMO, but why is it not referenced as Figure 1? I make a very bold claim that there are not such examples, and we might disagree, but without validation quantum chemistry can propose whatever. It had 70 years to show us how powerful it is. So when Alán (Aspuru-Guzik) was showing us that in 5 years AI can discover the best battery material, I think we should have a little bit of doubt if it is just AI or it is just playing with data. Otherwise we need to reshape our thinking of scientific problems.

Kurt Wüthrich: We are ready for coffee. We will reconvene at 11:25 for the general discussion on the subject.

General discussion

Ben Feringa: Thank you very much for your contributions. I am impressed in what the field already can do. So, this is about logics and using these models and calculations. I do not know about other labs but in my lab a lot of discoveries are made by mistakes, serendipity. So, in general I have a question: is there a possibility that you can say something about allowing us to make mistakes using these machine learning, this artificial intelligence, etc. Do you calculate this in? Because often mistakes lead to the most marvelous discoveries, new insights, etc. I would like to hear your comments on this.

Eugene Shakhnovich: I would like to comment on that. I think, at least in our case, indeed these machine learning approaches generate a lot of mistakes that are extremely useful. During our coffee break I talked with several people. For example, Thomas had a good question about how many false negatives and/or false positives are generated? Then, at least for molecular design purposes, false negatives exceed false positives. The question is that we learn them. Then we can generate iterative databases of very similar molecular structures that classify these falses and trues. That allows the next round of learning to be more efficient. This iterative learning by mistake is really important in these set-up approaches, these mistakes are featured as a bug in this case.

Lee Cronin: I am going to answer Ben Feringa for looking for serendipity and creativity. I think we are mischaracterizing part of the discussion. I think Alán Aspuru-Guzik, Bartosz Grzybowski and other panel members discussed this quite well. If we do not consider strong AI, we are considering aggregating data and coming up with new insights of data we already have. He has an interesting point: collecting high quality data is really the bottleneck in discovery science in chemistry. Bartosz said very nicely that we do not trust reaction databases. So, one thing were I am very excited about is that you need to start making self-driving labs, if you like, that collect data in real time; NMR-data, MS-data, what is happening when you are chucking the reagents together into a flask, the integrating of the data collection, what really is going on in your experiment... And the way you analyze that, your predictive power. Then you can do not weight extrapolations but a little bit, because you can look into the uncertainty of your experiment. As a discovery scientist, which I would like to be one day, I would like to mix-up stuff in my pot and work out which I guess that are going to be most reactive. This is there is the mischaracterizing insome of the earlier discussion. You can capture serendipity in the system. We have an AI that did that, which discovered new molecules. And you can anticipate what is going on. What Bartosz (Grzybowski) is feeling is that organic chemists have learned to play chess very well. But they are doing what most chess players do very well. They learn all the opening moves. As molecules are being more complicated, we are still making relatively easy small molecule. But when we are coming to molecules that have a molecular weight of a couple of thousands, hundreds of steps, each reaction, each transformation is going to be individual decision how to make that work. There will be no generic. That is, I think, where we should start to push things. The answer to Bartosz, I do not know if he agrees, is that the organic chemist should be pushed to think bigger, make more complex molecules and use AI as a way to get there more quickly in the laboratory. Because, then you could save a graduate student 3 years if you could cut out 10 steps. My question to the panel and back to Ben Feringa, how can we use AI to maximize serendipity and labor cost?

Bartosz Grzybowski: So Lee spoke about my feelings, so I have to answer this. In an ideal world what Ben Feringa said about serendipity: if we had a complete mechanistic understanding of an organic reaction, how it happens, there would not be serendipity. Because, I believe, on this level it should be completely deterministic. But I agree that our knowledge is incomplete. It would be very interesting to map the space of chemical reactivity with which chemical groups are still poorly understood. Which types of chemistries are more prone to serendipities? I do not expect too much serendipities in Diels-Alder or Cope rearrangement, it is beaten to death or at least a little bit. But there are all kinds of multicomponent reactions which we do not know. If you change a little bit the ratio of the starting materials of the substrates, you get a different product. Maybe our effort in this should not just be bigger molecules, but unexplored places in chemistry: new mechanisms, new reaction types. This is where all the serendipity by definition almost is.

Alán Aspuru-Guzik: Can I add to that a little bit? As I was saying briefly to Kurt Wüthrich, there is an entire field of AI: curiosity-driven algorithms. Those algorithms for example play videogames and try to solve the videogame in different ways every time by frostbite learning. They are learning creative ways of solving those problems. One of the things I want to say to chemists is that they should send their students or go themselves to AI conferences and become friends with AI professors, AI scientists. Because, they are working on very cool algorithms, there is a lot of work on the interface, we are just putting them into chemistry. It is a big fruitful area. I think we can bring them back challenges of what is interesting for us and I think that AI is going to surprise us how well it is going to perform on doing this.

Laura Gagliardi: I want to ask a question. My perception of organic chemistry, and I am not an organic chemist, is that serendipity is also the exception along the way where you have to figure out the unique condition your reaction can occur. How do those AI discover these exceptions/situations? Do you think it is like playing a videogame?

Lee Cronin: Can I just answer that? It is about having the right analysis in there. Alán Aspuru-Guzik's example is very good. The game has been written and that game runs on "Silicon", a program that runs the game, that well posed but big problem. One of the things we have in organic synthesis is what Bartosz Grzybowski, and again I will not speak for him, was talking about is that organic synthesis in some parts (Cope rearrangement, Diels-Alder, some couplings) are well defined. However, if we go into the space where things are ill defined, then you have the chance to make a discovery in almost all steps. In general, if you say to an organic chemist go and discover a molecule, they will divide by zero. Because there is no "let's discover a molecule". If you then write down a target, they will generate roots to that target and have accidents on the way and make discoveries. The best way to do that, to maybe answer your question of the AI, is give it a target. It tries to do it in real space. So, you actually dream a molecule. You then try to make that molecule with a robot. And then you have an in-line assay to workout where things go wrong. And then you bring in the human. Because one thing we have to say, and maybe it is to simplistic for the generational gap, is that AI is not going to replace any chemist. AI is going to expand the number of molecules known dramatically and that means more chemists, more jobs, more money. So, I think having this close loop is where you will find the most serendipity.

Kurt Wüthrich: Well, I will take the risk to simplify things. There are statistics showing that chemists and physicists come up with novelties and big discoveries at an early age, when they make mistakes. And then they learn from these mistakes and that leads to breakthroughs. Now, how can artificial intelligence bring in that element of the student who makes mistakes, who reacts to these mistakes and discusses them in the group. And then learns and makes the next step and the next mistake, and so on. If you work with rats or mice in behavioral science, you can use the animals only once because then they have learned and will not make the same mistakes as before. So, in some way we are dealing here with a similar situation when talking about the age at which we perform innovative research.

Dean Astumian: It seems that one of the big differences between say playing games and chemical synthesis is in order to benefit from mistakes you actually have to make the molecule. Whereas in chess, AlphaZero learned to beat the world champion by playing two billion games and it did not actually have to move the pieces. Which would have been impossible in the time scales we are talking about. But with benefiting from chemical computers you actually, at some point, have to go into the laboratory and mix the stuff up and make the molecule *in vitro*. And see if it is in fact what was predicted or if it was a mistake and you get something really new.

Bartosz Grzybowski: So, I will try to connect the things of making mistakes and your comment Dean. To give you a timeline of one of these Nature products just for validation. Because, I am not interested in this particular molecule. My students are interested in validating the computer's prediction. We are talking about 1.5 year for three people, this is extremely expensive. What Ben Feringa asked about serendipity and discover, in an ideal world I would like to have it like mathematics with no mistakes and no probabilities, just deterministic. I go to the lab, I cook it and I have it. It would cut a year of my student's efforts. There is always a trade-off between trying to get something unexpected versus the desire to get all these programs to the level that they make no mistakes. Synthesis is very unforgiving to mistakes. If you have a ten-step synthesis, if any of them is wrong the entire sequences do not work. If we could minimize the errors and false predictions this would be great.

Joanna Aizenberg: I want to move the discussion to another domain, away from organic synthesis at this point. And ask questions what are the perspective of machine learning or artificial intelligence in inorganic material synthesis, in particular taking into account that inorganic materials, in most cases, rely on crystallinity, rely on order. However defects, domain boundaries and things that are natural outcomes of a synthesis would actually determine the performance. What is it that we can talk about, in terms of, using machine intelligence to design or improve our ways to make materials that, by definition, will have defects? And how could we control defect engineering?

Alán Aspuru-Guzik: There is work, by Elsa Olivetti in MIT, on natural language processing for all inorganic synthesis procedures. To try to understand when people are cooking an inorganic material, how do they reach these different phases. The difference, I think, between inorganic synthesis and organic synthesis is that inorganic synthesis is, as you say, defect driven, it is also very process driven. Depending how you anneal the material, how you reach it, you will end up with different polymorphs or amorphous phases, etc. What she did, what was very creative I think, was to go to the past corpus of all anecdotal evidence of the humans and use machine learning to predict what will differentiate these conditions. Kristin Persson and Gerbrand Ceder just published another paper where they did that. They learned even from the abstract from the papers, they tried to predict new synthesis for new materials just from the abstracts. There is a lot of conceptual learning that could happen. That shows you perfect inorganic crystals without the melting color. But why not adding the melting color with more information for say of surfaces or materials in boundaries or so. It requires more data, more thinking and more domain knowledge of the inorganic chemist and combination of computer scientist. I do not think it is impossible to get there. Those two examples are mentioned.

Joanna Aizenberg: Would you say it is not as ready as it is ready for organic materials?

Bartosz Grzybowski: So Joanna, I know when we were in graduate school, or you were a post-doc and I was a graduate student, we did with George Whitesides crystal engineering. It came to nothing, the predicted polymorphs; because it all depends on these very weak interactions. I think, especially in organometallic chemistry the moment I understand these electron back donations, these weird bonds like π^*-orbitals to d-orbitals, this is very hard to codify, these rules are very malleable. In organic chemistry you do have a C-C bond, etc. either a single or a double bond. In organometallic chemistry, to codify this in machine language, you need to think about the strengths of the interaction and these interactions with many small interactions. I think it is actually very very challenging. Matt Sigman is doing this stuff, doing catalyst predictions. Based on the premise that similar catalysts and similar ligands will have similar properties. But this is some sort of QSAR but is much more challenging then organic synthesis.

Lee Cronin: Let me just answer this quickly. We have an assistant who does that. The thing is, you do not do the computation *in silicon*. What you do is, you do the computation, and I will talk about this a little today, in the material. What you use deep learning to do is to choose the reagents. So, what happens is that the robot chooses the reagents, or the algorithm does, the reagents get mixed up, and you do the chemistry. And then you have a high throughput system that measures the physical properties. You have the defects there, and you measure the band-gap, you measure the lattice potential,.... Then that data is put into a database, and that updates your algorithm, and you keep going around. The answer I am trying to give you here is that to do this properly, you need a closed-loop system that combines thinking, making, and detecting. It is very difficult, as the panel just said, to *ab initio* put in defects. How do you do that? Just make the material. Why compute the defects when you can make and measure them? And having them in the loop and seeing the process of the synthesis as the computation. As long as you can then decode it and use that to discover new materials, this is the way I think we should go. This is called hybrid machine intelligence for synthesis. But you have to have a closed-loop. It is only just happening.

Mark Ellisman: I am a bit shy and a bit inarticulate. So I hope you will bear with me while I ask a question and then make a comment. The question concerns the talk given about synthesis of natural products and applying deep learning to find a better way. This meeting has some session on water. I wonder if in any of the natural product works you are considering, this current view that cytoplasm, even in plants, is rather organized, and free water is more limited. So, some of the reactions that you might end up favoring might be better simulated in the context of an environment where you modulated water with PEG or something like that, whatever works *in vitro*. The second is a comment and a caution. It has to do with concerns about the maintenance of the so-called scientific method. Which is that when one publishes one's work, you want to be sufficiently disclosed, so that it is replicable. I am worried a little bit coming from computational science for decades that, as we increase the complexity with which we make predictions, simulations, or scientific work based on computing, that we are not rigorous enough with regard to the provenance required. Particularly in AI, so that the work can be precisely replicated. So that is a comment, and I hope it becomes a thread through the meeting. But the first one is the real question.

Bartosz Grzybowski: I will answer your question regarding natural products. Chemists synthesize natural products in completely different solvents, not in water. I understand that cells make it in certain conditions with enzymes and all that, and Eugene Shakhnovich could say much more about this. Instead, chemists operate in dichloromethane, toluene, and all these kinds of stuff, so the problem does not exist. Now in terms of integrity and data. In natural product synthesis, all of this field historically has been based on faith. Because someone reports that they made a natural product in x% of yield and provide the yield of individual steps and the spectra, but there is no proof of this. I think, in computer science and AI it is actually much better because journals now require that you disclose the code. So, you show the gut of the program. For myself, I will just tell you that the program I just showed you there is actually used by many people because it is brokered by Sigma Aldrich, so it is supposed to be global reach. You are absolutely right that the validation is super important. But compare to chemistry and synthetic chemistry. When x from a famous university in America publishes the synthesis of a natural product, the only thing I can do is to have faith that what they report is true. Because I cannot request a sample of this material, I cannot check the procedures, it is only what they describe. There was a recent paper in 2018 when a group from the US published a paper in which they claimed that their reaction worked on the thousandth try. They tried one thousand times to carry out this reaction, and they claimed that on the thousandth attempt it worked. So how reproducible this would be? And it appeared in a top-level journal. So, be the judge.

Eugene Shakhnovich: Let me jump in on this issue about natural products. I think one very interesting task for organic design is to include enzymes into the whole story. Because, what we are trying to do and others as well, is to try to use enzymes for particular challenging steps. And probably a fusion of what Bartosz and others are doing is some sort of thing which comes naturally from biology. Because it is, of course, not surprising that biology and cells invented enzymes. I think we are making ourselves a big disservice by not including enzymes in these steps. Given that we have a huge database of enzymatic reactions and cells, we probably can start using some of these algorithms to learn particular specifics of reactions and to see what kinds of enzymes can be optimized for that. This is a co-optimization of reactions and enzymes. There are significant successes in protein design in the last ten years or so. Which is something of great interest and which will help us practically, in terms of understanding how biology works to create natural products.

Kurt Wüthrich: About 25 years ago, Schultz and Lerner were awarded the Wolf Prize for the preparation of enzymatically active antibodies. This hype has largely disappeared. What struck me at the time was that organic synthetic reactions that can otherwise only be performed in dry organic solvents, could by these enzymatically active antibodies be performed in aqueous media. Maybe that has some bearing on your comment.

Bartosz Grzybowski: Very quickly. This would be great, but the largest database of enzymatic reactions, for example, BRENDA has about 6 to 7 thousand examples. To compare with, the Reaxys has 40 million organic reactions. It would be great to have more data about how to use enzymes. What databases like BRENDA give you is 'ok this enzyme can do this particular molecule and maybe three other molecules', but you do not have any information about the scope of these reactions. How do you get to the level of data that we have been collecting in Chemistry since the French revolution, actually because the Reaxys database goes back to the French revolution, is unclear to me.

Thomas Cech: We heard many impressive examples of either new products or new synthetic pathways that were achieved through AI or machine learning. But I did not hear much discussion. You always show a validated spectacular example at the end, but out of the final list of compounds or pathways that were derived from the AI, what percentage of those were, in fact, successful and how many did you not show us because they were unsuccessful?

Bartosz Grzybowski: This I guess is a question directed to me. I have a lab of four organic synthetic students in Warsaw. So far, they cooked 12 pathways. The longest one was 16 steps, the shortest one was 4 steps. So far so good, all of them work.

Alán Aspuru-Guzik: We design a material with AI and make and measure it. For example, flow batteries or a better example perhaps is organic light-emitting diodes. We predicted 40 top-performing organic light-emitting diodes, out of which 10% of them, 3 or 4 were top performers. Then we learned, as the comment made before, from the errors from the others to improve our model. Roughly in organic material design, I would say the hit rate is about 10%. This is also roughly what happened to us in the drug candidates that I mentioned before.

Eugene Shakhnovich: We have very similar statistics of about 10% in designs which are supposed to be potent for the target. The complete inactive would be 10%, but there was a range of compounds which were active against the target, meaning that there was some binding affinity registered, but way below than predicted, like 100-fold below than predicted. But again, we use this to build up the negative database and try to find features which discriminate the right ones from the wrong ones. We did a lot of that in the company I have been associated with pharmaceuticals in somewhat in the earlier steps that helped a lot in improving models. There was more synthetic capacity, these chemists generated 100 molecules which were extremely helpful to improve the models.

Bartosz Grzybowski: Regarding the figures mentioned by Eugene, I would like to add that they have maybe a slightly more difficult task because it is predicting drug ability or properties as a flow battery material. In organic synthesis, it is not surprising that the hit rate is 100%, assuming that every step is correct. Then the pathway should be correct, or we did something grossly wrong. So, I think, they have a task that is a little more challenging than ours. Because in our case if the induvial step is correct, the whole process will be correct.

Laura Gagliardi: I would like to go back to the inorganic materials and the debate of the defects. Lee and you gave a very simplified description of how to deal with these defects. For example, metal-organic frameworks are probably very complex because they have both organic and inorganic components. They also depend on the conditions at which you should do your experiment, the material deteriorates with humidity, and so on. There are so many unknowns. Sometimes the experimentalist cannot provide the theoretician even definitive structural data. So how can one deal with this?

Lee Cronin: Just answering the question. We have a robot that makes models for fun. In fact it is a random number generator. So, all it does, it takes structures and causes, add them together (Advanced Normalization Tools), and the more disorder you have, either you try to make more isomorphous materials at the same time but code the same crystal structure, the quicker they crystallize, or the probability function for crystallization changes and you make random numbers. So one of the things we are trying to do is to precisely understand the process conditions that give rise to crystallization. Fortunately, making models is actually really simple. Everyone just cooks them up pretty much in an autoclave. Actually, controlling that and doing that design of experiments is pretty simple. What I think a lot people do, is, after the cooking, looking at what is crystallized but do not look at what is in the mother liquor. What you must do is look at the mother liquor as well. We have a system where we are passing the mother liquor to NMR and MS-spec and look at how the ligands have changed and have been depleted and then update the database. We start to accumulate all the information. Models generate a great kind of industry in prediction of regular crystal structures, of course. It is great in applying graph theory and topology and alike. But there is a disconnect with putting in, as Joanna said, defects and getting other properties. The answer is really: 1-Record your process premises in a standard format that is machine-readable and human-readable; 2-Do multiple assays, not just crystallography (we have a little robot that puts the crystals in a tube with a rotating anode, and it gets the crystal structure in 5 minutes per crystal now). But then we must look what is in the mother liquor. Because then you find out what is decomposing, what else is going on. And that allows you, that dark network of unknown reactions allows you to make more discoveries.

Ben Feringa: Just adding to this. I hear a lot about designing structures, computational design, the needed structure of molecules, structures of materials. I fully agree with Joanna would it not be that we should start from a different perspective to design functions. What is the function that we need instead of what structure we need? What will then be the determining factors for the function, being defects, step-edges or whatever? The same, and that is my real question, and that goes back to the previous discussion: What is a molecule? Do we really understand what molecules are? Maybe Stefan Hell does because he looks at single molecules, and he knows what a single molecule is. But I am struggling with it. So, in our models and what we calculate and also in your transformations, do we really understand if our classic way of looking at a molecule, as an organic chemist looks at balls and sticks, is still valid? Or do we have to look at molecules in the situation where the molecule had a function, a property, or whatever as a part of a community of molecules? I would like to have your opinion about this.

Alán Aspuru-Guzik: The first thing is that all the inverse-design engineering models that we mentioned are for functions: Which is the structure that gives you the function? Your question sounds more philosophical, but from the machine learning perspective, I also mention that the molecular representation is an open problem. Trying to understand what is a molecule for a computer is interesting because depending on the features that you expose to the computer algorithm, the computer algorithm will learn in a different way. So, for a machine learning perspective, a molecule can be as simply as a string. We recently invented a new format called selfies, that replaces the smiles, that represents a molecule as a string. You can go from that simple representation all the way down to a three-dimensional wavefunction. All those things we call a molecule, we call different conformers or pulses of a molecule. So, it is a complicated question, but for machine learning it is crucial. Because how we give to the machine learning the representation of a molecule will represent the accuracy.

Eugene Shakhnovich: Let me quickly add. You know, these designs for which I told about, this drug by chemical design? The design is for the chemical function. The simplest function would be binding to a specific target. But, for example, recently several people have done design for the switching function. So, one way to look at compounds is to make isomers which can be photoswitchable, one active and one inactive. That combines quantum chemistry, of course excited state switching, and that requires a lot of machine learning because for that you cannot, for all the reasons mentioned by others, do straightforward quantum chemistry. So, I would say that function design is at the core of many designs, and I agree that it is really crucial to design for functions because, in most cases, this is what is going on. However, if we progress further, more and more complex functions will be designed. But binding to a particular protein, at least in my perspective, will definitely be a function.

Joanna Aizenberg: I want to continue that exact point. A molecule is wonderful, but if we talk about functions it is never one molecule alone, it is not only a collection of molecules but how they interact with their environment. There are probably ways to do it, on a molecule in the medium, on the molecule in the environment that definitely will affect its function, definitely will affect its properties. How one could integrate the molecule within the space, confined space, solvents, or whatever would affect its properties.

Bartosz Grzybowski: So, what we have seen here is part of our efforts also just to automate what used to be the focus on the structure. In the 20th century, people were making natural products to illustrate their methodology or, in some cases, to show they were better than their colleagues. I fully agree that the 21st century should be about the function, and the function should be beyond the molecule. If we take functions like self-replication, molecules do not self-replicate, systems self-replicate. DNA needs other proteins to replicate, and so on, maybe with the exception of some reactions. But in predicting the functional materials, it would be nice to ensure that all the components are synthesizable. So, one is the prerequisite of the other. There are methods to make sure that the components of the bigger thing that you are making can be made. There was a lot of effort in computational drug design when all was fine and dandy until people said that it is a great ligand, but I cannot make it, or I can make it in 25 steps. I think there is synergy between the two: the function-oriented and the structure-oriented, and the structure should enable the function.

Veronique van Speybroeck: So, I would also want to add on the comment of Joanna, and also on the issue of the environment. Indeed, if we talk about the function of a molecule or of a material, it is also important to evaluate it under operation conditions. That is mainly my question. Operation conditions could be temperature, the presence of gas molecules, let's say for an inorganic material we could have gas molecules in the pores. How do we incorporate all these complexities in these machine learning approaches? Can we progress there to include these operation conditions in relation to the function?

Alán Aspuru-Guzik: Maybe we can provide an example that connects all of the last three questions. Because we designed organic light-emitting diodes and needed to overcome all those problems. First of all, we needed to find a property that we were looking at the molecular level. It was temperature assisted delayed fluoresce; it is an optical property of the molecule's energy levels. Then eventually, it will have to be transferred to the environment which is a host molecule and a bunch of layers in a device that has to be tested in operating conditions. So, the way we did was the way that Todd Martinez was talking about, sequential AI. First of all, we figured out that the AI could predict the quantum properties of the molecule in the gas phase. Once we were able to do that, we used another very simple AI model to transfer the experiments in the bulk, in other words, in the host material. For example, there is a red-shift of the lines or broadening of the lines, that we had to learn. Then, when we predicted that, we proceeded in making some devices that actually literally ran with electrical current and emitting light. Getting the light from them helped us to inform us of the process all the way to the molecular level. So, it is a funnel where you tried to have different levels of machine learning and experiments working together. So, it is not from the beginning all the way to the end that you will predict everything, but at each level, you have to use different tools, different things. But eventually it involves experiments, it involves devices and everything.

Todd Martinez: One way to think about that, which is maybe simpler than what Alán was talking about and people are striving at, is in the context of the nanore-actors where you find reactions and then you try to refine to get barrier heights in order to put them into transition state theory, to put them into kinetics. There is an assumption in all of that, which you probably all see, which is that if I pull a reaction out and then calculate the barrier, that is a good representative of the barrier in the environment. It is the natural place to start testing basically simple models, transition state theory, that we all have an idea about. You can start asking to what extent it is true. So, to what extent is the frequency of reactions that occur under some particular method equal to the rate that is predicted from transition state theory for isolated molecules. We started doing this, and the answer is that it depends, it depends on the environment. So, the conclusion that one comes to is actually what Alán said, that there is a sequence of models. So, there is a simple view that there is a molecule, and I can think about the molecule itself with certain properties belonging to it. And then there are refinements on that, if I put it in different environments it will change. Which goes back to this question of trying to learn how the machine learns the coarse-grained models. So, if the machine should learn that, you should provide that, and then it will learn that the barriers have some dependence on the environment, and it starts learning. You can see how this all can work, but it is definitely useful to start from the idea, from 100 years or more of chemistry, that a molecule is an independent thing, that its properties are largely dependent on itself, and then view the rest of it as something that needs to be folded in as barrier heights or reaction rates, reaction energies, modified by the environment.

Andreas Walther: I have a question regarding when you look at functions integrated into a higher system. You made the example of the OLED where you optimize the optical properties of the materials, but to make an OLED all the layers play a role, like the electron injection layer, and also the processing plays a role. Do you also optimize with machine learning the manufacturing and would you also look at different components of the system which could actually make this optical material to perform better?

Alán Aspuru-Guzik: In that project, we did not as in a project where you have automation, and this is the reason why Lee Cronin and I are working together on robotic chemistry. I have a paper where we built perhaps the first thin film self-driven lab. It is a lab that actually makes thin films with spin coating to exactly get at that. And we optimized the whole transport layer of all the materials. So, we are going in that direction. So, we have a robot that drops spin casts, measures optically, measures electricity,.... So, the idea is to integrate all that with the synthesis and then eventually close the loop in a laboratory system as Lee Cronin is addressing. So yes, we need to get there and to do machine learning as well.

Bartosz Grzybowski: On this, I like to be a little polemic. If you would like to automate 100%, the cost of automation is a nonlinear function of the percent of automation. So, to automate a little bit of your process it is cheap, but if you want to automate every single screw or maybe every single TLC or finding a system to run a column this is enormous. Even mister Elon Musk, 30% of his Teslas are suffering from this drive to automate everything. So, they need to go back, and a human being needs to fix it. But do we really need to automate 100% of it? Maybe streamlining 50% of it would be enough. So, if you ask me about organic synthesis. I do not think a robot will do all these multi-step syntheses because the robot would need to know such a spectrum of solvents, systems, conditions for running a column, etc. This would be God himself right? So, maybe it is enough to have a dedicated graduate student.

Raimund Ober: As an outsider, I would like to ask a question about the underlying mathematical models that you are using. If I understand correctly, there are two extremes: one is the deep learning models that do not look at any of the physical properties but use a general neural network-based model. The other one is that you have been discussing which is based on physical principles. And both of them seem to have their pros and cons. The deep learning model does not seem to take into account, at least to me, the knowledge that the chemistry field probably has acquired over dozens if not hundreds of years, at least not in mathematical terms. Whereas the physical model tries to do that but then has difficulties because some of the more advanced connections are very difficult to model in physical terms, which we of course struggle with, and for that we like the idea of deep learning. The question is ideally, as an outsider, I would think that a hybrid model would be the proper thing, where you take into account the kind of physical model that you are convinced that is correct, and supplement that with deep learning, with the flexible mathematical parametrization of the rest of the phenomena that you are uncertain about. I am just wondering if any of such attempts have been made in the field to try to come up with such a hybrid model?

Bartosz Grzybowski: Because the talks are only 10 minutes, the reactions rules are knowledge bases, physical-based. All of the other things like prediction of pKa, which proton is going to react, and all that is AI-based. So, it is a mixture. There is a drive right now, people have attempted to learn all of the chemistry just from data. Just go to a database set and learn it all. I think that is utopian. You are absolutely right that just physical models will never get to that level of sophistication. So, it has to be a mixture and apply AI to subproblems for which the data is clean and abundant enough. But the cleanness of the data is very important for a given subproblem.

Eugene Shakhnovich: I would like to add that it is especially important in biology because of the huge dimensionality of space and, at least in all that we try to do, we start from some low dimensional subset of variables that determine a particular biological phenomenon. Then, on that, we map the computational changes into the AI and make some predictions. In biology, it is crucially important to have this type of hybrid AI. I do not believe in *ab initio* biology at least at this stage of development.

Thomas Hermans: Maybe a question for Bartosz Grzybowski and maybe Lee Cronin. So, a trend in the chemical industry is to go to flow reactors to do things in continuous flow versus batch. Is there a way to incorporate that into your AI? Let's say the layout of your tubes or when you add A to B and under what conditions, can that be incorporated into your AI?

Bartosz Grzybowski: So, flow chemistry, we just made the program with RPy with our collogues at MIT. So, if you know how to have in one tube 10 different chemicals, none of which will precipitate, none of which will clog the tube, then you are golden. But how to predict it rather than with trials and errors? I know that there are papers showing 6-steps synthesis in the flow, but I also know how long of trials and errors it takes to predict it. So, surprisingly there is no general model predicting if a molecule in a solvent is going to be soluble or will precipitate. There, are no such models, with the exception of a few solvents. So, this is not a problem of AI per se, but lack of data. And I know that Novartis is trying to predict crystallization and all that stuff, but they are doing this for tens of years and with very little effect, to my understanding.

Lee Cronin: Let me add quickly that the problem with flow is that flow is not built for intensification. I think someone (Raimund Ober) said earlier about one of the great things we should be doing is trying to integrate hundreds of years of chemistry and all the information in there. By doing things in batch you can do that. So, I think that the problem is more towards automating the batch, and then when your need process intensification, you then basically go bach-semiflow. And then you can do AI on that, you know how the process behaves, you can get feedback control. I think the reason why people are saying that flow is adopted by big pharma is the MIT Nevada's collaboration. Because there are no current processes end-to-end that use flow. People want them to, and make contributions, and do great work but exactly for the reasons that Bartosz Grzybowski pointed out. So, I think it is about making sure that that data you collect in batch is usable for everyone else, operable and usable.

Thomas Hermans: On the last comment. We are working on technologies to achieving an effective 100% plug flow which would then translate directly to batch. Once that is obtained, you can fully make the switch.

Kurt Wüthrich: Well, my watch tells me that from now on we can only accommodate outstandingly intelligent and important questions. Who wants to comment?

Winfried Denk: Let me ask a dumb question then in the hope that serendipity will save me from that. What can chemistry do for AI? Let's say chemists understand what understanding is, could they help the AI community to understand understanding as such?

Bernd Hartke: I have a general question. I have collected terabytes of data per day for many days. I have fed all this big data into my deep neural network and I expect that I can get extrapolations for thousands of new points. As a second step, can I now go back to the deep neural network and ask it to generalize, to give me a real insight and not just point predictions?

Alán Aspuru-Guzik: There is an entire field of AI called explainable AI that aims to do that. So, for a lot of the questions in the field that people are asking, that the AI practitioners are asking, it is important to go to the state of the art of AI to know where they are and, as the previous question was eluding to, help them learn.

Bartosz Grzybowski: Quickly I would like to add what chemistry can do for AI and what journals can do for AI. Do you know what the average yield that people report in scientific publications is? So, the yield is how well did the reaction work. It is 80%. Nobody reports yield below 40% or something. This is a big lie, right? There is an initiative right now called the journal of negative results and my country is very much represented there, many Polish are publishing there, but that maybe is a question of national character. If chemists could thoughtfully say that something failed, then all of these networks would work better if they had negative examples. But then again, cultural issue.

Kurt Wüthrich: I see there are no further hands up. So, we can recess for lunch. Thank you all for your exciting contributions.

Session 2

Modeling of Functional Materials

THEORETICAL MODELLING OF FUNCTIONAL MATERIALS

BERT M. WECKHUYSEN

Debye Institute for Nanomaterials Science, Utrecht University,
3584 CG Utrecht, the Netherlands

Introduction

Functional materials represent a wide range of physical objects in which the structuring of matter at different length scales leads to various physicochemical properties, and ultimately also to one or more functions at the macroscopic scale. These objects are intrinsically multi-scale and multi-component, ideated by humankind to help them to do something, and which are usually produced in large-scale facilities so that many people can ultimately make use of them to enjoy life.

Examples of functional materials are batteries, catalysts, sensors, adsorbents, membranes, fuel cells, and coatings, as schematically shown in Fig. 1, but there are many more materials surrounding us in our daily lives. Critical for the development of current and future functional materials is that we can not only ideate them but also control and structure matter at the micro-, meso-, and macroscale in a very precise manner so that the composed matter is exactly doing what we want.

Hence, scientists wish to precisely know about the composition–structure–function relationships of functional materials, simultaneously in space (e.g., crossing the different length scales) as well as time (e.g., understanding how they age so we can make them in such a way that they last forever). This quest for rational materials design is certainly not a new question. However, it is more easily stated than practically realized. It requires that we are able to explain how functional materials work (i.e., explanatory materials design) and how we can predict their function and create new and better functional materials (i.e., predictive materials design). For both scientific approaches, we will have to develop experimental and theoretical methods, and most effectively these two sets of methods have to be combined in an integrated fashion so they can cross-fertilize each other.

The main aim of this introductory article is to describe some of the ongoing developments in the field of functional materials design, in which I wish to stress the importance of combining experiment with theory, and *vice versa*. Only when the brightest scientists from different scientific disciplines are really working together will we be able to foster the field so that rational materials design not remains a pipedream.

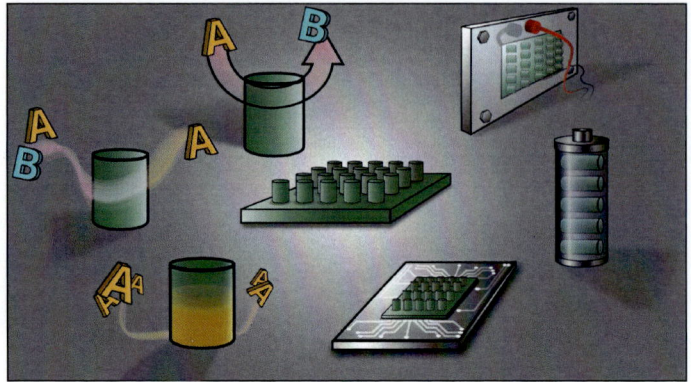

Fig. 1.   Examples of functional nanomaterials, including fuel cells, batteries, and solid catalysts, including the possibility of artificial intelligence (AI) and high-throughput (HT) design principles to accelerate, for example, the conversion of reactant A to reaction product B.

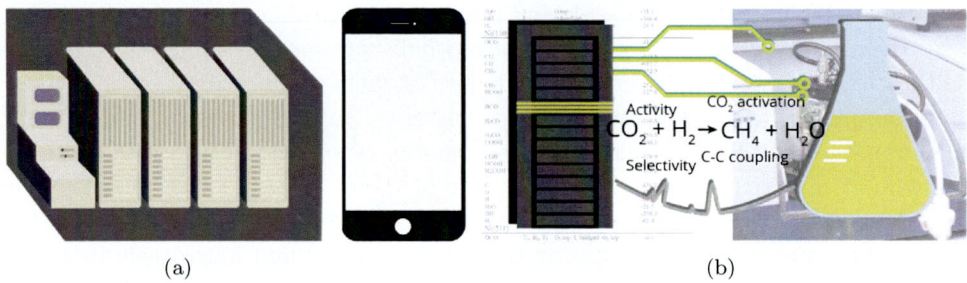

(a)   (b)

Fig. 2.   (a) Large hardware of the past to compute, e.g., functional materials, illustrating the increasing capabilities of mankind in modelling and computing complex systems; (middle) current smartphones have a memory that exceeds by far what one could do in the early 1950s. For example, the IBM Model 350 in 1956, a gigantic machine, could only store 4 MB, while an iPhone X nowadays takes many pictures of 3 MB in size. When someone has 10,000 pictures stored on his or her phone, it represents 30 GB. This GB information is stored in a device that now easily fits in someone's jacket pocket; and (b) showcase example, discussed in this proceedings article, namely, the Sabatier reaction in which CO_2 is hydrogenated into CH_4, thereby determining which reaction network is leading to the formation of methane, as well as if there are catalytic pathways to facilitate carbon–carbon coupling so that higher hydrocarbons could be made directly from CO_2.

As schematically illustrated in Fig. 2, thanks to the (r)evolution of computational power, many scientists now have access to (relatively) cheap and easily accessible hardware together to well-established software, such as VASP [1] or ADF [2]. As a result, we are currently able to compute many experimental data of molecules as well as more and more properties of functional materials. Examples include electronic and vibrational spectroscopic data. Indeed, almost everyone can now take a computer or any other hardware device with proper software to compute, e.g., an infrared spectrum of a molecule. However, before this is done, we have to geometry optimize the 3D structure of this molecule. For molecules, obviously, we are well aware about their molecular structure, but for functional materials, such as solid catalysts, this information is often not known or at best only partially known. The reason is that functional materials cannot (always) be described as a crystal with a 3D long-range structure, as, e.g., the surfaces of functional materials often contain a lot of imperfections, which are very important, as these imperfections or defect sites may even represent the active sites of the functional material under study. Furthermore, these imperfections are often present in minute amounts, making them also from an experimental point of view difficult to study. Hence, the clear need to combine theory and experiment.

Figure 3 illustrates the complexity of a functional material by taking a Ni/SiO_2 catalyst as a showcase. This material is able to catalyse the so-called Sabatier reaction, named after Nobel laureate Paul Sabatier, which is the hydrogenation of CO_2 into CH_4 [3–5]. The catalyst consists of Ni metal nanoparticles deposited in the pores of an amorphous high-surface-area SiO_2 support. These supported Ni metal nanoparticles can have different sizes and hence different Ni metal faces, namely, Ni(100), Ni(211), and Ni(111), which are exposed in different relative ratios to the

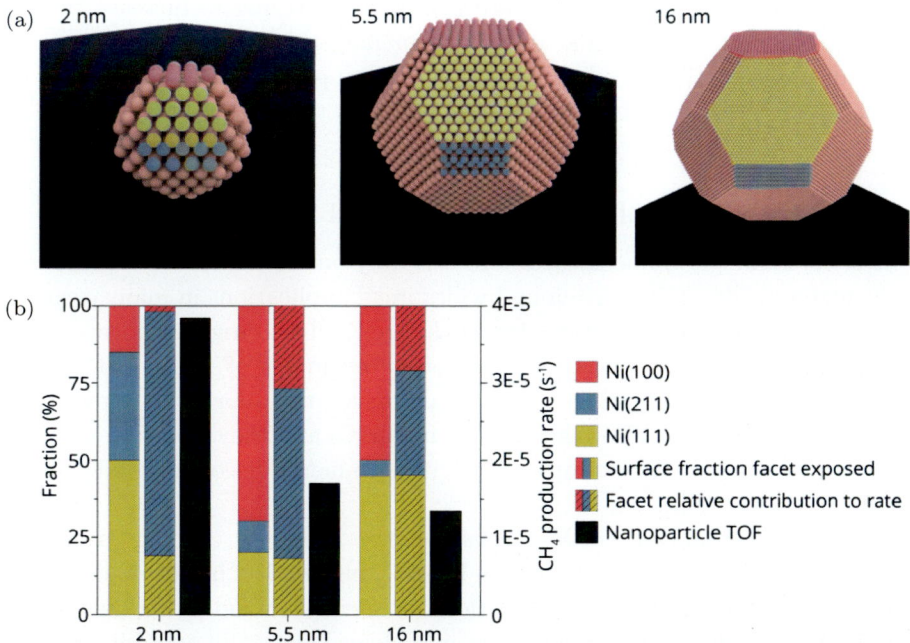

Fig. 3.   Illustration of the chemical and structural complexity of a solid catalyst. As a showcase, a Ni/SiO_2 material is discussed, which is able to catalyse the hydrogenation of CO_2 into CH_4. (a) This catalyst consists of Ni metal nanoparticles dispersed on a high-surface-area SiO_2 support (black) and depending on the size of these nanoparticles, different metal facets are relatively exposed to the outer surface. (b) Fraction of the different metal facets for the Ni metal nanoparticles of 2, 5.5, and 16 nm, as well as the turnover frequency (TOF) for the CO_2 methanation reaction. The graph also shows the relative fraction of surface facets exposed, as well as their contribution to the CO_2 methanation rate.

surrounding atmosphere. Figure 3 also shows supported Ni metal nanoparticles of 2, 5.5, and 16 nm, and their relative contribution of the Ni metal faces to the CO_2 methanation rate. Interestingly, full theoretical modelling of 2-nm Ni metal nanoparticles on a quantum level somewhat represents the limit of what is currently possible to compute given the high number of Ni metal atoms constituting this metal nanoparticle. Furthermore, theoretical modelling of metal–support interactions, or even the additions of additives, such as promoters and poisons, is far from trivial and often we have to put limitations on the size as well as on the chemical and structural complexity of the computer models to make the theoretical calculations practically feasible within the timespan of, e.g., a PhD work.

Summarizing this point, although we wish to make theoretical models as realistic as possible, current increased capabilities of calculations remain insufficient to cope with the full complexity and dynamics of functional nanomaterials. Hence, the models are still rudimentary and only reasonable approximations, enabling the determination of specific physical and chemical properties [6]. It is clear that the outcome of these (still limited) theoretical calculations is largely determined by the precision with which we can fully describe the functional material under study. One may then not wonder that we are continuously trying to increase the precision of the models of functional materials, making them, e.g., larger on the one hand, while on the other hand mimicking as best as possible the behaviour of these materials under true *operando* conditions [1, 7, 8].

Towards Computational Design of Functional Materials

Zeolites, crystalline aluminosilicates, are certainly one of the most versatile microporous materials [9]. They have found widespread applications in many areas, such as gas adsorption and separation, ion exchange, and heterogeneous catalysis. In other words, zeolites are one of the prototype examples of functional materials, which people have been tried to continuously make and tailor to their own wishes. Interestingly, nature can also make zeolite materials and they can, for example, be found near volcanoes. An example is stilbite, a mineral discovered and named by Axel Fredrik Cronstedt.

Since the inception of the Materials Genome approach, aiming to accelerate both materials discovery and innovation [10, 11], much effort has been directed into the computational search for new functional materials, including zeolites. Seminal examples include the work of the groups of Michael Treacy [12] and Jihong Xu [13], who have developed large databases for hypothetical zeolite framework structures. These databases consist of many millions of potential zeolites to be experimentally made. Surprisingly, there are only ∼230 different zeolite framework structures synthesized in the lab, indicating that many more zeolite-based materials could be explored for their potential catalytic, adsorption, and ion exchange properties [14]. In other words, there has been over the years a continuous search for a practical

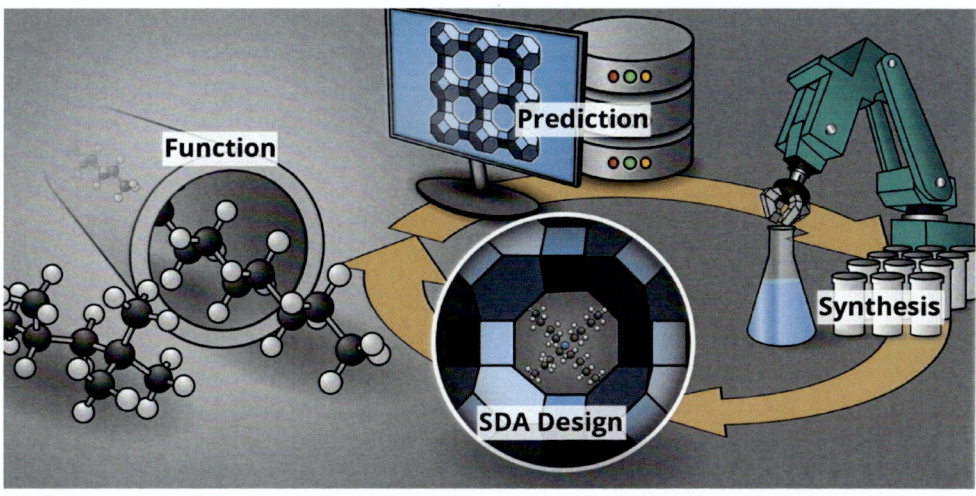

Fig. 4.   Layout of the combined experiment theory strategy to synthesize new zeolite materials based on large databases of hypothetical zeolite framework structures. By evaluating various structure-directing agent (SDA)–framework interactions, the most suitable SDA–structure combination can be selected and evaluated in the actual materials synthesis in the laboratory.

synthesis recipe for synthesizing one of these few million zeolite framework structures waiting to be discovered (see Fig. 4).

A very recent example of such effort is by the group of Bong Hong, who explored a library of diazolium-based cations as organic structure-directing agent (OSDAs) to synthesize a new zeolite framework structure, PST-30 [15]. PST-30 is a 2D zeolite with both 10-membered rings (MRs) and 8-MRs. To make this possible, they compared the stabilization energy of a series of OSDA cations in several preestablished hypothetical frameworks. More specifically, they found that by using one of the 32 candidate OSDAs, namely, 13DMP-C_4^{2+} (1,1'-(1,4-butanediyl)*bis*(2,5-dimethyl-1H-pyrazol-2-ium), under highly concentrated excess-fluoride conditions, it was possible to crystallize PST-30, as computationally predicted. Furthermore, careful materials characterization revealed that the 18-hedral cavity of PST-30 was occupied by one pyrazolium moiety of 13DMP-C_4^{2+}, as proposed in the computational screening analysis of the OSDA library explored. This elegant study demonstrates the usefulness of computational screening in the search for a suitable OSDA to direct the synthesis of pre-established zeolite framework structures.

Another interesting showcase is the synthesis of Boggsite with the framework structure BOG. BOG is a zeolite with a 3D channel system with intersecting 12-MRs and 10-MRs [16, 17]. It is a rare, naturally occurring zeolite, difficult to make in the laboratory. It was Avelino Corma and co-workers, who were the first to make Boggsite synthetically [18]. However, this was not done in the same chemical composition, as the synthetic variant did contain Si and B instead of Si and Al, although Al could be incorporated in the framework structure after ion exchange.

Here, again, by exploring a wide range of OSDAs, now based on phosphazene-derived organic cations, it was possible to synthesize BOG. By using molecular modelling the minimization of the energy of zeolite–OSDA systems was explored and potential OSDAs were used to synthesize the targeted zeolite framework.

Towards Computational Understanding of Materials Functionalities in Action

We not only wish to use computational methods to help materials scientists to make new or better functional materials but also to properly understand their function as function ultimately determines the performance of an engineered material. In the case of solid catalysts, it is very important to have control over three properties, namely, catalyst activity, stability, and selectivity. In particular, the final one seems not to be a trivial property to control, and requires that chemists understand the different reaction pathways a catalyst may facilitate. By doing so, they may find clues on how the selectivity towards a particular reaction product can be steered. To provide such insights, catalyst scientists use both theory and experiment. In the case of experiments, often *operando* spectroscopy and microscopy are used to determine (potential) reaction intermediates and how their presence is influenced by experimental variables, such as temperatures and pressures. Similarly, theory may provide energy barriers for the different reaction pathways. In what follows, we will illustrate this approach for the catalytic hydrogenation of CO_2 to CH_4 over Ni-based materials, a reaction which has been investigated in detail in our research group [3, 19, 20].

Experimental work by using, among other catalysts, a series of Ni/SiO_2 materials, varying in their Ni metal nanoparticle size in the range from 1 to 7 nm, shows that CO_2 methanation is a structure-sensitive reaction [19]. Structure sensitivity implies that catalytic activity is function of the precise structure of the metal nanoparticle. Hence, there is a dependency of the CO_2 TOF as a function of the Ni metal nanoparticle size, as illustrated in Fig. 3. *Operando* transmission FT-IR spectroscopy was used to relate the observed methanation activity to the presence and evolution of (surface) reaction intermediates. For this purpose, CO_2 methanation reaction was evaluated using different conditions, namely, $200°C$, $300°C$, $400°C$ and 1, 5, 10, and 20 bar. The highest measured TOF was found at $400°C$ at 20 bar with catalyst containing Ni metal nanoparticles with a size $\sim$2.5 nm. The relative amounts of the three main reaction intermediates, surface CO^*, gaseous CO, and surface $HCOO^*$ were found to be a function of the reaction conditions as well as of Ni metal nanoparticle size. As shown in Fig. 5, the formation of CO^* and $HCOO^*$ originates from two different reaction mechanisms, the direct CO dissociation (also denoted as the Carbide Pathway) and a hydrogen-assisted CO dissociation (also denoted as the Formate Pathway). The main product formed for each catalyst material is CH_4 with the formation of gaseous CO up to 10% (although it is also a function of the Ni particle size). Also, slight amounts of, e.g., ethane were found. From the conducted *operando* spectroscopy and activity measurements, we learned that CO_2 hydrogenation is a structure-sensitive reaction and at least two reaction mechanisms are actively taking place. The rate-determining step is most likely based on the ease with which CO* is hydrogenated and on the amount of available H* on adjacent Ni sites. The ratio of different products formed also depends on the Ni particle size, which means that product selectivity is also structure-sensitive.

In order to better understand these observations, an in-depth theoretical study was conducted [20]. Density functional theory (DFT) was performed of all possible reaction intermediates on four different Ni facets, namely, Ni(111), Ni(100), Ni(110), and Ni(211). In this way, sets of stable geometries of each reaction intermediate were obtained, which were used to study each elementary reaction step — by calculating the energy of its transition state — of the three potential reaction mechanisms (the already mentioned Carbide Pathway and Formate Pathway, as well as the Carboxylate Pathway) on the four Ni metal facets. Based on these calculations, the following observations could be made. First of all, the reason that surface CO* species were detected is that this intermediate is very stable on each Ni facet. However, a subsequent reaction step with CO* is much more difficult, due to the significantly higher energy barriers, as demonstrated in Fig. 5. Also, the rate-limiting step in the energetically most favourable reaction mechanism was found to be the hydrogenation of CO* towards HCO*. The formation of HCOO* by CO_2 hydrogenation is energetically feasible, but the conversion of HCOO* either back to CO_2^* or further to HCO* is energetically more demanding due to the higher energy barriers. Therefore, HCOO* formation will be relatively fast compared to any further reaction with this intermediate, which results in a longer lifetime of HCOO* on the Ni surface. Hence, it could be observed by *operando* FT-IR spectroscopy. Second, the formation of gaseous CO can be explained by the desorption of surface CO* from any Ni metal facet. Although adsorbed CO* was found to be very stable on each Ni facet, CO* desorption is entropically very favourable due to the increase in degrees of freedom. The reaction rate for CO* desorption was calculated using the Arrhenius equation with the CO* adsorption as activation barrier. It was found that under reaction conditions, it is indeed possible for surface CO* species to desorb.

As mentioned above, with *operando* FT-IR spectroscopy and on-line gas chromatography, e.g., ethane was detected. Based on DFT calculations, it was then found that both C* and CH* are the most stable carbon intermediates. Thus, it was assumed that there will be a relatively higher coverage of the C* and CH* surface species than the less stable CH_2^* and CH_3^* surface species. The energetics for the coupling reaction between C* + CH* was calculated on the four Ni facets. The most facile coupling reaction was found on Ni(111) with an activation barrier of 82 kJ/mol. The barriers on the other Ni facets were twice as high. Thus, ethane formation is most likely to take place on Ni(111). The potential formation of ethanol

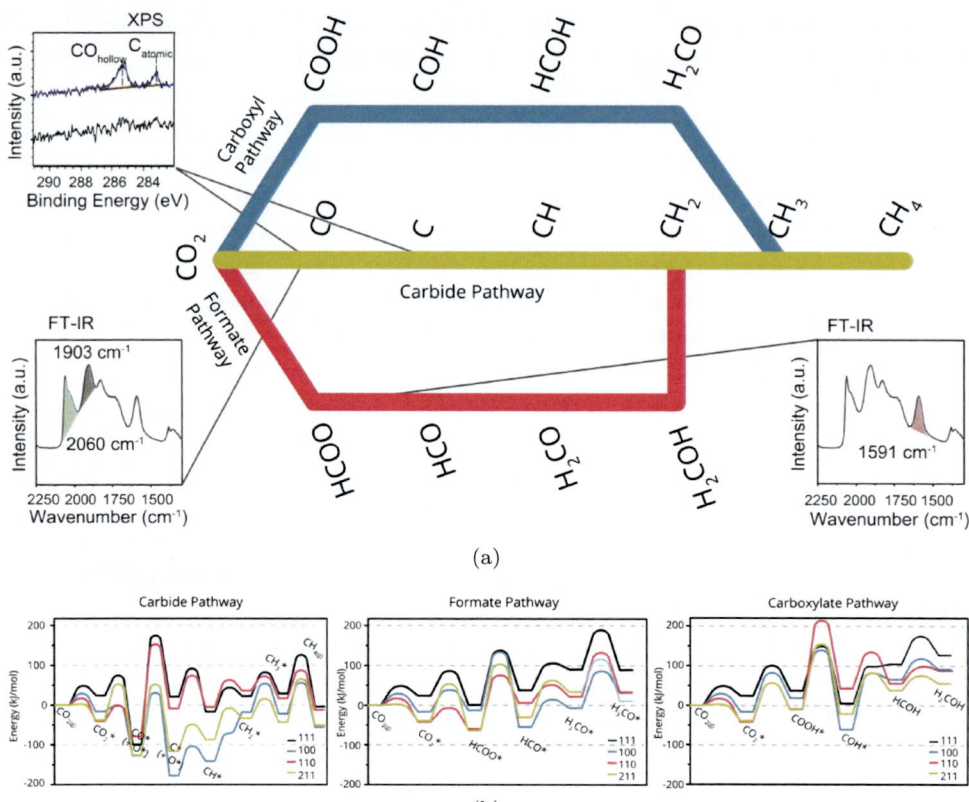

Fig. 5. (a) The three potential reaction pathways for catalytic CO_2 hydrogenation, known as the Sabatier reaction, for a Ni-based catalyst, including the experimentally observed surface reaction intermediates, namely, CO* and HCOO*, as observed with *operando* FT-IR spectroscopy, and CO* and C*, as observed with X-ray spectroscopy (XPS). (b) Energetics of the three pathways, i.e., Carbide Pathway, Formate Pathway, and Carboxylate Pathway, over different Ni metal facets (i.e., (111), (100), (110), and (211)), as determined by DFT calculations [20].

was also evaluated with the elementary reaction in which CO* inserts into C*. The CCO* formation was found to be endothermic. Also, the backward energy barrier was found to be significantly lower than the formation of CCO*. This makes it unlikely that ethanol is formed. Similarly, methanol was also not experimentally detected. The reason for this is that the formation of carboxylate-intermediates is less stable compared to the intermediates formed in the Carbide and Formate Pathways. Furthermore, methanol formation via hydrogenation in the Formate Pathway (i.e., $H_3CO^* + H^*$) is an endothermic process, therefore it does not occur on a Ni surface.

Based on this combined experiment and theory approach, it can be concluded that CO_2 hydrogenation over Ni likely proceeds via the Carbide Pathway with hydrogen-assisted CO* dissociation. CO* dissociation was found to be most facile via COH* on Ni(100), Ni(110), and Ni(211) facets and via HCO* on a Ni(111) facet. CO_2 hydrogenation was found to be energetically the least demanding over the Ni(110) facet with a rate-limiting step of 110 kJ/mol. However, a combination of the four Ni facets (= allowing the surface intermediates to migrate between every Ni facet) results in a mechanism with the overall lowest energy profile with a rate-limiting step of 99 kJ/mol. Finally, we have found that the adsorption of $CO_{2(g)}$, the very first step in CO_2 methanation, is already structure-sensitive. This elementary step is most facile on a Ni(211) facet (i.e., the stepped surface), because on this Ni facet, the reaction is exothermic and has the lowest energy barrier. The Ni(111) facet (i.e., the terrace) would be the least favourable for this surface reaction because each calculated CO_2^* was found to be endothermic. Also, the highest energy barrier for CO_2 adsorption was found on the Ni(111) facet.

This showcase example illustrates the clear win–win when theory and experiment are properly combined. However, it is also an example of elucidating structure–property relationships, while one would really like to move from descriptive to predictive functional materials science, where we explore property–structure relationships. It should also be clear that the discussed materials complexity of a Ni/SiO_2 catalyst only considers the different Ni metal facets, and not yet, e.g., the support oxide. It furthermore excludes the presence of additives, such as promoter elements (e.g., K), as well as alloying elements (e.g., Co or Cu). Clearly, theoretical modelling has to further develop so we can fully describe large, structurally and chemically complex systems, i.e., metal (alloy)–promoter–support systems, and under real(istic) reaction conditions, in which we incorporate both pressure and temperature, and in the case of liquid-phase catalytic reactions also solvent molecules. We are not yet there, but the path forward is clear and many developments are expected to take place in the upcoming decades. One of the main challenges is to couple the knowledge from one scale (i.e., time and length) to the other scale (e.g., micro-, meso-, and macroscale) and translate them into macroscopic properties, namely, activity, selectivity, and stability. A final remark has to be made here that the described experimental and theoretical developments for catalysis equally apply for research in, e.g., batteries and fuel cells.

Acknowledgments

Utrecht University (in the frame of the Strategic Theme Sustainability), the Netherlands Organization for Scientific Research (NWO, in the frame of the Gravitation program, MCEC, Multiscale Catalytic Energy Conversion), and the Advanced Research Center Chemical Building Blocks Consortium (ARC CBBC) are gratefully acknowledged for their financial support. I wish to thank Charlotte Vogt, Ellen Sterk, Katarina Stanciakova, Thomas Hartman, and Eelco Vogt, all from Utrecht University, for fruitful discussions and valuable input related to the writing of the article.

References

[1] C.R.A. Catlow, V. van Speybroeck, and R.A. van Santen, *Modelling and Simulation in the Science of Micro- and Mesoporous Materials* (Elsevier, Amsterdam, 2018).

[2] G. te Velde, F.M. Bickelhaupt, E.J. Baerends, C. Fonseca Guerra, S.J.A. van Gisbergen, J.G. Snijders, and T. Ziegler, *J. Comp. Chem.* **22**, 931 (2001).

[3] C. Vogt, M. Monai, G.J. Kramer, and B.M. Weckhuysen, *Nat. Catal.* **2**, 188 (2019).

[4] J.B. Senderens and P. Sabatier, *Compt. Rend. Acad. Sci.* **82**, 514 (1902).

[5] P. Sabatier and J.B. Senderens, *Compt. Rend. Acad. Sci.* **134**, 689 (1903).

[6] J. Sauer and H.-J. Freund, *Catal. Lett.* **145**, 109 (2015).

[7] F. Meirer and B.M. Weckhuysen, *Nat. Rev. Mater.* **3**, 324 (2018).

[8] I.L.C. Buurmans and B.M. Weckhuysen, *Nat. Chem.* **4**, 873 (2012).

[9] E.T.C. Vogt, G.T. Whiting, A. Dutta Chowdhury, and B.M. Weckhuysen, *Adv. Catal.* **58**, 143 (2015).

[10] A. Jain, S.P, Ong, G. Hautier, W. Chen, W.D. Richards, S. Dacek, S. Cholia, D. Gunter, S. Skinner, G. Ceder, and K.A. Persson, *APL Mater.* **1**, 011002 (2013).

[11] S. Chibani and F.X. Coudert, *Chem. Sci.* **10**, 8589 (2019).

[12] M.D. Foster and M.M.J. Treacy, A database of hypothetical zeolite structures (2010). http://www.hypotheticalzeolites.net.

[13] Y. Li, J. Yu, and R. Xu, Hypothetical zeolite database. http://mezeopor.jlu.edu.cn/hypo.

[14] http://www.iza-structure.org.

[15] D. Jo and S. Bong Hong, *Angew. Chem. Int. Ed.* **58**, 13845 (2019).

[16] J.J. Pluth and J.V. Smith, *Am. Minerol.* **75**, 501 (1990).

[17] https://europe.iza-structure.org/IZA-SC/framework.php?STC=BOG.

[18] R. Simoncas, D. Dari, N. Velamazan, M.T. Navarro, A. Cantin, J.L. Jorda, G. Sastre, and A. Corma, F. Rey, *Science* **330**, 1219 (2010).

[19] C. Vogt, E. Groeneveld, G. Kamsma, M. Nachtegaal, L. Lu, C.J. Kiely, P.H. Berben, F. Meirer, and B.M. Weckhuysen, *Nat. Catal.* **1**, 127 (2018).

[20] C. Vogt, M. Monai, E.B. Sterk, J. Palle, A.E.M. Melcherts, B. Zijlstra, E. Groeneveld, P.H. Berben J.M. Boereboom, E.J.M. Hensen, F. Meirer, I.A.W. Filot, and B.M. Weckhuysen, *Nat. Commun.* **10**, 5330 (2019).

OPERANDO MODELLING OF FUNCTIONAL NANOPOROUS MATERIALS

VERONIQUE VAN SPEYBROECK

Center for Molecular Modeling, Ghent University,
Technologiepark 46, 9052 Zwijnaarde, Belgium

My View of the Present State of Research on Modelling Functional Nanoporous Materials

Nanoporous materials, having pores with dimensions less than 100 nm [1], are very tractable for usage in many application fields such as gas separation, ion exchange, catalysis, energy storage, conversions, etc. [2–4] Ideally, one would be able to design materials with properties tailored to specific needs including high surface areas, nanosize and quantum confinement effects, ordered porosity, and high adsorbing and sensing abilities [5, 6]. Ultimately, one would obtain absolute control in building structures at the atomic scale to obtain the desired functional behaviour. This is an ambitious goal which poses major challenges both for theoreticians and experimentalists. Even if theoreticians would be able to predict how the material should be composed at the nanometre scale to obtain the desired functional behaviour, experimentalists would face the challenge to synthesize these nanometre-controlled materials and to characterize them with the highest possible resolution.

Inherently, one is confronted with a problem of attainable length and timescales both in experimental and theoretical research. From an experimental characterization point of view, there is the ambition to develop spectroscopic methods that push the boundaries of attainable spatial and temporal resolution to systematically smaller scales [7]. In contrast, computer models use a bottom-up reductionistic approach, starting from atomistic models which should be representative for the material's function observed experimentally. Up to a decade ago, theoreticians typically started from rather simplistic representations of nanoporous materials, where only a fraction of the material was accounted for, neglecting in many cases its periodic nature, its topology, and the inherent presence of defects or any spatial heterogeneities [8]. In 2019, the field evolved substantially. More realistic models are used, but the theoretically attainable length scales of maximum a few nanometres are still far from experimental crystal sizes.

Another level of complexity is introduced by the fact that materials have an inherently very dynamic nature under realistic working conditions. This naturally brings us to the adjective "operando", which is used both in theoretical and experimental research and refers to evaluating the material under operating conditions.

Around 2000, the terminology "operando spectroscopy" was launched within catalysis research, originating from the need to develop spectroscopic approaches able to give information on how an operating catalyst is working [9–11]. Major experimental advances have been made since then [12]. Likewise, the field of computational modelling is shifting progressively towards operando models, which capture the intrinsic dynamic behaviour under working conditions. Reaching this goal is certainly not possible by a single method alone. Instead, a range of models based on molecular dynamics (MD) methods, microkinetic models, and machine learning algorithms are currently explored [8, 13–15]. This brings us to currently attainable timescales within the field of modelling functional nanoporous materials. Typical molecular motions and vibrations occur on the picosecond timescale, whereas many processes related to activated processes are not captured in this short time domain. Various methodologies are now being explored to simulate such rare events [16, 17]. Looking from an even more holistic perspective, one needs to be aware that materials have an intrinsic lifecycle. In catalysis research, this is referred as the birth, life, and death of a catalytic solid [18, 19]. Current computational methods are certainly not evolved to an extent that they can capture such *in vita* conditions. In summary, there is still a spatial and time resolution gap to be bridged between theory and experiment.

My Recent Research Contributions to Modelling of Functional Nanoporous Materials

Within my group, we have developed and applied a series of methods to study functional nanoporous materials under operando conditions, which mimic experimental conditions as closely as possible. To reach this objective, we have developed a series of methods that allow one to map the *free energy surface* under realistic conditions of temperature, pressure, etc. Hereafter, I will illustrate the approach for the description of *phase transformations within soft porous crystals under the influence of external stimuli and for the description of the active site and its catalytic function in nanoporous materials*. Both processes are examples of rare events, as they are typically activated compared to the thermal energy available in the system. The situation can be compared to a topographical map when driving through the mountains (Fig. 1). During a typical MD simulation of a few hundred picoseconds, one would sample only the valleys of the mountain landscape. However, to map the activated processes, one should be able to also describe mountain passages with sufficient probability. Therefore, we use MD techniques which enhance sampling in these lower probability regions. Many of these techniques originate from biomolecular simulations, where also slow dynamics processes such as protein folding are important. These methods may be used in combination with a first principle description of the potential energy surface (PES) or with a classical force field for the molecular interactions. The choice is determined whether chemical bonds are

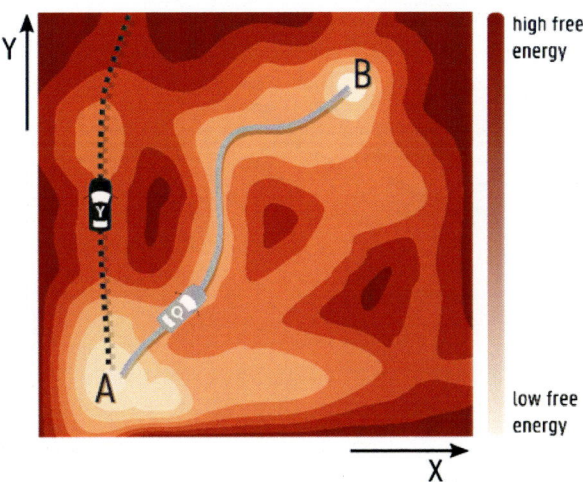

Fig. 1.   Exploration of a hypothetical free energy surface, by driving the sampling in various directions.

formed or not during the activated process and has a profound consequence for the accessible timescales of the simulation.

Responsive behaviour of soft porous crystals refers to materials showing a bi-stable or multi-stable behaviour under influence of external stimuli such as temperature, pressure, guest adsorption, or even external fields while retaining their crystallinity. Such functional response is tractable for applications including controlled drug release, gas adsorption, separation, and sensing [20, 21]. To understand the window of operation and design nanoscale devices with a controlled functional response, intimate knowledge is necessary on the nanoscopic origin of the experimentally observed phenomenon.

Experimentally, framework flexibility can be followed and recognized by monitoring the response of the material upon exposure to the external stimulus [22]. However, from experiment, one cannot construct the underlying Helmholtz free energy surface associated with the transition, as the material undergoes various irreversible structural transitions between (meta)stable equilibrium states (Fig. 2). In contrast, using classical MD simulations performed with in-house developed force fields, it is possible to construct the Helmholtz free energy profile in terms of a collective variable steering of the transition [23, 24]. A simple illustration is shown in Fig. 2 for materials exhibiting topological flexibility, where the phase transformation is accompanied by substantial changes in the volume. Pressure versus volume profiles are constructed in a dedicated thermodynamic ensemble, where the volume is controlled but the cell shape may vary [25]. Using this operando modelling approach, various potentially interesting materials were discovered in close synergy with experimentalists. For mechanical energy storage applications, for instance, we predicted for which materials a large amount of work needs to be exerted to induce

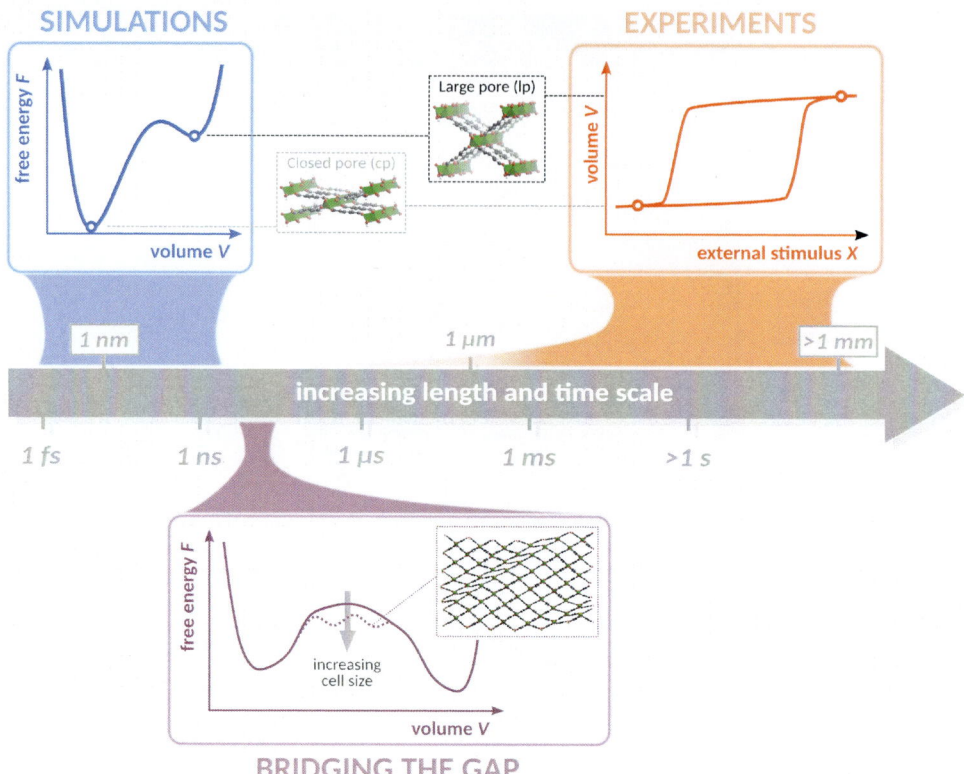

Fig. 2.   Illustrating the gap in accessible length and timescales between simulations and experiments and our current approach in bridging this gap.

the phase transformation [26, 27]. Similarly, materials to be used in nanoscale temperature switch devices preferentially morph between two phases in a desired temperature window. We discovered how a possible temperature switch could only be realized thanks to a delicate interplay between dispersion stabilization at low temperature and entropic effects at higher temperatures [28].

Despite the validity of the sketched thermodynamic approach, the kinetics and precise nature of the transition mechanism remain to be resolved. Recent experimental work showed how the material's ability to morph between various phases is critically affected by the crystal size. Downsizing the crystals from the micro- to mesoscale with primary crystallite sized between 10 nm and 1 μm substantially suppressed the responsive behaviour [29–31]. To shed light onto these phenomena, we are confronted with the spatial resolution gap between theory and experiment. Based on simulations with nanocells with sizes below 10 nm, a phase transition would only occur if the barrier between two (meta)stable states would completely disappear, as the smallest barrier in the thermodynamic potential of a single unit cell would translate into a huge barrier for the entire system. This mechanism is referred to as collective behaviour for the entire system. However, for a realistic material, it is rather unlikely that a crystallite would morph from one phase to the other in a synchronous way. To capture possible spatial heterogeneities during the phase transformations, we recently performed simulations on substantially larger models in the mesoscale range above 10 nm. Despite the fact that the systems are still substantially smaller than experimental crystallite sizes, important new insights into the transition mechanism were discovered. The mesoscale simulations revealed how a phase transformation nucleates locally within the lattice and creates an interfacial defect, where the two phases temporarily coexist and which propagates through the lattice. The barrier for this phenomenon was found to increase for crystals of smaller size, which corroborates with experiments. Further atomistic understanding of experimentally observed phenomena, such as the presence of disordered domains [32], will eventually require fundamentally new models (vide infra).

Operando modelling of the catalytic function in nanoporous materials is very important as many catalytic processes occur at higher temperatures or are influenced by the guest adsorption in the pores. The more "standard" modelling approach starts from a few points on the potential energy surface, such as reactants, transition state, and products. This is a huge oversimplification of the working catalytic material. In reality, the scene is much more complex: competitive pathways may be at work, which are essential for determining product selectivities, various guest molecules may be present in the pores of the material which may alter the nature of the active site or facilitate certain reaction channels, and reactive intermediates may change the operating temperature window [33–35]. Using enhanced sampling MD techniques based on a density functional theory description of the PES, we were able to map the free energy surface at realistic conditions. As such, a dynamic reorganization of catalytically active sites was observed for various processes (Fig. 3). For Brønsted acidic sites in zeolites, protic molecules in the pores of the material were able to capture the proton originally located on the lattice to form protonated reactive clusters [36, 37]. For Ni-zeolites used in ethene oligomerization, it was shown that ethene molecules reversibly mobilize the active site and exchange with the zeolite as ligands during reaction [38]. Such reactant-mobilized active sites have also been observed in the selective catalytic reduction of nitrogen oxides with ammonia in Cu-SSZ-13 zeolite and some other cases [39]. For other materials used for reactions taking place in milder conditions, such as metal-organic frameworks, the confined solvent plays an intriguing role and may facilitate ligand exchange processes or create new active sites [40, 41]. With the quest for conversion of new feedstocks such as biomass, new heterogeneous catalysts will have to be designed, with more complicated active sites, combining, for example, Brønsted and Lewis acid functions in close proximity [42]. Ideally, one could atomically design so-called single active sites at the surface of a solid catalyst, which gives the desired function and is robust in the desired operation window [43–45].

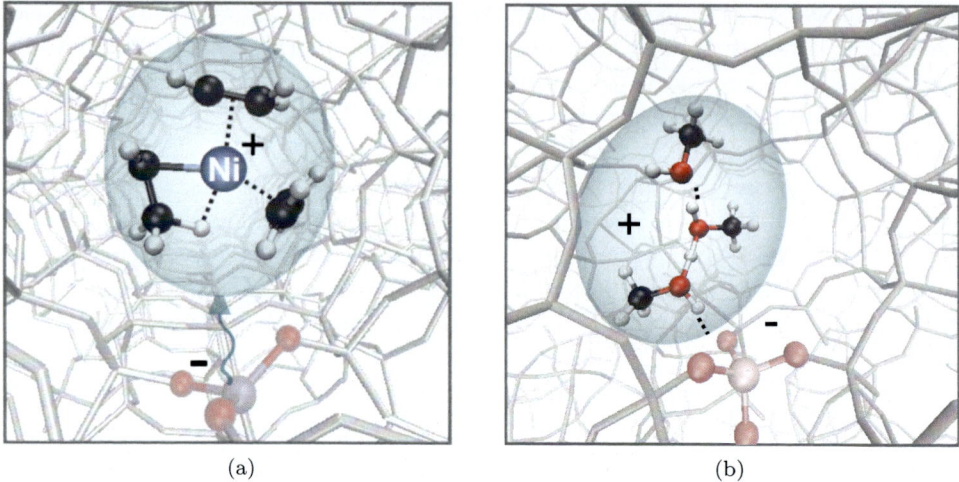

(a)   (b)

Fig. 3.   Illustration of dynamic reorganization of active sites in zeolites. (a) The active site for
ethene dimerization in Ni(II)-zeolite shows mobility upon coordination with ethene molecules.
(b) Formation of protonated clusters in Brønsted acidic sites in zeolites. Left figure adapted from
Ref. [38].

Outlook on Future Developments of Research on Modelling of Functional Nanoporous Materials

In my opinion, one of the great challenges within the field of computational
modelling of functional nanoporous materials is "How can we model functional
nanoporous materials at operating conditions within a similar spatiotemporal
window as attainable experimentally". This will require merging a reductionistic
bottom-up modelling with a top-down experimental approach. To this end, the
material needs to be modelled in an inclusive way with spatial heterogeneities and
needs to be brought under realistic operating conditions. To pursue this challenge, a
synergistic approach will be needed with experimentalists. Here, I see many analo-
gies with operando spectroscopy, where one tries to push the boundaries towards
better spatial and time resolution.

As mentioned in the introductory notes, the current limitations in computing
power and the currently available models strongly limit the accessible length and
timescales, and explicit atomistic simulations of, e.g., micrometre-sized crystals are
not yet feasible. To further advance the field, it will be necessary to define together
with experimentalists the targeted space–time windows. Next, fundamentally new
models will have to be developed to close the spatial and time resolution gap.
Inspiration may be sought from other fields like biochemistry, where much longer
timescales are being simulated [46]. To reach longer length scales, multiple strategies will have to be explored. Inspiration might be sought in coarse-grained techniques, where one abandons the atomistic description of the system and switches to

dynamical evolution of larger beads grouping various atoms. Similar models have been successfully used in biomolecular systems. Fundamentally new algorithms, for instance, based on machine learning potentials, might also be interesting. In any case, inspiration from various fields will have to be merged and a synergistic approach between various communities will be necessary.

Acknowledgments

All current and former members of the Center for Molecular Modeling are acknowledged for their contributions and discussions, which allowed us to advance in the field of modelling nanoporous materials. All funding bodies which have supported this fundamental research over the years such as the Fund for Scientific Research, Flanders (FWO), the Research Board of Ghent University (BOF), and the European Research Council under the European Union's Horizon 2020 research and innovation programme (consolidator ERC grant agreement No. 647755-DYNPOR (2015–2020)) are acknowledged. Ghent University is acknowledged for providing the framework to conduct fundamental research.

References

[1] M. Thommes, K. Kaneko, V. Neimark Alexander, P. Olivier James, F. Rodriguez-Reinoso, J. Rouquerol *et al.*, *Pure Appl. Chem.* **87**, 1051 (2015).

[2] V. Van Speybroeck, K. Hemelsoet, L. Joos, M. Waroquier, R.G. Bell, and C.R.A. Catlow, *Chem. Soc. Rev.* **44**, 7044–7111 (2015).

[3] Y. Cui, B. Li, H. He, W. Zhou, B. Chen, and G. Qian, *Accounts Chem. Res.* **49**, 483–493 (2016).

[4] R.E. Morris and P.S. Wheatley, *Angew. Chem.-Int. Edit.*, **47**, 4966–4981 (2008).

[5] J.-L. Brédas, K. Persson, and R. Seshadri, **29**, 2399–2401 (2017)

[6] M. Wächtler, L. González, B. Dietzek, A. Turchanin, and C. Roth, *Phys. Chem. Chem. Phys.* **21**, 8988–8991 (2019).

[7] I.L.C. Buurmans and B.M. Weckhuysen, *Nat. Chem.* **4**, 873–886 (2012).

[8] V. Van Speybroeck, K. De Wispelaere, J. Van der Mynsbrugge, M. Vandichel, K. Hemelsoet, and M. Waroquier, *Chem. Soc. Rev.* **43**, 7326–7357 (2014).

[9] M.A. Banares, *Catal. Today* **100**, 71–77 (2005)

[10] B.M. Weckhuysen, *Chem. Commun.* 97–110 (2002).

[11] B.M. Weckhuysen, *Phys. Chem. Chem. Phys.* **5**, 4351–4360 (2003).

[12] M.A. Bañares, M.O. Guerrero-Pérez, and A. Urakawa, *Catal. Today* **336**, 1 (2019).

[13] A. Bruix, J.T. Margraf, M. Andersen, and K. Reuter, *Nat. Catal.* **2**, 659–670 (2019).

[14] L. Grajciar, C.J. Heard, A.A. Bondarenko, M.V. Polynski, J. Meeprasert *et al.*, *Chem. Soc. Rev.* **47**, 8307–8348 (2018).

[15] K. De Wispelaere, S. Bailleul, and V. Van Speybroeck, *Catal. Sci. Technol.*, **6**, 2686–2705 (2016).

[16] C. Abrams and G. Bussi, *Entropy* **16**, 163–199 (2014).

[17] R. Demuynck, S.M.J. Rogge, L. Vanduyfhuys, J. Wieme, M. Waroquier, and V. Van Speybroeck, *J. Chem. Theory Comput.* **13**, 5861–5873 (2017).

[18] F. Meirer, S. Kalirai, D. Morris, S. Soparawalla, Y.J. Liu *et al.*, *Sci. Adv.* **1**, 12 (2015).

[19] B.M. Weckhuysen, *Natl. Sci. Rev.* **2**, 147–149 (2015).

[20] S. Horike, S. Shimomura, and S. Kitagawa, *Nat. Chem.* **1**, 695–704 (2009).

[21] A. Schneemann, V. Bon, I. Schwedler, I. Senkovska, S. Kaskel, and R.A. Fischer, *Chem. Soc. Rev.* **43**, 6062–6096 (2014).

[22] L. Vanduyfhuys, S.M.J. Rogge, J. Wieme, S. Vandenbrande, G. Maurin *et al.*, *Nat. Commun.* **9**, 9 (2018).

[23] L. Vanduyfhuys, S. Vandenbrande, T. Verstraelen, R. Schmid, M. Waroquier, V. Van Speybroeck, *J. Comput. Chem.* **36**, 1015–1027 (2015).

[24] L. Vanduyfhuys, S. Vandenbrande, J. Wieme, M. Waroquier, T. Verstraelen, and V. Van Speybroeck, *J. Comput. Chem.* **39**, 999–1011 (2018).

[25] S.M.J. Rogge, L. Vanduyfhuys, A. Ghysels, M. Waroquier, T. Verstraelen *et al.*, A. Ghysels, *J. Chem. Theory Comput.* **11**, 5583–5597 (2015).

[26] P.G. Yot, L. Vanduyfhuys, E. Alvarez, J. Rodriguez, J.P. Itie *et al.*, *Chem. Sci.* **7**, 446–450 (2016).

[27] S.M.J. Rogge, J. Wieme, L. Vanduyfhuys, and S. Vandenbrande, G. Maurin *et al.*, *Chem. Mater.* **28**, 5721–5732 (2016)

[28] J. Wieme, K. Lejaeghere, G. Kresse, and V. Van Speybroeck, *Nat. Commun.* **9**, 10 (2018).

[29] Y. Sakata, S. Furukawa, M. Kondo, K. Hirai, N. Horike *et al.*, *Science*, **339**, 193–196 (2013).

[30] S. Krause, V. Bon, I. Senkovska, D.M. Többens, D. Wallacher *et al.*, *Nat. Commun.* **9**, 1573 (2018).

[31] S. Wannapaiboon, A. Schneemann, I. Hante, M. Tu, K. Epp *et al.*, *Nat. Commun.* **10**, 346 (2019).

[32] M.J. Cliffe, W. Wan, X. Zou, P.A. Chater, A.K. Kleppe *et al.*, *Nat. Commun.* **5**, 4176 (2014).

[33] P. Cnudde, K. De Wispelaere, L. Vanduyfhuys, R. Demuynck, J. Van der Mynsbrugge *et al.*, *ACS Catal.* **8**, 9579–9595 (2018).

[34] I. Yarulina, K. De Wispelaere, S. Bailleul, J. Goetze, M. Radersma *et al.*, *Nat. Chem.* **10**, 804–812 (2018).

[35] K. De Wispelaere, B. Ensing, A. Ghysels, E.J. Meijer, and V. Van Speybroeck, *Chem.-Eur. J.* **21**, 9385–9396 (2015).

[36] K. De Wispelaere, C.S. Wondergem, B. Ensing, K. Hemelsoet, E.J. Meijer *et al.*, *ACS Catal.* **6**, 1991–2002 (2016).

[37] S. Bailleul, S.M.J. Rogge, L. Vanduyfhuys, and V. Van Speybroeck, *ChemCatChem* **11**, 3993–4010 (2019).

[38] R.Y. Brogaard, M. Komurcu, M.M. Dyballa, A. Botan, V. Van Speybroeck *et al.*, *ACS Catal.* **9**, 5645–5650 (2019).

[39] C. Paolucci, I. Khurana, A.A. Parekh, S.C. Li, A.J. Shih *et al.*, *Science* **357**, 898 (2017).

[40] J. Marreiros, C. Caratelli, J. Hajek, A. Krajnc, G. Fleury *et al.*, *Chem. Mater.* **31**, 1359–1369 (2019).

[41] J. Hajek, C. Caratelli, R. Demuynck, K. De Wispelaere, L. Vanduyfhuys *et al. Chem. Sci.* **9**, 2723–2732 (2018).

[42] P. Sudarsanam, E. Peeters, E.V. Makshina, V.I. Parvulescu, and B.F. Sels, *Chem. Soc. Rev.* **48**, 2366–2421 (2019)

[43] S.M.J. Rogge, A. Bavykina, J. Hajek, H. Garcia, A.I. Olivos-Suarez *et al.*, *Chem. Soc. Rev.* **46**, 3134–3184 (2017).

[44] C. Copéret, A. Comas-Vives, M.P. Conley, D.P. Estes, and A. Fedorov *et al.*, *Chem. Rev.* **116**, 323–421 (2016).

[45] J.M. Thomas, R. Raja, and D.W. Lewis, *Angew. Chem. Int. Ed.* **44**, 6456–6482 (2005).

[46] B.E. Husic and V.S. Pande, *J. Am. Chem. Soc.* **140**, 2386–2396 (2018).

MODELLING EMERGENT PHENOMENON USING CHEMICALLY ENCODED COMPUTATIONS

LEROY CRONIN

School of Chemistry, University of Glasgow, Glasgow, G12 8QQ, UK

Introduction

As chemical, materials, and biological systems become more complex, the demands on the computational approaches to model these systems increase. The result is a computational arms race between the resources required to instantiate the model at the correct level to give useful predictive power and the size of the most tractable problem. Not only will computational resource always be limited for most problems as they become large but most problems will also assume that the underlying physics and chemistry can be captured discretely. Another route to investigate very complex systems is to compute the problem in a chemical substrate. This means that the problem becomes coded in physically which can evolve to the solution [1].

My View of the Present State of Research on Modelling Using Non-conventional Computation

The field of unconventional or natural computing [2] has emerged during the last two decades as a route to exploit the physical properties of certain systems that are naturally encoded to solve certain problems. Examples could be solving a maze [3] by physically drawing the structure in a microfluidic chip and then finding the shortest path to the centre of the maze by attaching electrodes to the start and end and discharging a 20–30-kV potential into a low-pressure helium atmosphere. There are many other systems that have been used to explore big problems including DNA origami as self-assembling circuit boards [4], or the exploration of a materials formulation space [5]. The implementation of unconventional computing using a chemical system, for instance, the computing capability of chemical reactions for image processing, shows that the abstract of problems into physical space is possible [6]. In chemical unconventional computing, reaction diffusion/excitation waves are often employed to implement computing systems. For example, various types of logic gates can be implemented in a similar way to a collision-based billiard-ball computer [7]. These systems well demonstrate the computational capabilities of chemical systems to carry out specific computations in a distributed fashion by exploiting the massively parallel nature of chemical systems (i.e., Avogadro-scale units for parallel computation). While these approaches show promise, they are not programmable.

My Recent Research Contributions to Modelling Functional Materials

We have taken two approaches to explore the modelling of complex material systems in different substrates. In the first example, we have shown how it is possible to encode the self-assembly of gigantic inorganic molecules into a chemical self-assembly experiment by controlling the physical template, pH, and the reduction potential. Under these conditions, we show how a simple inorganic salt can spontaneously form information-rich autocatalytic sets of replicating inorganic molecules that work via molecular recognition based on the $\{PMo_{12}\}$ Keggin ion, and $\{Mo_{36}\}$ cluster [8, 9], see Fig. 1.

These small clusters are able to catalyse their own formation via an autocatalytic network, which subsequently form a template of the assembly of gigantic molybdenum blue wheel (Mo_{154}-blue), $\{Mo_{132}\}$ ball containing 154 and 132 molybdenum

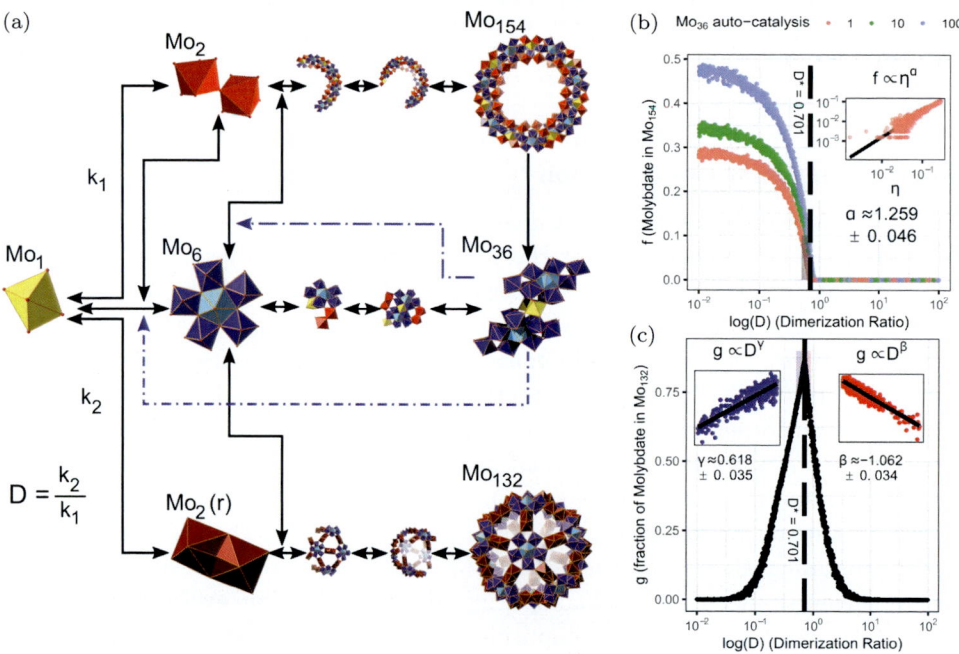

Fig. 1.   Scheme of stochastic model. (a) Schematic representation of the stochastic kinetic model. (b) The model exhibits a critical transition to the formation of the giant nanostructure $\{Mo_{154}\}$, the scaling relation near the critical point is shown in the inset. (c) Near the critical point, the formation of the other giant nanostructure $\{Mo_{132}\}$ is maximized, showing critical scaling near the transition, with different scaling coefficients on either side of the transition.

atoms, and a new $\{PMo_{12}\} \subset \{Mo_{124}Ce_4\}$ nanostructure. Kinetic investigations revealed key traits of autocatalytic systems including molecular recognition and kinetic saturation. A stochastic model confirms the presence of an autocatalytic network driven by molecular recognition, where the larger clusters are the only products stabilized by information contained in the cycle, isolated due to a critical transition in the network.

The second example uses a chemical processor that utilizes individually addressed but fully interconnected cells of a chemical oscillating Belousov–Zhabotinsky (BZ) reaction, as the data-processing medium [10]. The system can be programmed to achieve flexible chemical computation by addressing the individually controlled stirrers and adjusting the stirring speeds, see Fig. 2.

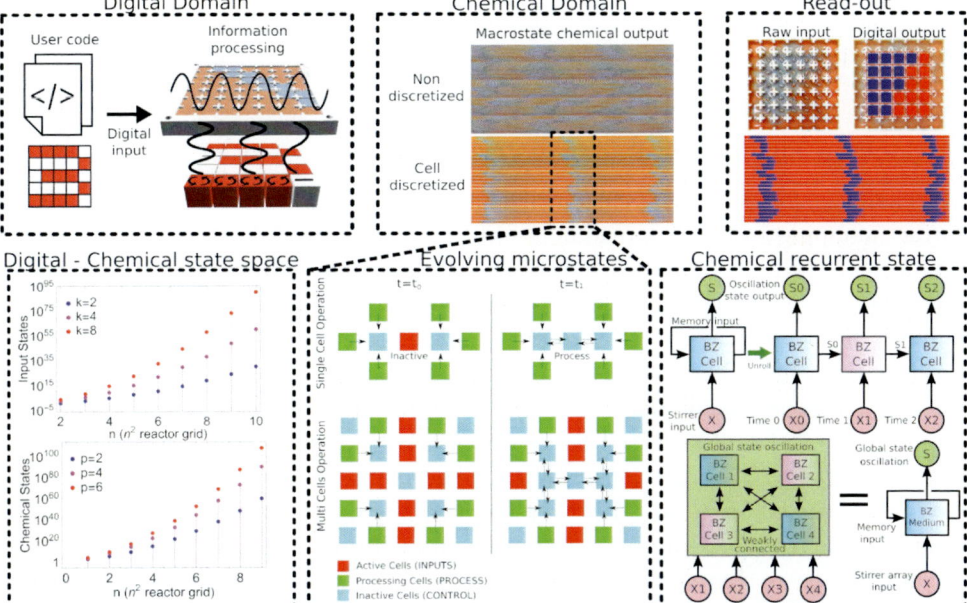

Fig. 2. Chemical processor paradigm. The user inputs the code via a 5×5 grid of cells. This grid of cells controls the stirrers on/off and also the speed. (Chemical Domain) Based on the selected grid, oscillating waves will appear in the platform. If the cells are discretized, these oscillations will localize. (Read-out) Using a camera and image processing, the states of the chemical processor are read by a digital computer. (Input Vs Processing Space) Input States plot shows the scaling of the number of input states with the number of BZ cells on the experimental platform at different PWM stirring inputs ($knxn$, where k can be 2, 3, and 4). Chemical states plot shows scaling of the number of chemical states (defined by initial phase and frequency of BZ oscillations, see SI) with the number of BZ cells on the experimental platform at a different number of measurable oscillation frequencies ($2nxnpnxn$ where p can be 2, 3, and 4). (Evolving microstates) Because the cells are weakly connected, their oscillations will convolve, and be able to perform complex computations by controlling the stirring speeds into (a) active cells — fast stirring — for inputs, (b) process cells — slow stirring, and (c) inactive cells — no stirring (see SI). (Chemical recurrent state) Because the BZ oscillations have memory, the global state of the medium not only depends on the input but also on the state of previous iterations.

Our chemical processor architecture relies on data storage and processing via electron transfer between molecules of $[Fe(Bpy)_3]^{2/3+}$ as a catalyst for the BZ reaction and a read-out where the oxidized regions containing Fe(III) are blue, and the reduced states containing Fe(II) species are red. The output from the array is produced by recording a video of the BZ medium to monitor the oscillation states of the reaction in the individual cells.

To achieve programmability, we designed a platform that controls the inputs as oscillations of the BZ reaction at local sites in a grid ("cells") by externally controlling the oscillations in each cell with a magnetic stirrer, where cells are triggered when a stirrer is turned "on" or when an "off" cell is surrounded by "on" cells and the chemical oscillations transfer to it. As a result, programming the platform can be used to exploit the chemical states arising from interactions between spatiotemporal excitation patterns. Moreover, the stirring patterns can be individually changed at any time based on user input; therefore, the BZ processor described here can be programmed at any point during its execution.

Outlook on Future Developments of Research

The challenge for the area of chemically encoded computations will be designing the correct mapping function to read in the problem, and then allow the read-out of the solution. In the first example, I showed how by doing a combination of kinetic experiments and a stochastic model based upon the assembly of the cluster using simple building blocks, it was easy to predict the assembly via a series of interconnected catalytic networks. The source of the information encoding in this case comes from the self-replicating small inorganic templates which themselves follow a template of the larger cluster architecture. These templates prevent a combinatorial explosion in the number of products, ensuring that the physical clusters assembled are the read-out of the computation. In this respect, the template acts as the "program" to run the computation. However, a key problem with this view is understanding when a computation is complete and expanding the concept to other materials. These could include the coupling of template replication to catalytic assembly to explore an assembly of other types of nanostructures to avoid combinatorial explosion.

In the second example, I described how a reaction–diffusion system, placed into a digitally addressable encoder array of stirrers, can be used to encode representations of abstract problems, but the ability to map these into real material problems will depend upon controlling the nearest neighbour interactions. If this can be done, then many interesting problems from condensed matter physics can be mapped into these 2D arrays such as the Ising model. Mathematical problems that are combinatorially hard can be solved if the grid size approaches 100×100 or 10,000 cells. In this case, the system might be able to beat silicon for some types of problems and even match that of quantum computers if "chemical supremacy" in the combinatorially addressable chemical computer can be achieved.

Acknowledgments

This work is funded by the UK EPSRC, US DARPA, and EU ERC.

References

[1] J.M.P. Gutierrez, T. Hinkley, J.W. Taylor, K. Yanev, and L. Cronin, *Nat. Commun.* **5**, 5571 (2014).

[2] S.L. Harding, J.F. Miller, and E.A. Rietman, *Int. J. Unconv. Comput.* **4**, 155 (2008).

[3] A.E. Dubinov *et al.*, *Phys. Plasmas* **21**, 093503 (2014).

[4] J. Elbaz, O. Lioubashevski, and F. Wang, *Nat. Nano.* **5**, 417 (2010).

[5] L.J. Points, J.W. Taylor, J. Grizou, K. Donkers, and L. Cronin, *Proc. Natl. Acad. Sci. USA* **115**, 885 (2018).

[6] L. Kuhnert, K.I. Agladze, and V.I. Krinksy, *Nature* **337**, 244 (1989).

[7] E. Fredkin and T. Toffoli, *Int. J. Theor. Phys.* **21**, 219 (1982).

[8] H.N. Miras, G.J.T. Cooper, D.-L. Long, H. Bögge, A. Müller, C. Streb, and L. Cronin, *Science* **327**, 72 (2010).

[9] H.N. Miras, C. Mathis, W. Xuan, D.-L. Long, R. Pow, and L. Cronin, *Proc. Natl. Acad. Sci. USA*, **117**, 10699–10705 (2020).

[10] J.M.P. Gutierrez, A. Sharma, S. Tsuda, G.J.T. Cooper, G. Aragon-Camarasa, K. Donkers, and L. Cronin, *Nat. Commun.*, **11**, 1442 (2020).

EXCITED-STATE AROMATICITY FOR THE DESIGN
OF NEW FUNCTIONAL MATERIALS

MIQUEL SOLÀ

Institut de Química Computacional i Catàlisi (IQCC), Universitat de Girona,
C/Maria Aurèlia Capmany, 69, 17003 Girona, Catalonia, Spain

Introduction

The controversial concept of aromaticity [1] has been traditionally connected to benzene and derivatives. In the last few decades, the idea of aromaticity has been enriched by new fascinating aromatic compounds and, nowadays, the concept of aromaticity has been extended to almost the entire Periodic Table. The main feature of aromatic molecules is that they exhibit high electronic delocalization in closed 2D or 3D circuits that results in significant energetic stabilization, low reactivity, tendency towards bond length equalization, and characteristic spectroscopic and magnetic features [2].

The concept of aromaticity is used mainly to explain the properties and reactivity of certain molecules in their ground states. Its application to rationalize the properties and reactivity of molecules in excited states is scarcer. The classic example of excited-state aromaticity is the aromaticity of the lowest-lying triplet excited states of conjugated monocyclic compounds of $4n$ π-electrons predicted by Baird in 1972 [3] and confirmed by the identification of the planar triplet ground states of $C_5H_5^+$ and $C_5Cl_5^+$ [4, 5]. Computational confirmation of the Baird rule came from different studies [6–13]. Some of them also proved that the singlet excited state with the same configuration as the lowest-lying triplet excited state is aromatic too [14–16]. Excited-state (anti)aromaticity can explain the photochemistry of many compounds [17]. The excited-state intramolecular proton transfer (ESIPT) [18, 19] reactions are one of these situations. A particular case of these ESIPT reactions corresponds to compounds that have a phenol group which is hydrogen bonded to a nitrogen or oxygen atom of the same molecule. The two possible isomers for these species are the enol and the keto forms that are related through an intramolecular proton transfer process. In the ground state, the enol form is the most aromatic and the most stable and the proton transfer to yield the keto form is thermodynamically unfavourable. In the lowest-lying $^1\pi\pi^*$ excited state, the enol form is antiaromatic and the proton transfer that generates the keto form allows the ring to get rid of this antiaromaticity [20, 21], which explains why the ESIPT becomes kinetically and thermodynamically favoured in the excited state. Excited-state (anti)aromaticity is important not only in photochemistry but also in the design of new materials that act as molecular motors [22, 23], high-spin organic molecules [24–26], photoluminescence materials [27], or in photovoltaic cells. In this latter case, the stabilization of a triplet charge transfer state (CTS) state relative to a singlet CTS can be helpful by slowing down the charge recombination process and improving the efficiency of the cell [28, 29]. To stabilize some triplet CTS, one can benefit from the aromaticity of the triplet states.

Recent Research Contributions to Excited-state Aromaticity

Stable high-spin molecules, with two or more unpaired electrons of parallel spin direction ($S = 1$ or greater), are difficult to produce. These materials are not only of theoretical interest but high-spin molecules are also intensely sought as nanomagnets for applications in spintronic devices [30], magnetic refrigerants [31], or as contrast agents for magnetic resonance imaging [32] to name some applications. Recently, several larger polycyclic and macrocyclic high-spin compounds that are influenced by Baird-aromaticity in their ground states have been generated [33, 34]. In the quest for stable organic molecules with triplet ground states, we analysed a series of 48 different pentafulvenes (see Fig. 1) [35]. Pentafulvenes are influenced by their exocyclic substituents in an opposite manner in the T_1 and S_0 states (Fig. 1(a)). The usual ground state S_0 is stabilized by electron-donating groups (EDGs) that favor the resonance structure having a Hückel aromatic cyclopentadienly anion with six π-electrons. On the contrary, the lowest $\pi\pi^*$ triplet state is stabilized by electron withdrawing groups (EWGs) that benefit from the resonance structure with a triplet cyclopentadienyl cation, which with four π-electrons is aromatic according to Baird's rule. Therefore, fulvenes are influenced by their substituents in an opposite manner in T_1 and S_0 states. A substituent that enhances/reduces fulvene aromaticity in S_0 reduces/enhances aromaticity in T_1. We analysed [35] which substituents and which positions on the pentafulvene core are the most influential for designing compounds with low or inverted singlet-triplet energy gap (E_T).

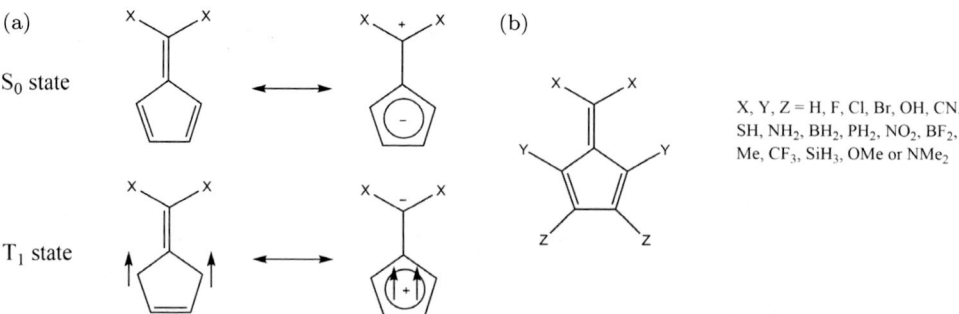

Fig. 1.   (a) Aromatic resonance structures influencing the S_0 and T_1 states of pentafulvene. (b) The set of disubstituted pentafulvenes, having either two X, Y, or Z substituents.

Calculations carried out at the UM06-2X/cc-pVTZ level show that, in the parent pentafulvene, the triplet state is higher in energy than the singlet ground state by 33.9 kcal/mol. EWGs in the exocyclic carbon atoms reduce this energy difference, which becomes 25.9 kcal/mol for $X = $ CN. EDGs located in the Y position (see Fig. 1(b)) also lead to a stabilization of the triplet with respect to the singlet ground state ($E_T = 19.9$ kcal/mol for $Y = $ NH$_2$). Finally, Z substituents have a minor effect on the singlet–triplet energy gap [35]. The combination of EWGs as X (CN, NO$_2$, or CF$_3$) with EDGs as Y (NH$_2$, NMe$_2$, OH, OMe, or SH) brings down E_T considerably ($E_T \leq 12$ kcal/mol). Interestingly, the triplet state becomes the ground state for $X = $ CN and $Y = $ NH$_2$, the singlet state being higher in energy by 3.5 kcal/mol. Finally, protonation at the cyano N atoms of 6,6-dicyanopentafulvenes increases the π-electron-withdrawing capacity of cyano groups generating a fulvenium dication with a triplet ground state (T$_0$). The five-membered ring (5-MR) of all species with a triplet ground state is clearly Baird-aromatic according to HOMA, FLU, and NICS calculations.

In the case of polycyclic conjugated hydrocarbons (PCHs), we found that the topology of the PCH is important to determine the E_T value [36]. Dibenzobiphenylene isomers A3LL and *cis*-A3BB of Fig. 2 constitute an example. The T$_1$ triplet state energies relative to the singlet S$_0$ ground state at the (U)B3LYP/6-311G(d,p) level are 2.75 and 1.06 eV, respectively. In comparison to A3LL, the singlet–triplet energy gap in *cis*-A3BB is smaller by about 1.7 eV. The reason for this big reduction is displayed in Fig. 2. For the A3LL isomer, the two most external 6-MRs have a Clar π-sextet. In the T$_1$ state, the central biphenylene unit could have a Baird π-octet. In the Clar π-sextet model [37, 38], π-sextets, π-octets, etc., should be disjoined and for this reason this Baird π-octet is not formed as indicated by the ACID plot. However, in the *cis*-A3BB isomer, we have the same Clar π-sextets in

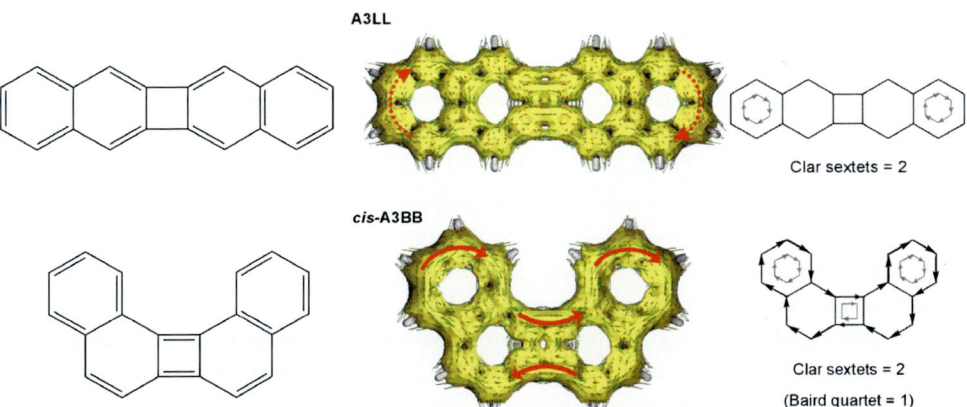

Fig. 2.   Resonant structures of A3LL and *cis*-A3BB isomers in their ground state, ACID plots and schematic drawings of the ring-currents in their T$_1$ states. Black arrows represent stronger currents, while grey arrows represent weaker currents. Adapted from Ref. [36].

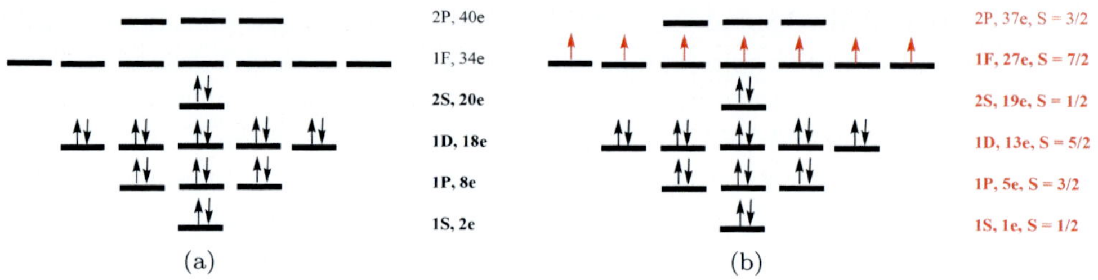

Fig. 3.   The (a) closed-shell and (b) open-shell jellium magic numbers. Occupancies of a system with 20 e- for the closed-shell and of 27 e- for the open-shell electronic structure are considered.

the outer 6-MR but, in addition, we have a disjoined π-quartet in the 4-MR. The presence of this additional Baird π-quartet provides an extra stabilization for the T_1 state of this kinked isomer as compared to the linear one.

Finally, we have also described a new way to design high spin species in metallic clusters. The observed experimental abundances of a number of metal clusters have been explained with the spherical jellium model. This model assumes a uniform distribution of positive charge corresponding to the cluster atomic nuclei and their innermost electrons in which the interacting valence electrons move. The energy levels of valence electrons for such a model are $1S^21P^61D^{10}2S^21F^{14}2P^61G^{18}2D^{10}$ ... (see Fig. 3), where S, P, D, F, and G letters denote the angular momentum and numbers 1, 2, 3 indicate the radial nodes [39, 40]. Metallic clusters with 2, 8, 18, 20, 34, 40, etc., valence electrons are particularly stable because these numbers correspond to closed-shell electronic structures in the jellium model.

The Baird rule shows that monocyclic annulenes in the lowest-lying triplet state are aromatic (and particularly stable) with $4n$ π-electrons. With this number of electrons, they have an electronic structure with the last shell half-filled with two electrons of same spin. Inspired by this Baird rule, we decided to investigate if such a rule can be extended to atomic clusters. If the last energy level of valence electrons for the jellium model is half-filled with same-spin electrons, the system should have some extra stability and should have aromatic character [41]. As seen in Fig. 3(b), this situation is reached for the magic numbers of valence electrons of 1 ($S = 1/2$), 5 ($S = 3/2$), 13 ($S = 5/2$), 19 ($S = 1/2$), 27 ($S = 7/2$), 37 ($S = 3/2$), 49 ($S = 9/2$), etc. We checked this hypothesis in ten beryllium clusters: Be_4, Be_9, and Be_{10} are closed-shell systems, whereas Be_3^{+1}, Be_6^{-1}, Be_7^{+1}, Be_9^{-1}, Be_{10}^{+1}, Be_{13}^{-1}, and Be_{14}^{+1} follow the open-shell jellium model with a same spin half-filled last energy level. All open-shell species are in their ground state, except for Be_3^{+1} and Be_6^{-1}. Our results [41] show that those clusters whose last energy level of valence electrons is half-filled with same-spin electrons in the jellium electronic structure are aromatic (*open-shell jellium* aromatic) and present an extra stability, as proven by their larger atomization energy. This new set of magic numbers proposed may be a powerful tool for researchers who work in the quest for stable single high-spin molecules for their use as single-molecule-based magnets [42, 43].

Outlook on Future Developments of Research on Excited-state Aromaticity

In my opinion, from a theoretical point of view, there are at least two aspects that require further investigation. First, it is important to examine whether the Soncini-Fowler rule [44], which generalizes the Hückel and Baird rule to high-spin states, is followed by organic and inorganic compounds. Preliminary results [16] did not provide enough evidence for this extended rule of excited-state aromaticity. And second, the aromaticity of highly twisted species in the ground and excited states should be analysed in detail. According to Rappaport and Rzepa [45], annulenes with an even number of twists and writhes are aromatic in their ground states if they have $4n + 2$ π-electrons and they are aromatic in the lowest-lying triplet states if they have an odd number of twists and writhes. However, there is a loss of $2p_{\pi}$–$2p_{\pi}$ overlap with increasing number of twists and writhes. So, it seems that after a given number of twists and writhes, the electronic delocalization should not be effective enough to generate aromatic species.

Acknowledgments

This work has been supported by the Ministerio de Economía y Competitividad (MINECO) of Spain (Project CTQ2017-85341-P) and the Generalitat de Catalunya (project 2017SGR39 and ICREA Academia 2014 prize). Co-authors of works discussed in this chapter are gratefully acknowledged.

References

[1] M. Solà, *Front. Chem.* **5**, 22 (2017).
[2] Z. Chen, C.S. Wannere, C. Corminboeuf, R. Puchta, and P.V.R. Schleyer, *Chem. Rev.* **105**, 3842 (2005).
[3] N.C. Baird, *J. Am. Chem. Soc.* **94**, 4941 (1972).
[4] R. Breslow, H.W. Chang, R. Hill, and E. Wasserman, *J. Am. Chem. Soc.* **89**, 1112 (1967).
[5] M. Saunders, R. Berger, A. Jaffe, J.M. McBride, J. O'Neill *et al.*, *J. Am. Chem. Soc.* **95**, 3017 (1973).
[6] F. Fratev, V. Monev, and R. Janoschek, *Tetrahedron* **38**, 2929 (1982).
[7] H. Jiao, P.V.R. Schleyer, Y. Mo, M.A. McAllister, and T.T. Tidwell, *J. Am. Chem. Soc.* **119**, 7075 (1997).
[8] V. Gogonea, P.V.R. Schleyer, and P.R. Schreiner, *Angew. Chem. Int. Ed.* **37**, 1945 (1998).
[9] E. Steiner and P.W. Fowler, *J. Phys. Chem. A* **105**, 9553 (2001).
[10] E. Steiner and P.W. Fowler, *Chem. Commun.* 2220 (2001).
[11] A. Soncini, P.W. Fowler, and F. Zerbetto, *Chem. Phys. Lett.* **405**, 136 (2005).
[12] P.W. Fowler, E. Steiner, and L.W. Jenneskens, *Chem. Phys. Lett.* **371**, 719 (2003).
[13] S. Villaume, H.A. Fogarty, and H. Ottosson, *ChemPhysChem* **9**, 257 (2008).
[14] P.B. Karadakov, *J. Phys. Chem. A* **112**, 7303 (2008).
[15] P.B. Karadakov, *J. Phys. Chem. A* **112**, 12707 (2008).
[16] F. Feixas, J. Vandenbussche, P. Bultinck, E. Matito, and M. Solà, *Phys. Chem. Chem. Phys.* **13**, 20690 (2011).

[17] M. Rosenberg, C. Dahlstrand, K. Kilså, and H. Ottosson, *Chem. Rev.* **114**, 5379 (2014).

[18] A.P. Demchenko, K.-C. Tang, and P.-T. Chou, *Chem. Soc. Rev.* **42**, 1379 (2013).

[19] A.J. Stasyuk, M.K. Cyrański, D.T. Gryko, and M. Solà, *J. Chem. Theory Comput.* **11**, 1046 (2015).

[20] C.-H. Wu, L.J. Karas, H. Ottosson, and J.I.-C. Wu, *Proc. Nat. Acad. Sci.* **116**, 20303 (2019).

[21] B.J. Lampkin, Y.H. Nguyen, P.B. Karadakov, and B. VanVeller, *Phys. Chem. Chem. Phys.* **21**, 11608 (2019).

[22] J. Sturala, M.K. Etherington, A.N. Bismillah, H.F. Higginbotham, and W. Trewby *et al.*, *J. Am. Chem. Soc.* **139**, 17882 (2017).

[23] B. Oruganti, J. Wang, and B. Durbeej, *Org. Lett.* **19**, 4818 (2017).

[24] N.M. Gallagher, A. Olankitwanit, and A. Rajca, *J. Org. Chem.* **80**, 1291 (2015).

[25] M. Mauksch, and S.B. Tsogoeva, *Phys. Chem. Chem. Phys.* **19**, 4688 (2017).

[26] Y. Qiu, L.J. Fischer, A.S. Dutton, and A.H. Winter, *J. Org. Chem.* **82**, 13550 (2017).

[27] S. Shokri, J. Li, M.K. Manna, G.P. Wiederrecht, and D.J. Gosztola *et al.*, *J. Org. Chem.* **82**, 10167 (2017).

[28] M. Izquierdo, B. Platzer, A.J. Stasyuk, O.A. Stasyuk, A.A. Voityuk *et al.*, *Angew. Chem. Int. Ed.* **58**, 6932 (2019).

[29] B. Liu, H. Fang, X. Li, W. Cai, L. Bao *et al.*, *Chem. Eur. J.* **21**, 746 (2015).

[30] M. Murrie, *Chem. Soc. Rev.* **39**, 1986 (2010).

[31] M. Evangelisti and E.K. Brechin, *Dalton Trans.* **39**, 4672 (2010).

[32] B. Cage, S.E. Russek, R. Shoemaker, A.J. Barker, C. Stoldt *et al.*, *Polyhedron* **26**, 2413 (2007).

[33] W.-Y. Cha, T. Kim, A. Ghosh, Z. Zhang, X.-S. Ke *et al.*, *Nat. Chem.* **9**, 1243 (2017).

[34] W. Zeng, H. Phan, T.S. Herng, T.Y. Gopalakrishna, N. Aratani *et al.*, *Chem.* **2**, 81 (2017).

[35] S. Yadav, O. El Bakouri, K. Jorner, H. Tong, C. Dahlstrand *et al.*, *Chem. Asian J.* **14**, 1870 (2019).

[36] R. Ayub, O.E. Bakouri, K. Jorner, M. Solà, and H. Ottosson, *J. Org. Chem.* **82**, 6327 (2017).

[37] E. Clar, *The Aromatic Sextet* (Wiley, New York, 1972).

[38] M. Solà, *Front. Chem.* **1**, 22 (2013).

[39] W. Ekardt, *Phys. Rev. B* **29**, 1558 (1984).

[40] W.A. de Heer, *Rev. Mod. Phys.* **65**, 611 (1993).

[41] J. Poater and M. Solà, *Chem. Commun.* **55**, 5559 (2019).

[42] C.A.P. Goodwin, F. Ortu, D. Reta, N.F. Chilton, and D.P. Mills, *Nature* **548**, 439 (2017).

[43] B.M. Day, F.-S. Guo, and R.A. Layfield, *Acc. Chem. Res.* **51**, 1880 (2018).

[44] A. Soncini and P.W. Fowler, *Chem. Phys. Lett.* **450**, 431 (2008).

[45] S.M. Rappaport and H.S. Rzepa, *J. Am. Chem. Soc.* **130**, 7613 (2008).

MODELLING METAL–ORGANIC FRAMEWORKS AND OTHER FUNCTIONAL MATERIALS WITH ELECTRONIC STRUCTURE THEORIES

LAURA GAGLIARDI and CHRISTOPHER J. CRAMER

Department of Chemistry, Chemical Theory Center, and Supercomputing Institute, University of Minnesota, Minneapolis, Minnesota 55455, USA

State of the Art of Electronic Structure Theories for MOFs and Other Materials

Metal–organic frameworks (MOFs) are porous crystalline materials with an increasing number of possible applications, ranging from gas storage and separation, to catalysis, chemical sensing, photocatalysis, and drug delivery. The solely experimental exploration of the enormous chemical space spanned by MOFs, with their modular construction from a variety of nodes and linkers, is impractical and computational modelling is indispensable in combination with experimental efforts to guide the design of novel materials with specific properties. In terms of electronic structure theories, the Kohn–Sham density functional theory (KS-DFT) is widely applied in materials science because it is affordable for large systems and it provides semiquantitative accuracy. KS-DFT has been, and continues to be, the most used electronic structure tool for modelling functional materials, thanks to its broad generality and versatility. For some systems, however, it has been shown that chemical accuracy cannot be achieved without resorting to a wave function-based theory. Typical situations in which KS-DFT can give non-negligible errors include the description of dispersion forces, magnetic properties, and properties in general in correlation-dominated materials, such as transition-metal-containing systems characterized by locally open shells on the metals.

Our Recent Contributions Exploring the Characterization of Functional Materials with Wave Function-based Electronic Structure Theories

We have explored phenomena related to MOF-based gas separations, catalysis, and magnetic behaviour [1, 2]. We usually start our investigations by performing KS-DFT calculations to acquire initial insight into the structure of these materials and their fundamental electronic properties. One key questions is, for example, what is the spin multiplicity of the metals in the nodes (often, but not always, metal oxides) and also what is the overall spin of the material. Some of the most relevant

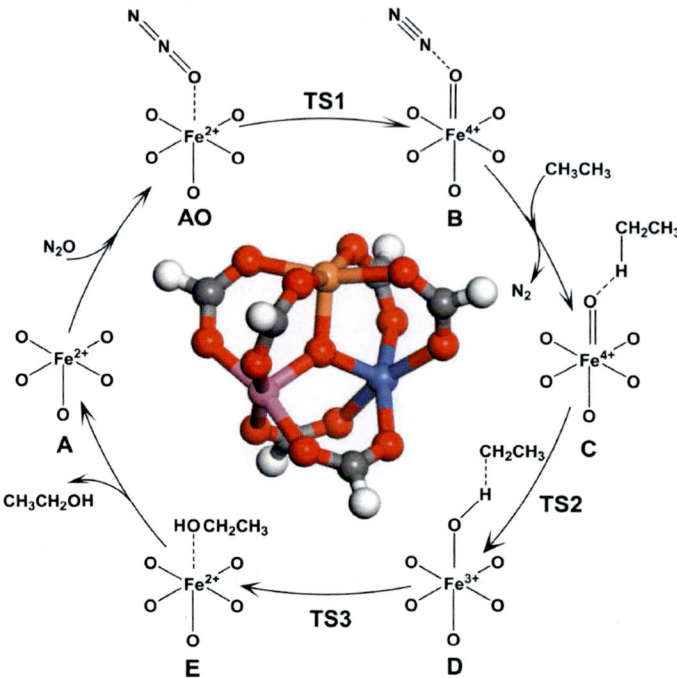

Fig. 1.   Model cluster $M_1(III)M_2(III)Fe(II)(\mu_3\text{-}O)(HCOO)_6$, where M_1 and M_2 can be Al, Cr, or Fe. Shown here with ethane to ethanol catalytic cycle. Reprinted with permission from Ref. [3]. Copyright 2018, American Chemical Society.

challenges for KS-DFT are the need for broken-symmetry solutions because of the multireference (MR) character, self-interaction error (SIE; sometimes referred to as delocalization error), and the limited degree of universality in specific density functionals. We faced these challenges in the study of single non-heme Fe(II) ions present in the nodes of MIL-100, MIL-101, and MIL-808 MOFs with promising catalytic potential for C–H bond activation [3]. We explored the reaction profiles converting methane and ethane to the corresponding alcohols ethanol and methanol, and we explored how the chemical composition of the MOF node might modulate the catalytic activity. Three metal sites are present in each node, one of which was held constant as Fe(II), while the other two were varied over Al(III), Cr(III), and Fe(III). Most work was conducted with DFT, using the M06-L functional, but multireference calculations based on a complete active space wave function, followed by second-order perturbation theory, CASPT2, were also performed for some cases.

Both M06-L and CASPT2 predicted high spin on all Fe and Cr centres, but the degree of difference was not consistent across methods. The primary use of the CASPT2 calculations was to justify performing the DFT reaction coordinate calculations on the high-spin surfaces. One reason why MR methodologies like CASPT2 tend not to be widely used is that they are very expensive and they are not as "black-box" as DFT. The user has to carefully select the active space as well as inspect the outcome of the calculation to ensure that it is physically meaningful. If the system under consideration has regions that are strongly correlated with one another, large active spaces may be needed, and the calculations can become prohibitively expensive. Such high computational cost together with the need for specialized expertise has slowed the mainstream use of MR theories in the computational study of MOFs and of functional materials more generally.

Another modern challenge in electronic structure theory is to predict a quantitatively accurate band gap and band structure for strongly correlated materials. Even within the weak correlation regime, the KS-DFT band gap cannot be interpreted as the fundamental band gap owing to the derivative discontinuity of the exchange-correlation energy. This band gap problem makes KS-DFT less appealing as an accurate band structure method for materials. Including Hartree–Fock (HF) exchange in hybrid functionals, or the on-site interaction (U) in DFT+U, may improve the performance in certain situations. Moving beyond DFT, GW has been widely considered as the method of choice for band gap/band structure predictions for weakly correlated systems. Recently, density matrix embedding theory (DMET) [4–6] has been proposed as a computationally cheaper alternative while offering similar accuracy for lattice models.

Outlook on the Further Evolution of Electronic Structure Theories for Functional Materials

An active field of research aims to reduce the cost of MR calculations and automate the choice of active space. Alternative options to CASPT2 for recovering electron correlation at low cost are in development, including multiconfiguration pair-density functional theory (MC-PDFT) [7].

Another challenge with MR methods is that even when they can be used relatively efficiently to obtain single-point energies, geometry optimization is even less black box in character, so usually one first generates potential energy surfaces with KS-DFT and then computes single-point energies with MR methods, which can obviously be problematic if the DFT potential energy surface is inaccurate. Recent analytical gradient implementations [8] for MC-PDFT may enable accurate geometry optimizations at this particular MR level. Regarding automating choice of active space, work has begun to appear in Refs. [9, 10].

Another important point meriting consideration is that common practice involves following a reaction along a single spin-state potential energy surface, but chemical reactions, especially in catalysis, may involve two different spin states, and KS DFT can suffer deficiencies in predicting the relative energies of such alternative states. Moreover, spin–orbit coupling may be an important factor controlling the propensity to move from one spin-state surface to another, and KS DFT is typically poorly suited to computing such coupling as at least one spin state is often ill described by a single determinant. To summarize, DFT is usually an economical *first* choice for modelling reactions and other phenomena of catalytic materials, but we believe that the field has advanced sufficiently that computational modelling should more regularly move a step beyond DFT to consider the use of more flexible wave function-based electronic structure theories, including MR methods.

In closing, we note that for prediction of the band structures and band gaps of highly correlated materials, extensions of DMET methods to the solid state by taking advantage of the local nature of Wannier functions and the translational symmetry of crystals are ongoing and they offer significant promise.

Acknowledgments

This work was supported by the Inorganometallic Catalyst Design Center, an EFRC funded by the U.S. Department of Energy, Office of Basic Energy Sciences under Award DE-SC0012702, and by the U.S. Department of Energy, Office of Basic Energy Sciences, Division of Chemical Sciences, Geosciences and Biosciences under Award DEFG02-17ER16362. Computer resources were provided by the Minnesota Super-computing Institute at the University of Minnesota.

References

[1] V. Bernales, M.A. Ortuño, D.G. Truhlar, C.J. Cramer, and L. Gagliardi, *ACS Cent. Sci.* **24**, 5–19 (2018).

[2] C.A. Gaggioli, S.J. Stoneburner, C.J. Cramer, and L. Gagliardi, *ACS Catal.* **9**, 8481 (2019).

[3] J.G. Vitillo, A. Bhan, C.J. Cramer, C.C. Lu, and L. Gagliardi, *ACS Catal.* **9**, 2870 (2019).

[4] G. Knizia and G.K.-L. Chan, *Phys. Rev. Lett.* **109**, 186404 (2012).

[5] H.Q. Pham, V. Bernales, and L. Gagliardi, *J. Chem. Theory Comp.* **14**, 1960 (2018).

[6] R. Pandharkar, M.R. Hermes, C.J. Cramer, and L. Gagliardi, *J. Phys. Chem. Lett.* **10**, 5507 (2019).

[7] L. Gagliardi, D.G. Truhlar, G. Li Manni, R.K. Carlson, C.E. Hoyer, and J.W.L. Bao, *Acc. Chem. Res.* **50**, 66 (2017).

[8] A.M. Sand, C.E. Hoyer, K. Sharkas, K.M. Kidder, R. Lindh, D.G. Truhlar, and L. Gagliardi, *J. Chem. Theory Comp.* **14**, 126 (2017).

[9] S. Keller, K. Boguslawski, T. Janowski, M. Reiher, and P. Pulay, *J. Chem. Phys.* **142**, 244104 (2015).

[10] J.J. Bao, S.S. Dong, L. Gagliardi, and D.G. Truhlar, *J. Chem. Theory Comp.* **14**, 2017 (2018).

TOWARDS IN-SILICO DESIGN OF FUNCTIONAL MATERIALS

KATARINA STANCIAKOVA and BERT M. WECKHUYSEN

Debye Institute for Nanomaterials Science, Utrecht University,
3584 CG Utrecht, The Netherlands

Our View of the Present State of Research on Modelling of Functional Materials

In the coming decades, society has to deal with enormous social and environmental challenges of which we ourselves are at least in part the cause. Our success crucially depends on how quickly and efficiently we can come up with sustainable solutions that preserve our standard of life, while being not environmentally harmful. In that respect, the development of new materials and chemicals that are sustainable and fulfil the principles of green chemistry, in which circularity (e.g., carbon-to-carbon concept) is a key aspect, is of the highest importance. A lot of effort is currently invested into the discovery of biomass-derived functional materials (e.g., green coatings) or new materials for CO_2 capture and storage (i.e., CCS), electrochemistry to not only convert water into H_2 but also to make more complex organics (i.e., electrochemical synthesis of organic molecules), as well as into the energy storage, as reviewed in some recent articles [1–4].

The rational design of new materials is not possible without the detailed knowledge of chemical processes on all scales, down from the atomic structure all the way up to macroscopic level. The state-of-the-art *operando* and *in-situ* measurements generate time-dependent dynamic data of functional nanomaterials, such as solid catalysts, batteries, and fuel cells that are relevant for the realistic and even real conditions, thus providing us with the most detailed view on the chemistry of these materials. On the contrary, the routine utilization of such measurements to systematically study existing as well as new materials is not yet possible, mainly due to the large pool of options one would have to explore and limited material, time, and human resources. Here, miniaturization as well as high-throughput approaches could be helpful, but still it is unclear how far we can push these developments in the decades to come. Hence, we will not be able to move towards the rational design of materials without the help of *in-silico* methods that have proved to be indispensable in the search for fundamental understanding during the past decades. This is well illustrated in the review of Catlow and co-workers on their application in the field of zeolite chemistry [5].

Thanks to the rapid evolution of computational power, we have already reached the accuracy required to reproduce experimental data, such as UV/VIS, IR, or NMR spectra. On the one hand, we are continuously increasing the precision of our models, while on the other we build complex models mimicking the behaviour of the materials under *operando* conditions on different time and various length scales [6]. However, the ultimate strength of computer simulations is in their predictive nature and the continuous development and application of new algorithms, such as machine learning, allows us to fully access their potential. Several successful examples already exist where modelling and high-throughput materials screening were used to guide the design of new materials, e.g., in a search for the best electrocatalyst for the hydrogen evolution reaction (HER) [7] or to engineer germanium-containing zeolites [8].

In what follows, we will show how the systematic computational research performed on top of advanced experimental studies allows one to go from description to explanation and eventually towards making predictions.

Our Recent Research Contributions to the Functional Design of Solid Catalysts

As a showcase example of functional design of solid catalysts, we will describe zeolites that are one of the most versatile microporous materials in the field of inorganic chemistry. They find widespread applications in many areas, such as gas adsorption and separation, ion exchange, environmental protection, as well as catalysis. Examples of the latter include fluid catalytic cracking (FCC), hydrocracking, and aromatization. The envisaged future applications of zeolite-based materials include the production of chemicals from oxygen-abundant alternatives, such as biomass and waste, including but not limited to alcohols, ketones, and aldehydes. One of the biggest advantages of zeolites is the option to tailor their catalytic and physical properties during and after the zeolite/their synthesis. The most common methods to prevent diffusion limitations, control the zeolite acidity, and improve their stability are post-synthesis desilication and/or dealumination that are usually harnessed by steam or acid treatment. In our earlier work, we have already shown that dealumination does not affect the structure of zeolites uniformly and a certain preference in the active sites that are being removed exists [9]. Therefore, if performed in a controlled manner, water-induced zeolite dealumination can be a tool to tune the stability and efficiency of the zeolite catalysts for current and future processes, which follows the principles of green chemistry.

In recent computational work [10], we have applied density functional theory (DFT) to systematically study the initial stage of water-induced dealumination in the well-known and widely used zeolite ZSM-5. This zeolite material has the framework code MFI and consists of a micropore structure in which two types of 10-membered ring (MR) channels, namely, a straight and a zigzag channel, are intersecting. The structure of ZSM-5 is illustrated in Fig. 1. By including two explicit

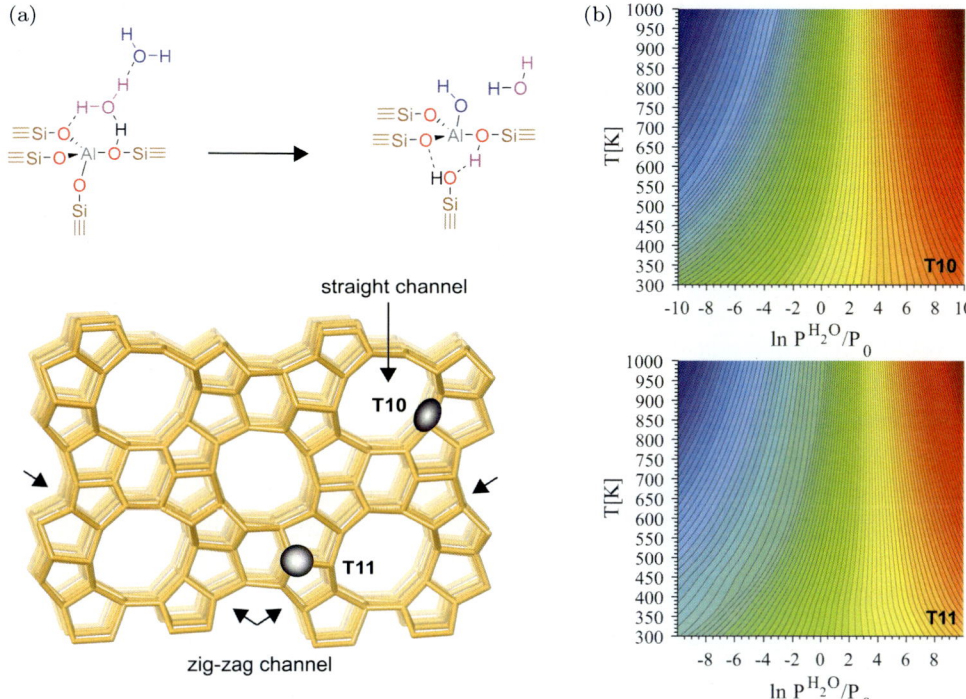

Fig. 1. In recent work, we have modelled the initial stage of the dealumination reaction (i.e., the first Al–O bond breaking) in zeolite ZSM-5 (a). Different catalytic site locations were considered, namely, the T10 (in the 10-membered ring (MR) zigzag channel) and the T11 (in the 10-MR straight channel) position. Reactivity diagrams (b) in which blue regions correspond to the lowest reaction rates and red regions correspond to the highest reaction rates for dealumination reveal that the relative reactivity of the sites depends on the reaction conditions. Adapted from Ref. 10.

water molecules into the zeolite model, we have shown that microsolvation alters the dealumination reaction landscape as well as its mechanism. More specifically, we have identified a dealumination mechanism, which is thermodynamically and kinetically preferred and involves the spontaneous breaking of the Al–O(H) bond due to coordination of the water molecules at the Al atom in the *anti*-position to the Brønsted acid site proton. At temperatures relevant for zeolite steaming, we find that the regioselectivity of Al sites during dealumination is not determined by the stability of the Al–O(H) bond, but rather by the accessibility and the solvation of the active Al site.

Based on these modelling results, we suggest that pressure-controlled dealumination can be used as a post-synthetic treatment to precisely tune the Al distribution and synthesize hydrothermally more stable and reactive zeolite-based catalysts, as illustrated in Fig. 1 [10]. However, the validity of this hypothesis yet must be confirmed by a.o. experimental spectroscopic and catalytic data, thereby reinforcing the concept that combining theory and experiment is a powerful approach, which has to go hand in hand to solve intricate problems in the functional materials science era.

Outlook on Future Developments of Research on Design of Functional Materials

The ultimate challenge for computational chemistry is the *ab-initio* development of new and up to now unknown functional materials from scratch. In that respect, the utilization of high-throughput benchmark methods will play a key role in the rational design of such functional materials. This can be achieved either via the development of new concepts that would be able to capture the essence of the desired material properties in terms that are simple enough to be computationally feasible, while maintaining the accuracy, or by use of algorithms, such as machine learning or neural networks.

Besides, with the constant growth of the computational power, the continuous growth in the accuracy of the models can be expected. With much more precise computer simulations in our hands, we can more accurately predict important physicochemical properties, such as adsorption, diffusion, or chemical reactivity, the optimization of which is crucial for the *in-silico* design of functional nanomaterials. Pairing experiments, preferentially performed under *operando* conditions, with advanced theory on real(istic) models will become routine and will provide more fundamental insights into the structure as well as function of materials.

Still, there is a lot of room for improvement and surely many developments are expected to take place in the upcoming decades. One of the main challenges is to couple the knowledge from one (i.e., time and length) scale to the other (e.g., micro-, meso-, and macroscale) and translate it into the terms of macroscopic behaviour, including chemical reactivity and stability [11]. A great example here is our understanding of catalytic reactions taking place in a porous solid. From an atomic point of view, the catalytic site can be defined by a group of a few atoms, but as we zoom out, the confinement effect, the presence of defects exceeding the unit cell lengths, and many other phenomena, including molecular transport limitations, start to affect the overall reaction landscape. Moreover, the structure of a catalyst material does not only depend on the reaction conditions alone but surely evolves with time due to its adaptive nature and physicochemical processes, such as catalyst ageing [11, 12].

Finally, special effort must be made for the improvement of the direct multiscale methods, such as QM/MM or MM/CG, that combine the strengths of the accuracy of computational methods with the efficiency that the calculations can be performed. Within the multiscale methods, we are still not able to sufficiently model the presence of explicit solvent molecules that would be able to freely diffuse over the whole system, which is the bottleneck for the realistic modelling of the dynamics of chemical reactions taking place in many functional materials or biological systems. In that respect, the development adaptive multiscale methods [13] that make use of machine learned potentials trained to the specific quantum chemistry method can be a feasible option. Clearly, materials and life scientists should inspire each other. By doing so, they will help in developing new frameworks of thinking to theoretically model complex systems in the most accurate and efficient manner.

Acknowledgments

Utrecht University (in the frame of the Strategic Theme Sustainability), the Netherlands Organization for Scientific Research (NWO, in the frame of the Gravitation program, MCEC, Multiscale Catalytic Energy Conversion), and the Advanced Research Center Chemical Building Blocks Consortium (ARC CBBC) are gratefully acknowledged for their financial support.

References

[1] P. Gallezot, *Chem. Soc. Rev.* **4**, 1538 (2012).
[2] D.M. D'Alessandro, B. Smit, and J.R. Long, *Angew. Chem. Int. Ed.* **49**, 6058 (2010).
[3] J.K. Nørskov, T. Bligaard, J. Rossmeisl, and C.H. Christensen, *Nat. Chem.* **1**, 37 (2009).
[4] F. Cheng, J. Liang, Z. Tao, and J. Chen, *Adv. Mater.* **23**, 1695 (2011).
[5] V. Van Speybroeck, K. Hemelsoet, L. Joos, M. Waroquier, R.G. Bell, and C.R.A. Catlow, *Chem. Soc. Rev.* **44**, 7044 (2015).
[6] S. Bailleul, S.M.J Rogge, L. Vanduyfhuys, and V. Van Speybroeck, *ChemCatChem* **11**, 3993 (2019).
[7] J. Greeley, T.F. Jaramillo, J. Bonde, I. Chorkendorff, and J.K. Nørskov, *Nat. Mater.* **5**, 909 (2006).
[8] Z. Jensen, E. Kim, S. Kwon, T.Z.H. Gani, Y. Román-Leshkov, M. Moliner, A. Corma, and E. Olivetti, *ACS Cent. Sci.* **5**, 892 (2019).
[9] L. Karwacki, D.A.M. de Winter, L.R. Aramburo, M.N. Lebbink, J.A. Post, M.R. Drury, and B.M. Weckhuysen, *Angew. Chem. Int. Ed.* **50**, 1294 (2011).
[10] K. Stanciakova, B. Ensing, F. Göltl, R.E. Bulo, and B.M. Weckhuysen, *ACS Catal.* **9**, 5119 (2019).
[11] A. Bruix, J.T. Margraf, M. Andersen, and K. Reuter, *Nat. Catal.* **2**, 659 (2019).
[12] J.S. Lim, N. Molinari, K. Duanmu, P. Sautet, and B. Kozinsky, *J. Phys. Chem. C*, **123**, 16332 (2019).
[13] J.M. Boereboom, R. Potestio, D. Donadio, and R.E. Bulo, *J. Chem. Theory Comput.* **12**, 3441 (2016).

SESSION 2: MODELING OF FUNCTIONAL MATERIALS

CHAIR: B. WECKHUYSEN
AUDITORS: F. DE PROFT[1], Y. OLIVIER[2]

[1] *Research Group of General Chemistry (ALGC), Vrije Universiteit Brussel (VUB),
Pleinlaan 2, 1050 Brussels, Belgium*
[2] *Unité de Chimie Physique Théorique et Structurale & Laboratoire de Physique du Solide,
Namur Institute of Structured Matter, Université de Namur,
Rue de Bruxelles, 61, 5000 Namur, Belgium*

Discussion among the panel members

<u>Bert Weckhuysen:</u> I would like to organize it as follows: first, I have prepared for every speaker a question, and then we go one after the other. After that, I would like to invite the different speakers to comment on each other and I have some meat to throw here, where you can discuss. Then we have a break, a coffee break; we come back and then we discuss a little bit with the whole audience here. I have a first question to Veronique Van Speybroeck: how generic is the mobility of the active sites and how can we engineer it?

<u>Veronique Van Speybroeck:</u> Thank you for the question. As I have already also explained in my statement, we have first seen these for Bronsted acidic sites, but it's seems much more generic than what I've shown here. I've shown also the example of the nickel zeolite, but we have also seen it recently for palladium zeolite. I think that it's interesting that we also see that these active sites basically can be engineered. For example, we have also seen the analogies with homogeneous catalysis and I think that this is also interesting that, depending on the conditions, like for example, the metal can detach from the material and resembles what we see in homogeneous catalysis and do that back and forth depending on the conditions. The ambition would be to try to find these conditions to indeed engineer it: it can be temperature, guest molecules. I think it's a rather generic effect. How to precisely engineer it? It has been seen in quite some examples, but to see it as a general feature to engineer it, I think it's quite still ambitious. Probably we also have to learn from what we have seen and try to also do smart analysis of the data that we have seen there and try to also learn from that. And therefore also, some techniques maybe that we heard in the morning might be interesting. I think we also might be inspired from other fields there.

<u>Bert Weckhuysen</u>: Thank you very much, Veronique. I would like to ask a question to Lee. You showed your chemical computer, your chemically encoded computations. I was very much intrigued by it, but I was just wondering... you showed it but what can we really learn and do in complex fields, like chemical-biology, material systems, synthesis, etc ?

<u>Lee Cronin</u>: That's a very good question, I think the challenge we had when we were trying to just do a chemical computation was just for the hell of it right, to start with. To say could we actually do the chemistry, do the computation in the chemistry rather than have a display screen? Because one of the problems is you could just get the chemistry, you would just output the result of something you put inside, so doing the solar autometer. I didn't show the rule 250, but there is a number of rules in there. We can show the chemical computer acts formally like a Turing machine. So, that means you should be able to take any calculation instantiating chemistry, but it would be very very slow if you try to do a calculation in a solar autometer. So that didn't feel very productive. Then we realized there is a whole host of problems and actually some of these kinds were inspired by a collaboration we've got with Alán Aspuru-Guzik funded by Darpa. To minimize energy in systems, there are lots of materials sciences systems, Monadic systems and also Nicola Spalding who was in ETH when I was talking about this once gave me a physics model, so she's very good, wants to take cosmological structures and maps them into materials science problems and uses the same mathematics. So, the short answer I believe is yes, we can read out. The energy minimizations are really good example, when we have a grid that is 100 by 100, that's 10000 stirrers. My group is kind of scared and hiding. That will do energy minimization faster than any silicon on the planet. And that's kind of cool because that means it's almost like a Monte Carlo method you wouldn't be able to out, so if you compare with an i9, so an i9 processor, with 10 billion transistors, that chemical computer would output it faster. And that's what we're going for.

<u>Bert Weckhuysen</u>: May I interrupt? When do you think you can make it?

<u>Lee Cronin</u>: We could make the 100 by 100 tomorrow, I think my group won't, so I think we will have to get a company to do it and once we've shown some kind of interesting applications exploring B-SAT problems and others that are suited to this type of architecture, I think we'll get a lot of assistance making it. So we could build it tomorrow, but the point is we've shown it works on a 7 by 7 so maybe going to 10 by 10 the problem we're going to have is when you go to a bigger grid is making sure you keep the oscillations synchronized, so there is an error correction problem. Not the same as quantum error correction problem but nevertheless an error correction problem. So, we need to prove what the limits are there. So technically tomorrow, but then there is going to be an error correction.

Bert Weckhuysen: Berend, I have a question for you, I was intrigued by the learning from the failed experiments, but I'm experimentalist, so I also relearn a lot from failed experiments. Can you explain me where does it come from and what is this good idea all about?

Berend Smit: It's a little bit what you observed, that if a PhD student or a postdoc start doing the synthesis, they get better and better with it, but then the problem is if they leave, all the knowledge gets lost. What we try to do is trying to quantify what it is, the chemical intuition which we feel not to turn to have a large effect on a property. The other potential this has is that you can do that across platforms, across different groups, or harvesting intuition on different type of reactions within the same field from different groups. The idea is basically advanced statistics to learn from what are the most sensitive things that happen if you go towards the successful reactions.

Bert Weckhuysen: What are the bottlenecks?

Berend Smit: The main bottleneck is that people don't publish the failed data in a systematic way. So, encouraging groups to report all their data in way you can access them, you can look at that. If you think about how much data is out there and if you can harvest them in a very intelligent way with electronic notebooks or other kind of things, there is a lot of things that are possible.

Bert Weckhuysen: Thank you very much. Then I turn to Miquel, you used the Baird rule for determining these species. I was just wondering if this rule is applicable to other situations to other species systems.

Miquel Solà: Yes, in principle, I showed you examples basically for the organic world, but also in the inorganic world we have several species, for instance, like Al_4^{2-} is an aromatic species that is aromatic in the sigma radial component, sigma tangential and π component with 2 electrons in each of these components and is Hückel aromatic for the 3 components. When you go to the Al_4^{4-}, you put two more electrons and in the singlet you have sigma tangential and sigma radial aromaticity, but the π is an antiaromatic molecule. However, if you go to the triplet then it becomes aromatic also in the π system. In fact, this molecule is more stable in the triplet state than in the ground state. There are other examples, for instance, $^5Ta_3$ is a molecule that has delta aromaticity in the Hückel mode, but it has π and sigma aromaticity in the Baird mode. It's an interesting example in which you have both aromaticities in the same molecule because you have them in the different components. And then there is all the field of photochemical reactions, there are many photochemical reactions that depend critically on the Baird aromaticity. There is for example the excited state intramolecular proton transfer, these ESIPT reactions. There is a phenol ring that transfers the electron to a nitrogen or something like that, in the ground state you get the keto form which is much less stable than the phenol form because the keto form is less aromatic, but when you go to the excited state, it is the other way around. In the excited state the phenol is antiaromatic and then it goes to the keto in order to relieve this antiaromaticity. There are many photochemical processes that can be understood taking into account this Baird aromaticity. There is a chemical review in 2014 by Henrik Ottosson and he shows many examples in which Baird aromaticity is fundamental.

Bert Weckhuysen: Laura, you showed molecules and periodic systems, chemistry in materials. How can we bridge these two fields?

Laura Gagliardi: Well, this is a challenge, but I think we have to work all together in that direction, because for example I see in the US the Department of Energy has really two domains, either you are a chemist or a materials scientist. I don't know, it's probably historical. I think that the challenge of scientists and engineers is to bridge this gap because molecules are important but some properties, some achievements can be made only with materials. But it's very important to try to understand materials really at the atomic and molecular level, so in terms of computations I think we can really try to combine different methodologies on different time scales and length scales, going from coarse grained simulations to atomic simulations and look at the details. But also, experimentally, it's important for example when one studies catalysis on these complex materials to try to really characterize the local structure and also in operando. So, we know what the structure is during the catalytic process, and we can eventually model these systems more realistically and predict novel materials with novel properties. But it's a challenge.

Bert Weckhuysen: Thank you. The last question of the first round is for Katerina. You showed this aluminum distribution and control in zeolites, ZSM5, but it's theory. So, I was just wondering: what about experiments? Do they confirm or is it in line? Is it plausible what you proposed?

Katerina Stanciakova: In this respect, I would say our knowledge is partial. From experiments we know that there is reduced active behavior if we apply steaming, under certain conditions, the aluminum preferentially moves from sinusoidal channels. The issue is the exact determination of these conditions, because it is really difficult to determine aluminum distribution exactly, namely because when we are doing some measurements usually, we are characterizing our sample on average. Then if we have trouble to determine aluminum distribution, it will be very difficult to find the conditions. However, I think that current developments in the field, such as synthesis of zeolite-oriented membranes which have all channels oriented in a certain way, can help to really improve our knowledge.

Bert Weckhuysen: Thank you. So, we had a first round of questions and I would like now to stir the discussion with one thing which I found a lot in different lectures that was: the importance of defects or I would say the flows of new heaven material which make and can be a perfect structure. And then you put it in to a reactor or in a device and then it actually develops, it's a kind of genesis in it and that goes towards a state where you have then defects or active sites. So, what is the importance, the relevance of defects, active sites and also what is the relevance of this whole genesis towards the complexity of the material? I thank almost every one of you who have somehow tackled it from one of the other sides. I was just curious. Who wants to comment on it? Who would like to start?

Lee Cronin: When it comes to materials complexity, I always find it really interesting that we can do a number of things, we can write down the Shannon information associated with the crystal structure. I really like what you were saying, thinking about design and experiments looking at electronic lab notebooks. So, if you have got two processes when you want to make a material. Is a material complex because the crystal structure is complex, or is it complex because the number of operations you have to do in the lab are long? This is really important. I think what we are doing in materials science and chemistry in particular is we're not thinking about the equivalence of those or the inequivalence of those. So for me the most important thing is how complex the material is, into how many unit operations it maps in my robot or for my PhD student, they're not the same, they make errors in different ways. So, I think if I can encourage us to think in that way, I will design an experiment so reproducibility becomes easier, and it's about what you can measure.

Berend Smit: If you look at our work on zeolites in metal organic framework, we always presume perfect crystals, infinitely large known surfaces. One of the reasons for that is the fact that the crystal structure is known, so we very much know what the structure is, and we have some confidence that what we do has some sense in reality. If you go to defects, then very often you can make a model where you get some inspiration of defects, like bonds are breaking or how the surface should look like. And we can even generate very sensible numbers, but it gets increasingly difficult if you ask the question: how is our defect structure, does that make any sense? So, the bottleneck I see here is often the characterization of defects, how can you make sure that the defect you invent on the computer has any sense in reality? And I see that as a big bottleneck for any defect.

Laura Gagliardi: May I also comment? I fully agree with what Berend says. On top of that, sometimes our experimental colleagues engineer defects on purpose to achieve certain reactivity, certain catalysis. And so again we would like to create reasonable defects in our models, but I think the problem is really complex. These defects evolve in the course of reaction, so maybe we start with a system that is realistic at the beginning but then during the reaction it changes. And how do we know?

Veronique Van Speybroeck: I would like to comment on that. This is a little bit what I wanted to emphasize also in my statement. Basically, there are three things, so from a modelling point of view we start bottom-up, from the molecular structure and we go to larger length scales and time scales. The question is where do we meet the experiment? They go from top-down zooming in the structures. When do we receive the information that is exactly comparable to what we model is a major question. And then indeed the question is things evolve operando, so how to merge that? For me it's a major challenge. We have to collaborate certainly between various fields of theory but also with experimentalists in close synergy, where we really learn from each other how they can go to spatial temporal resolution and operando and where can we meet each other. So that for me is one of the main challenges for future works, I'd say.

Bert Weckhuysen: As an experimentalist, I can further stress this point. Years ago I was involved in something here in Antwerp harbor. There was a reactor with a catalyst, and they loaded the same catalyst and then they gave the C3 or C4. And it took x days to C3 to come to its most active state. For C4 it was y and y was longer, it took longer time. Then the whole issue was that it shows first that the perfect material which was made, the so-called perfect, you put it in the reactor. It was not the active phase actually. At the beginning, it was a very lousy thing and it goes up. So, the active phase or defect phase, or whatever you called it, was formed in the reactor while it was running and by just switching one C to another, you get a totally different activation phase. At the end, the number of active sites was almost the same, but still it was a number which was very much below 100%. So, this kind of first operando type of measurements experimentally and then trying to link that to because we still only have often a signature of what the active sites are. We don't know what it actually is. We have certain assumptions and then it brings me to the question for the theoreticians: "ok look at these fingerprints, this is the behavior, this is what might my brain think could be the active site". And Laura made a very nice comment saying that the experimentalists do not have the experimental inputs for us to start to model or make it. So that brings me to how can we now have partial information from one site and partial information from the other sites, from experiment and theory, and how can we go toward a unified vision of this active phase in batteries, catalysts, whatever.

Laura Gagliardi: Actually, no. We have this center in energy frontier research center of catalysis of MOFs and it combines experimental and computational and sometimes the best we can do is find, design and model what is reasonable. It doesn't mean that it's real, but it's reasonable. We say, ok, if the catalyst has this structure and this is what happens. But for example, we deposit small metal oxide clusters on the node of a MOF and we don't know, we say they have atomic precision, but we don't know if it's one metal, two metals, three metals. They are small clusters, they are not nanoparticles, and at some point, I think actually the reality is that it's neither one nor the other. It's rather a coexistence of different nuclearities of clusters. We say ok, this model we decided to set up is reasonable, maybe it's not the real one but it's a good approximation to reality. And I think one has to be honest in this respect, and say this is the best we can do with the knowledge that we bring together.

Lee Cronin: One thing I find really interesting about when we're actually going through the process of making a model, doing some theory, is we're having to make some assumptions about what reality looks like. It's probably not late enough in the day to start questioning everything about our existence. But there are some situations where we simply don't know enough about reality to simulate it. This is the case in complex materials, we just heard. Catalyst aging is a classic one. In biology, animals, in understanding information transfer in biological systems, and there is going to be a point where we can't actually configure the simulation digitally to mirror what we see. So, then I wonder if we should start thinking about having hybrid computations. So, I give you the example of mayonnaise. Why simulate the self-assembly, the production of mayonnaise when you can just video its production with a high-speed video camera and then do image recognition and figure out what's going on at the phase boundary? So, I'm wondering, maybe what my co-panelists may think about this idea of closing the loop between experiment and theory before you make the next decision. It allows us to outsource the boring bad part of the computation into the substrate, and I don't know whether materials scientists are thinking about this more at the moment. In biology you have to do a bit more as it becomes messy.

Miquel Solà: Yes, maybe just to add one thing about the fact that we are using models and because of that of course this is always an approximation. Just remember a sentence of Dewar who said that we should not ask if a model is true or not, we should ask if they are useful or not. This it I think the point, at the end. If we perform simulations and they are useful to the experimentalists, I think we should continue, if not maybe...

Bert Weckhuysen: One final question which actually is directed to Laura and Veronique. Actually, if I recall well, you both stressed electronic structure, but at the end we are now discussing a lot about defects, structural aspects and not so much electronic aspects. What is now the state on electronic types of calculations going to electronic structure?

Laura Gagliardi: So, yes, the structure, the geometric structure is very important, but it may be related to the electronic structure, because for example there is a certain type of interaction between the ligand and the node that can be either guided by electronics or by steric effects. So, let's divide them up. I think the workhorse is DFT for electronic structure calculations and I think it works well

99% of the times but it's the opposite of a lottery. The time where it does not work you don't know why and then what do you do? Actually, you don't know if it works or not. In many cases, because we don't have benchmarks and that's a problem. We also need experiments to benchmark our calculations if we cannot do higher level calculations. So, one can say, ok, I construct this micro toy models and I run these super expensive calculations and I think this is one of the tools that we need, but we need to go from there to machine learning. As I was saying in my talk, where we machine learn vs DFT we will do as well as DFT in most of the cases, so one should use all these different approaches.

Veronique Van Speybroeck: So, we have seen first also in your talk that was very interesting that you describe very nicely that for very specific active sites you need very d electronic structure methods. The question is: you cannot expand it to very large size systems. So, should we remain to an atomic description for the whole system? For example you already mentioned that we have QM/MM, but maybe we have to make some part of the system in a coarser way and then try to learn from the cluster models where we have these more advanced electronic structure methods? Think also about fundamentally new models to even incorporate that complicated electronic structure, which is very important we know for also some MOFs but also other structures. The question that I have is indeed: can we still do it with all atomistic models? I believe sometimes we have to work with various levels of models where we can learn from. So you mentioned QM/MM, but maybe we even have to go to coarser models, to do that in the future. I don't know.

Laura Gagliardi: I fully agree with you and a multiscalar approach is needed where to combine all these different scales.

Bert Weckhuysen: I think I will close, and I think at least I need a coffee. So, let's have a break of half an hour and we reconvene at 16:20. Thank you.

General discussion

Bert Weckhuysen: Welcome back, we have now one hour to discuss this afternoon session and now it's up to you. We are looking forward to some questions and actually, I've found one volunteer and I thought that Joanna could start.

Joanna Aizenberg: I want to come back and a little bit expand on the question that I asked before about defects. They were somewhat addressed by your discussion and by examples that you presented. But I now want to add a little bit to that. When we are talking about real world materials, how about understanding of impurities? And let's say impurities as a word has a negative function. But how about additives? And can you actually give us ways of predict, control materials when we add active elements that help us or account for impurities that may actually affect the performance more so than even the carrier material itself?

Lee Cronin: So, we have built a robot that basically takes the literature and tries to make the literature. I'm just going to pick an example, not a defect, just organic molecules, just to illustrate an answer to our question, to talk about impurities and robustness and there is nothing that acts on. What does the robot do? It reads the literature and gives you a recipe, a list of things to buy from Millipore, Aldrich whatever, plug them in and you run the process. But then what it does is in parallel, it doesn't just do once, it has 6 instances of the reaction and we vary all the parameters in terms of the time, the temperature, the order of addition and also the errors in the addition. The other thing we've been doing is looking at different reagent grades, so five nines, three nines, 98%, garbage you get off the floor, and to try building a robustness index. Isn't it that she's building on some of the nice work that Frank Glorius has been publishing in his own catalytic systems to see if he can come up with a general robustness score for synthesizing a molecule? Now, it comes to your question. It would be great if we could almost have like a GitHub, I don't know if people are familiar with GitHub but it's like a programming hub that Microsoft has now bought sadly and it will break it by next year but it works at the moment. People write codes and you can copy people's code and inversion it and fork it, let's just say you put a new math you know in your GitHub. I download the code, modify, write down what I've modified and then remake it. What then needs to be in this area is once we have good essays, we can also put back in and say that the material functions as expected, catalytic activity, maybe band gap, maybe some other essay. So, I think it's important to basically have those things put together, so the way to cover an impurity is to basically make sure you have got a big database, and you have got a good design of experiments, you vary not just process parameters per reagent quality, and also you can add mess and not clean things. So, I don't know if that kind of answer's part of your question. I think personally simulating impurities is hard and I would say why to do it? Why not just make the materials with impurities in it and see how messes it up the assay.

Joanna Aizenberg: Lee you focused on impurities, I had two sides of that, another one is additives, where you actually add something that improves your reaction. In Nature, chaperones are all over the place where you have an active molecule and something else helping it. So how about additives?

Lee Cronin: I think the same answer goes for that. One person's impurity is the other person's additive, so you have to work out what it is. I think if you could some-how have the right essay to measure the improvement as you add in the additives. Because there are some things in formulation science for instance where people add things and they don't' know why. I think your question is really well posed. I think it's interesting and I think we lack the proper design for experiments to capture that and I'd say we should be capturing this and maybe some of my colleagues are working in the robotic area and also in the kind of integration area which could challenge the modelling side to maybe account for that.

Laura Gagliardi: I can give two examples, one is in catalysis, these catalytic processes that occur at the node of MOFs are really defect-guided. The catalytic performance depends on the nature of the defects and so from the experimental point of view, we would like our colleagues to fully characterize these defects, to tell us exactly how the defects look like, otherwise we are just trying different possibilities and maybe we have agreement in vibrational frequencies, we can say ok, this is the structure that is really present in your sample. And I don't know how simple it is to characterize defects. Also, the other thing that I think is important is the concentration of the defects. If I construct a finite model with my defect but in the real material how many of the defects do you have? This is an example and I was telling before this example: I showed about iron MOFs, the idea is that you have these 3 iron centers in your node and one iron out of three is reduced to 2+ and this is the one where the catalysis occurs. In theory, we say that one third of the iron centers, so 33% is catalytically active. Our colleagues, the experimentalists, made the material following all the prescriptions in the literature and then at the end of the day they did the catalysis and only 4% of the irons were catalytically active. We don't know why, I mean, because perhaps there are defects in these materials and I think it's very important to be honest, both the experimentalists and the theorists, as I was saying and then I stop in the MOF community. And I don't want to criticize anybody, but there is this tendency of saying "I made the next material with the largest surface area and the best catalytic behavior" but they have no idea what it is and what is responsible for the catalysis.

Berend Smit: We have looked at the possibilities also in MOF UO66, where you can replace or dope Ti with Nb or other metals just to improve the optical properties. What we then tried to do is to look at the range of metals you can replace and see which one was working. Then actually Nb worked very well and we tried to synthesize it; at the end of the day, we basically had to use the computations to justify that experimentally we made the material. Because the question is which Nb or which Ti is replaced by Nb and how do they distribute. But none of the conventional analytical techniques could give us any clue that our model at the molecular level made any sense, except that the optical properties match. And this is the essence of my point: you can try these models, you can do anything, you can calculate anything you like, but it's very difficult to justify that such defects, such doping is actually what you have in the real world. And there we never succeeded in that and that is always my hesitation.

Veronique Van Speybroeck: Maybe I can also add on that, regarding also the catalytic active sites like Laura mentioned. It can indeed be modified by adding something in the neighborhood, we have also seen a recent example where we added in

Bronsted acidic zeolites a calcium species in a pro-synthetic way and that improved to a major extent the selectivity and the lifetime of this material. But then the question is, of course, how is the calcium sitting in there? So that's very difficult also from experiments to know that and then we saw that we had to go back and forth. It could be mononuclear, binuclear, trinuclear species, and this would come as some sort of extra elements which are not necessary so close to your active site, but they also contributed function. This is something that we need to acknowledge that basically, if you also, for example, in catalysis or maybe in other materials, it is not really to look locally at what happens there but somewhere a little bit further in the neighborhood, it can also affect the function. And then my believe is that I am more and more convinced of that, that we have to go back and forth with the experiment. So it's not like we are going to predict something and then you have to go back to the experimentalists and see whether they can find some sort of validation, another spectroscopic method, you go back to your model and you can learn from each other. That is how we try to tackle this because it's not so trivial to do this modelling on these kinds of complicated systems.

Bert Weckhuysen: I would like to add two things. If you say additive and you think about catalysis, then I always say an additive you add to something and at the beginning it's a promoter and if you add too much it becomes a poison. You can often see a kind of maximum and that is very funny because it's like you are in the kitchen and you have salt, you add a little bit salt and you like your dish and if you add too much you dislike it. From an experimental point of view, it's so difficult as already mentioned. Often the additives have a very low Z or a very high Z, atomic number. What you have is that all the methods you tried to use failed; transition metal ions are actually easy to study because it has a d-d or whatever so you can follow it. But when you have bismuth and you want to do X-rays, spectroscopy, then you need to have certain things in the synchrotron and it is complicated because you cannot get light through something because it's too much absorbing, so it's a pain. Often additives are interesting, low amounts, but also a pain from a spectroscopic, analytical point of view.

Kurt Wüthrich: It is. I mean again I'm an outsider to your field and it has not become clear to me yet what is really simulated. Are you simulating binding affinity, are you simulating binding specificity, are you simulating turnover rates? What is it that you are really simulating and how do you compare to experiments? Do we have experiments that give us these sub-steps of the overall catalysis efficiency?

Katerina Stanciakova: I think this is exactly the point that very often we, computational people, we can calculate absorption energies, affinity and so on. But then when we talk to experimentalists, we find it difficult to find overlap. I can tell you examples of computing heat of absorption and measuring it in the lab, and I think it's really important that we try to put the knowledge together and each side always tries to well define what it is really doing. And I think it needs to be improved.

Laura Gagliardi: What we can calculate are structures, geometric information and vibrational frequencies. Sometimes, what we can do is compute the vibrational frequencies, since they are also measured. They don't know the structure. We know the structure that corresponds to certain ones. Then, the match between theory and experiment can help in the assignments of these modes and also in principle one can from structural information that comes from the calculations try to fit these in a NEXAFS spectrum. And again, I try to match the experimental and the computed structure. And then, for catalysis, what we can compute are the reaction steps, i.e., the energy it takes to break a bond, what is the barrier for that bond, what is the relative energies of two intermediates along the reaction. And then, if we want to compare that with what is measured, e.g. turnover frequencies, we need to really use these energies into a microkinetic model, so that this can be directly comparable with experiment.

We have this rate constant and making an error in a binding affinity of 1kcal/mol in the computation will give a rate constant off by several orders of magnitude. So I think, at the end of the day, what we are good at, is not looking at absolute values, but trends. So, the absolute value of the binding energies, I don't think we can determine spot on what is measured, but when we compare systems where we change the transition metal or where we change the inorganic component, where we create more defects, we can look at trends. This should resemble what is measured.

Miquel Solà: Just to add that, in homogeneous catalysis, it is relatively easy to calculate the barriers and the stabilization of the intermediates and then from here there are different models, for example the Kozuch-Shaik model to obtain turnover frequencies and turnover numbers. So this is in principle something that you can compare exactly with experimental values and see if your predictions are correct or not.

Kurt Wüthrich: Is this being done systematically to calibrate your parameters?

Miquel Solà: What people do is sometimes systematic. But in probably most of the cases it is not. It is something that should be done.

Kurt Wüthrich: What is the status of the force fields that you use? We heard this morning that, as soon as you go away from organic synthetic chemistry, we have problems with the force fields. What is the status of this part of your efforts?

Miquel Solà: In my case, I'm talking about DFT, not using FFs. So it is in principle a solution of the Schrödinger equation with DFT.

Laura Gagliardi: It depends on the systems, so for biological systems, proteins, FFs are pretty good, I would say. For systems with transition metals and also in catalysis, the problem is that you are breaking and making bonds, so the FFs, unless you have a reactive FF that really describes the change in electronic structure quantum-mechanically, you don't capture that. So one needs lots of experimental data to parametrize good FFs, and for some systems there are more experimental data than others. In catalysis, in modelling these functional materials that were mentioning today, we are not so ahead. One could do ab-initio simulations but then, like Car-Parinello or ab-initio MD, but the problem with them is that the timescale of those simulations is very short, you can run them for like picoseconds. Instead one might want to look at phenomena at larger timescales so the outcome is relatively informative.

Ben Feringa: I have 2 questions, if the chairman allows me, one is this. I was intrigued by some presentations where it was mentioned that you could calculate now, or simulate the movement of atoms at the surface, that means that heterogeneous and homogeneous catalysis are coming closer to each other?

I was earlier this week in Switzerland at this nano-surface conference with the whole STM community and one message I took home from this conference is that it is extremely difficult to predict how molecules will adsorb surfaces, let alone, in what pattern they form. So can you really see the dynamics at the surface and predict how the molecule will then adsorb or how it will interact with its neighbors. Because that intrigues me a lot, since it will guide us to make better catalysts of course, or surface assembled structures. So the real issue is how to predict what rearrangement you get?

Veronique Van Speybroeck: Regarding the adsorption, yes indeed we use these MD simulations to study the adsorption at the surfaces. It then becomes very important to do this also at really true conditions. E.g. we have seen that we have an adsorption of an alkene at a zeolite, than a lot of people would predict they form alkoxides, so a covalent bond with the surfaces, but we see, in some processes, if you go to really high temperatures, they behave completely different, and they become rather mobile species in the pores of the material. So that is something we have learned, that, regarding this adsorption, that, yes, we can model it, but one needs to do it at the conditions where the system is normally performed, in its true nature where is does its function.

It is not good to do static or 0 K calculations, because these species are in a lot of cases not representative for the true intermediates at operating conditions. And it is very difficult then to compare with experiments, because how are you then comparing with experiments? Because you then also need this data from experiments. You often do not have adsorption data of illusive intermediates for example at these conditions.

Ben Feringa: Yes, and then the dynamics. You get information from spectroscopy, I suppose?

<u>Veronique Van Speybroeck</u>: Yes, there we try indeed also to go back and forth with operando spectroscopy. They can also follow these conditions, what intermediates you have. But sometimes you are confronted with intermediates which are very illusive, with which you can do a lot of things and it becomes very very difficult. We try to simulate this indeed on the fly on the surface. This is basically what we try to do.

<u>Katerina Stanciakova</u>: What I see as a problem, when you want to couple these to experiments is that, if you have catalyst particles like platinum or nickel, it is very difficult to really know which type of facet you should use for modelling of your adsorption. And this really increases the complexity and then you are on one side, either you simplify your model to static calculations, or you just cannot test anything. Another problem is that your catalyst might change in time and there might be a restructuralization. A huge question is how we incorporate this into our models.

<u>Bert Weckhuysen</u>: I would like to come back to Veronique. I mentioned earlier about her system and comparison with this homogeneous catalysis. There is another system out, a paper, a few years ago, by an American team. They published in Science copper zeolites, and the copper zeolites are used in diesel exhaust catalysts and they interact with NO and ammonia. And also their system is a system where copper is mobile, interacts, and has two cycles, which has a lot of correspondence with homogeneous catalysis. And people are still arguing in the field, is it now mononuclear or binuclear copper? If the dynamics are taking place at a speed where the event is taking place but your spectroscopy measurement is slower, than you will see a bit of both. The second thing is, you stated about that you went to this conference, it was STM I belief?

<u>Ben Feringa</u>: (in background, not understandable) . . . especially also the dynamics you know about movements on surfaces and so on.

<u>Bert Weckhuysen</u>. Did they measure it at 100°C or 200°C?

<u>Ben Feringa</u>: No, either at low temperature or ambient conditions. So that is also a big gap there between what they see, and what you see.

<u>Bert Weckhuysen</u>: Almost all experiments which I have seen with TEM, certain scanning probe methods where you can increase temperature, there I must say: or it is very difficult to measure it due to the instability, or it is also because there are somehow molecules on the move. One can argue, for example with electron microscopy, that there is beam damage: you induce energy into the system. So than you don't know what you are measuring. Is it perturbing?

Ben Feringa: My second question, and this is probably directed to Lee Cronin, you mentioned about chemical computation and showed us this example, very fascinating. But when you think about computation, you think about processing speed. When I thought about what you were mentioning, shouldn't you call it, instead of chemical computation, chemical information processing for function? I give you an alternative and I want to hear your reaction.

Lee Cronin: You can call it whatever you want Ben. I think I understand what you mean. There are 2 points. The first one is, why put all the resources in computing what is really complicated, complex, hard whatever, rather than make them and measure them in context? That is the first point, I could almost imagine a hybrid simulation whereby I am taking isomorphic behavior in a material and using that to compute the best type of material, making a different material class. That would be a really interesting idea topologically.

Now, let's turn to a chemical computer. We started with the first iteration, which is just having stirrers in the center which turned the local oscillations on and off and you had a global patterning. And all you did there is, you encoded patterns, and that is fine because you had 2^{25} discrete input states and it is about the parallelization rather than the net speed. So if you compute a lot of discrete states, and then show how they are processed, then you can do a computation. And that is why we have nearest-neighbor things in there. Now, you are right, if I can now map the computation to, say, a material problem simulating defect migration, or something like this, then you'll turn chemical information processing, what for functional materials is pretty good. Because that is exactly what is does. And it kind of reminds me of some of the stuff George Whitesides is doing: this soft robotics, rather than outsourcing like, he had this elastomer that would pick up an egg, and one uses the polymer to form around the egg, you don't need any feedback control, and all the computation that you'd do to not cross the egg, if you have a metal terminator, or a human being, is done in the elastomer, okay. So that is a really nice way of outsourcing the computation. So I agree with you to some degree.

Ben Feringa: That is great Lee, this is the first time we agree on some point, no? I'm joking.

Lee Cronin: Let me make one final point. I am really inspired by quantum computing and the small niche of very important algorithms that would be implemented in quantum computers, and I want to ask the question: could we do analog computing in a chemical system, digitally read in, digitally define the code and read out, and then use, in the end, not just oscillations but go down to the single molecule limit, because an Avogadro of molecules is potentially an Avogadro of computing power, which beats any CPU on the planet.

Ben Feringa: I have some idea, I think we can do it. You know, we work on switches.

Lee Cronin: So that's twice we agree.

Andreas Walther: I have a question on the catalysis and on secondary influences. I think you said that catalysis on enzymes can be quite nicely explained, but there is an example when you do direct evolution of enzymes, you make a modification on the enzyme someway, you put a charged group somewhere really far away from the active center and it amplifies the catalytic activity, and it really doesn't have any chemical influence on the catalytically active center. And the second example that I would make, and I'm not sure how much this is influencing the modelling, but for what we know that if we put it in confinement, the water structure would change. So how much do you consider how the solvent for example in zeolites changes its structure, changes its activity, and how this actually does influence the catalytic process?

Laura Gagliardi: The first point is basically, going beyond the active site, and it is certainly very important for the reasons you just mentioned. So, from the modelling point of view, should we take that into account? Which means, if we, for example, want to use one of these layered approaches like QM/MM, we have to make the QM region bigger, because maybe it is not enough to treat this region beyond the active site classically, like point charges. So this means that we should always make sure that the size of our models are converged, that if we increase the active site, or the area just beyond, we don't capture new effects, that otherwise we wouldn't capture. And the same is true about solvents, so the question is, you make the material maybe in solvent, but then; I don't know, sometimes, during catalytic processes, the solvent is removed, so there is dehydration. But then, maybe some solvent molecules remain attached to your material, so the question is: do we have to treat them as non-innocent solvent molecules, so explicitly? Thus it is not enough to have an implicit solvent model and also maybe the solvent reorganizes dynamically, so it is not enough to treat it in a static way. For example, I think, in your modelling, Veronique, you treat the solvent dynamically and you treat it explicitly also, but these are very expensive calculations and sometimes the timeframe that you can capture, I don't know, is it always long enough to describe the phenomena of interest? So I hope I have answered your question.

Veronique Van Speybroeck: Regarding the water in these nano-porous materials, let's say so. We have done that in zeolites, where we have seen that it is very important. I have also shown it in one of the movies, so it really can change the active site so the active site can become mobile, or the transition state can be facilitated by presence of water. And in some processes, also water is produced during the chemical reaction and it can be beneficial so it is really necessary to take it into account. If we, for example, go to MOFs, like for example we have done a lot of modelling on zirconium-based MOFs, where the coordination number can change in a rather large way, depending on the presence of solvents like water, or other protic molecules, we can see, for example, that you can have an over-coordination. Then, for example, a linker opens, due to the fact that a protic molecule comes. These are really dynamic changes, that you see on the active site, due to the presence of the solvent. So, if you do not take it into account, you often do not have the right representation of your system, let's say. We have seen many examples, where it is really important. Of course, from a modelling type of view, if you do that in an explicit way; we do this; we take into account an explicit number of solvent molecules, the question is always, how many molecules do I have to take into a pore? It's not so trivial to have this comparison with experiments. For me it's not so evident. So we have some thermodynamic models that estimate how many molecules you need in the pores of your material but it is really not simple. But that it is important, is certainly the case.

Miquel Solà: Something about these allosteric interactions, this directed evolution that you mentioned. I have seen several papers, they do the mutation and they do MD and by changing the mutation, despite that the mutation is quite far from the active site, there is some change in the conformation of the active site, so it opens a channel, or there is a reactant that approaches the active site, there is something that is made by this change, that could be quite far away from the active site. So this is something that could be studied by MD.

Mark Ellisman: First, I was going to ask why there wasn't more electron microscopy in the analysis, but your answer was I think, the damage is your concern, that I still would wonder. But I was more motivated by something that Lee said, and I wrote it down: "information transfer in biological systems", was the quote that I think that he was referring to. So, I wonder, because this is some sort of a hybrid meaning, with biology and non-biology, for lack of a batter handle. If it isn't appropriate for at least some of the discussion now, to talk about biologically inspired advances in your field. For example, when you talk about local catalysis and implanting in complex materials such sites. If you think about what biology does, on a very small scale, it does self-repair. A little different than hidden healing of a semi-conductor device, it does it locally based on need, presumes synapses are made and broken intercellularly, you turn over molecules and replace them. So is there anything that comes to mind from such an inspired group, regarding how you would take advantage of what we are learning from biology, about how biological systems, actually remodel themselves across scales?

Lee Cronin: So, your question is: how can we be inspired by biology models across scales? I think, one thing I have been trying a lot is how we can review the problem as one big integrated problem. Because biology doesn't do anything other than act or persist in its fitness landscape, and I think, people increasingly are designing their algorithms and their robotics, and their materials separately. I wonder if they should come together more, and we should think about an integrated fitness function. Now, that's not exactly that, because one part of my other day job is that I try to make life forms. How does an object persist in time? I really get infuriated when the word 'self-repair' is used in the wrong way. You are quite right, biology does do self-repair because it uses resources, it mobilizes resources to actually affect the system, it's dissipative generally, and creates waste products. And I am wondering if you could use that kind of approach, that idea, when you are developing systems that will develop a, say, new type of quantum dot, and you are allowing some kind of selection process to occur at a much bigger scale.

But the simple answer to your question is: not enough people are thinking holistically because the problem is really hard to write down, and it comes back to the question: what are you measuring? And I think Joanna was mentioning this earlier, what are you actually measuring, what is your fitness function, what is your requirement? So, I think, once we've got better assays, say, a room temperature superconductor, well you might write down what that might look like, build a fitness function for it and then go search. How might we create complex materials, or materials that have more than a handful of parameters to define the outcome, is what we can learn from biology. And now we need to code it, someone you mentioned, like the Materials Genome today. Now we need to actually start using that encoding properly, rather than metaphorically. I don't know if that answers your question.

Mark Ellisman: It's a start, looking for more activation, if you like that. I was trying to catalyze thinking about where biology might inspire you to think differently than what we heard.

Bert Weckhuysen: I would like to answer the two questions you stated. First to follow up, in our field we are always inspired by the life sciences and the enzymes. We find them beautiful and, for example for zeolites, we have even combined them, 'zeozymes' we call them, zeolitic enzymes. And it is not just a word, people have also tried to see if they could make in a zeolite states of an active component, which would start to look like an active site in an enzyme. There are only a few examples of an 'entactic' state, where people have been trying to lock a certain copper coordination environment which was a rather unusual thing, which was held or kept by the structure or the structural component of the zeolite, the caging type thing. And people have also been trying to – in these cages – to see if they could get local pH gradients or concentration gradients which are different from the outside, that if you have a zeolite or a crystal, and you have locally something different than globally. Somehow that is happening. But I also have to admit, first, it is very difficult to measure and proof; second, there are not many examples that I am aware of, that really made it beyond saying: "oh yeah, I have copper with 'nitrogen-nitrogen-nitrogen oxygen' like galactose oxidase".

Your second question was about electron microscopy methods. Catalysis, in heterogeneous catalysis at least, there the temperatures and pressures are totally different from enzymes, well with exceptions from those found near volcanoes or whatever, and from there on all your methods start to be blurring, or you have to shoot energy in it and you can change energy states, that is what I already referred to. So yes, we do CryoTEM as well, but what does this tells us from what is really happening at that high temperature? That is still a question.

Alán Aspuru-Guzik: One point about bio-inspiration. At the Solvay conference of 2010, there were a lot of discussions about photosynthesis and artificial light harvesting. So you might want to look at the proceedings of this. There were tons of discussions, 9 years ago, when I was here, as a younger version of myself. So now, rather than dissipation and refueling, I think I have a question for Lee and his molecular computer. If you want to keep this thing going on, obviously you got lost of signal. How or are you going to refuel the chemicals such that the thing keeps just beating forever ?

Lee Cronin: You might have seen in the photograph pipes coming in and out, so actually the system is continuously being fed and emptied, so that's really important, because the BZ reaction in that kind of volume is about a liter, I guess. It starts to slow down after a few hours, it is very temperature sensitive. What it is actually doing, it is continuously feeding in / taking out so it goes forever. That's a simple answer to a good question.

Bartosz Grzybowski: So, two short questions. Do you think that there will be some examples of materials that are completely new, that computational methods can actually discover? Let me argue this way, that MOFs are based on arguably one of the simplest metallo-organic chemistry. From what Omar Yaghi is telling me when he was telling people it's possible, everybody told him it is impossible. In hindsight, you could argue that this could have been discovered by computer almost, once you know that a MOF can exist, it looks very logical and very simple. So are we going to see a new class of materials discovered by computer? And second part of the question: do you think we could have materials that actually think, store information and process them? We have 'memristors' that can store information. Do you think that, beyond what Ben Feringa is doing, his materials with switching capability, do you think, ultimately we can progress to materials that process information and learn how to think?

Laura Gagliardi: When MOFs were discovered, perhaps computers and computational models were not yet advanced enough to do this sort of high-throughput prediction. And probably now, yes, if you combine machine learning with some benchmark either from experiment or high-level theory, one could in principle discover new materials. I think there is always the gap between the lab and the computer, in the sense that maybe in the computer the material can exist, but then it is also a matter of making it and maybe, when you make it in the lab, there are competing reactions or the material could appear to be in a different phase compared to what you had predicted. I think that now we have so many computational tools, and also hardware tools, that this could happen.

Bartosz Grzybowski: So what should happen like so far every material is discovered. Be it, carbon nanotubes, buckyballs, Haber-Bosch catalyst 100 years ago. What kind of tools should we apply to discover completely new materials?

Laura Gagliardi: For example pillared materials where you have also for example Omar Yaghi who has talked about these multivariant MOFs, where you have, not just a linker and a node, but you have different links, so maybe different compositions of materials but also different phases, so maybe you want to make a perovskite and instead with the same composition, you make a different phase of the material with different properties.

Miquel Solà: C60 was discovered, let's say, first theoretically, it was a calculation of 1970 some Japanese, I don't remember what was the name, but it was first computationally and then experimentally.

Alán Aspuru-Guzik: And also, many solid-state materials have been predicted too, before they were made; inorganic solids, oxides, things like that.

Lee Cronin: Nicolas Bolden did invent some classes of solid-state materials that were then found, and then there are also topological insulators, so there is quite a few solid-state chemistry I think. I have to go and look carefully. But just to answer the second part of your question: 'do you think we will be out to invent materials that think'. That is a very big question for now, maybe the subject will have its own conference, one day. My argument is, you think biology produces abstracting machines, and we don't know how that works. No we would argue that one big problem is that we think that mathematics exist and it does in our heads, but does it in an objective reality? This is one of the arguments that I have at my students arguably in the bar, when we are trying to work out how we simulate things and do things.

But the short answer to your question is: the brain has a trillion neurons and between 1000-10000 connections per neuron, so there are more possible configurations in your brain than there are atoms in the universe so trying to make a thinking machine in silicon is hard let's say impossible, although that's a hard word to use. My feeling is that making materials that can compute may be the first step to how we could make non-living abstracting machines, if that makes any sense. Wouldn't it be great if a gel, a brain gel could invent mathematics, and that is kind of my next project after the computer.

Thomas Hermans: I have a question for Lee about the chemical computer. So, if you look at the energy dissipation that your system uses vs. an i9 (Intel Core i9 red) I mean, your laptop gets pretty hot so consumes chemical energy from your battery to your processor which is dissipated as heat. So how does your dissipation rate compare in terms of efficiency vs. an i9?

Lee Cronin: Thanks for that. I put that in the cover letter for the paper actually, which was hilarious. So, an i9, you can fry an egg on it, right, it does about 230 Watts full power, and we calculated that basically the BZ reaction is about a billion times less. However, you do have a silicon processor using the neural net with classification, so there is some processing, but we are doing it with a much simpler processor that only uses 50 Watts, so it is considerably less, and as the things scale up, when you got this 10000 element array, which will compete with an i9, or maybe even a supercomputer, I'm yet to adjust the scale, I just don't know the answer, I don't know how it scales yet. It might be literally trillions of times different. And that is kind of interesting. Your brain doesn't use as much energy but 20 Watts I think. Yet an i9, which is nowhere near uses 230. You can't fry an egg on your head.

Eugene Shakhnovich: Following up on this kind of biological sym applied to what you do with materials design. Of course, the most powerful idea of biology is evolution, and the question is, and there was also some mention about fitness function. So the question is, are you thinking about using some kind of evolutionary algorithms to design materials? Even *in silico* because, I'm old enough to remember kind of the hype about genetic algorithms, which reminded me a little bit of that, but in this case the real point of biological evolution is that it overcomes combinatorial complexity. The total number of possibilities is enormous but nevertheless, due to exponential expansion of favorable variance, it finds very quickly suboptimal but extremely efficient solutions. So, the question is can you couple what you do in terms of your objective fitness function, some sort of evolutionary scenario, where your algorithm will kind of exponentially expand into a suboptimal but high fitness solution. Is there something that you are contemplating to do, did this sort of ideas cross your mind?

Lee Cronin: I'll answer very quickly. I must apologize on the slide that I put up when I showed the robot with the nanoparticles and the formulation robot. Those both use different evolutionary algorithms in materials in a closed loop. So, what we do in that case, we use an evolutionary algorithm to produce nanoparticles and then, we climb up and down a fitness function. Then we use the physical seed, so the outcome of that evolution to the templates in the next generation and do an evolution again. So you couple a template in silico, and a physical evolution together and that allows you, when you then challenge the assembly; like I think this comes back to what Joanna was saying earlier, make the system dirty and tolerant of the conditions. You can use an evolutionary algorithm to produce new nanoparticles and clean quantum dots despite the fact that the system is error prone and full of junk.

Eugene Shakhnovich: And what is the role of natural selection in this kind of algorithms, what is the analogue of natural selection?

Lee Cronin: So, we put a mutation...

Eugene Shakhnovich: Yes, I understand, but the key-point is not mutation but selection.

Lee Cronin: Yes this is a big discussion, we use the environment to select, so what persist after time, physically persist, materially embodied evolution, which is what someone was talking about bio-inspired rather than just using a GA as a search, it is actually materially embodied.

Eugene Shakhnovich: okay, thank you.

Berend Smit: It's an approach we have been using once in a while. If you don't want to do brute force screening of libraries, you can also generate libraries with fitness function and you use GA and that kind of things to generate the library and then your objective function is some performance parameter of the material.

Laura Gagliardi: I actually wanted to comment on the question of Mark Ellisman about "what can we learn from biology?". All these catalysts by design are inspired by biology. Copper zeolites, the iron MOF, so first of all, we don't understand so well why things work. It is probably, as you mentioned, that this self-repairing is not so easy to reproduce *in silico* or when you engineer a material but also there is the fact that for example in these MOFs maybe you have these iron sites, that are like in the enzymes, but there are too many of them, they are too close and they interact among each other, but in a way that is not happening in the enzymes. So, we are trying to be inspired but we are not there clearly because the materials in the lab don't have the same features.

Mark Ellisman: So, you have world experts here on biophotonics, or photonics with Stefan Hell for example, with evolution and revolution and one opportunity that you have is, with some of these methods, to actually take a material and locally activate or inactivate in patterns under robotics. So, there are opportunities to use other forms of energy to modify devices that are modifiable. Light is just one example, if you have too many of your catalytic elements, and you had a photonic basis inactivating them, you could reduce your extra sites. I am not sure the scales would work, but there are opportunities that come from the kind of cross-fertilization that I think will happen more at dinner and over cocktails.

Bert Weckhuysen: But the thing is, if you look into photocatalysis heterogeneous type photo-catalysts, people try already these kinds of things to see how they can by triggering something, that something then follows. But making an artificial leaf, like you would have in biology, is still not an easy thing to make from an inorganic/organic material, and then having it operate for a while. People are inspired and try to do that. Second is, merely single molecules, the previous conference on catalysis, the Solvay conference, there people have been showing single molecule spectroscopy, things like what Stefan Hell is doing and so these methods are also already invading in our field to elucidate. There is one example of an American, his name is Peng Chen from Cornell university, he is trying to make an analogy between what is happening in heterogeneous and biological catalysis, he is educated at Stanford with Ed. Solomon, so there he had all that bio/inorganic type of thinking, and now he is trying with single molecule spectroscopy to see if he can make a bridge between these kinds of structures and enzyme type materials. These fields and their methods are coming together more and more.

I think there is time for one final question in this session before we accommodate what you stated at we can go for somewhere maybe a beer, have later other types of discussions about this session. Is there someone who has a question left? If not, I thank the speakers. I think we had a very nice overview of what the field can bring. Applause for them. And as you have seen in the program, for those who like to go to the reception in the city hall, it's at 7pm. We meet here in the lobby to go then together to the city. It will be a short walk and umbrellas will be available in the reception hall.

Session 3

Models and Experimental Data on Water Dynamics Complexity of Solid/Liquid Interfaces

MORPHING HARD AND SOFT BIO-INSPIRED MATERIALS BY REACTION-TRANSPORT DYNAMICS

C. NADIR KAPLAN* and JOANNA AIZENBERG†

*Department of Physics, Virginia Tech, Blacksburg, VA 02461, USA
†Paulson School of Engineering and Applied Sciences, Harvard University,
Cambridge, MA 02138, USA

Introduction

Engineering next-generation materials that can grow into efficient multi-tasking agents, move rapidly, or discern environmental cues greatly benefit from inspiration from biological systems. Therefore, continuum theories that describe how chemical reactions and transport processes govern the emergence of function and form in analogy with organisms play an essential role in the design of bio-inspired synthetic routes. By focusing on hydrodynamical and geometrical formulations, here we review the theoretical progress and outstanding challenges at the interface of materials chemistry and biology. For the future, the effective convergence of theory and experiment will result in optimized hard or soft bio-mimetic materials for applications ranging from bottom-up manufacturing to soft robotics.

Living systems are in many ways superior to inanimate materials: They constantly process information from ambient cues and thereby react selectively and flexibly to their ever-changing environment. The emergence of this complex pipeline that turns any organism into an "all-in-one automaton" relies on a cascade of intertwined processes from the molecular to cellular level [1]. At a molecular scale, the feedback between nucleic acids and proteins governs replication and metabolism, which, over eons, evolved into organizing an entire population of bio-molecules and achieving compartmentalization, homeostasis, growth, and reproduction of cells [2]. Either individually or by building superstructures ranging from tissues of eukaryotic cells to bacterial bio-films, this robust internal synchrony enables cells to exploit physicochemical processes for a range of tasks, such as bio-mineralization to develop intricate morphologies from hard materials, deformation of soft cellular or tissue material to undergo self-regulated motion, and transduction of chemical, mechanical, optical signals for communication, camouflage, or sporulation, to name a few.

Here, we discuss the continuum theories of reaction and transport phenomena in biological and bio-inspired systems, particularly mineralization of hard materials and complex sensing in soft matter, our recent work in this realm, as well as our view on the outstanding intellectual and computational challenges therein. Reaction and transport dynamics is accepted to be a common operation mode of life dating back to the seminal work of Alan Turing on pattern formation [3]. Most generally, chemical reactions together with solid and fluid transport, whether active transport or passive flow, are formulated in terms of sets of nonlinear partial differential equations (nPDEs) at a continuum scale. In this limit, the relevant length, time, and energy scales are much bigger than those of the microscopic constituents of a system, so that a coarse graining process naturally yields nPDEs. To date, continuum theories elucidated many chemical systems of interest via hydrodynamical or geometrical formulations. For the future, comprehensive theories that unify essential aspects of biotic systems will further advance our understanding of organisms and the synthesis of analogous bio-mimetic materials.

Mineralization of Complex Biological and Lifelike Forms: Present State of Research and Our Contributions

As a way of pattern formation of intricate 3D solids, bio-mineralization couples transport and reaction of species as well as their solidification at the growth front, where the length scales of the steep concentration gradients and the structure are separated. For instance, the molluscan and brachiopod shells record local shape changes during accretionary growth [4, 5] (Fig. 1). Briefly, a thin elastic mantle secretes and mineralizes calcium carbonate incrementally at the growth edge, thereby forming a hard shell over time to protect the soft body of the animal [4–6]. Although there exist structure-oriented theories, they simplify growth dynamics by focusing on prescribed elementary geometries [5, 7–9]. A deeper theoretical understanding of the growth of ornamented structures paves the way for the synthesis of arbitrarily complex brittle morphologies, which can serve as high-surface-area structures for catalysis [10].

Inorganic model systems, such as carbonate-silica precipitates (*aka* biomorphs), exhibit growth dynamics analogous to those of bio-mineralization. Biomorphs exemplify lifelike formation of inorganic composite morphologies, which are more complex than symmetric superstructures originating from crystal growth or droplet microfluidics [11, 12] (Figs. 1(a)–1(c)). Despite extensive experimental knowledge, a theoretical understanding has been missing owing to the geometric complexity of the emergent patterns, structural anisotropy associated with the long-range order of needle-like carbonate crystals that coexist with an amorphous silica background phase, and the length-scale separation in the system. In the last decade, two equivalent phase-field-based simulation methods that penalize spatial variations in the crystal orientation showed great promise in reproducing complex shapes reminiscent of coral-like composite precipitates [13–15]. Recently, we developed a geometrical theory that explains the accretionary growth and form of the biomorphs (Fig. 1(a)) and predicts new shapes that we then synthesized [16]. We employed the full range of patterns to build optical waveguides (Fig. 1(c)). The combination of morphology control and tailored optical properties brings us closer to acquiring versatile synthesis techniques from nature.

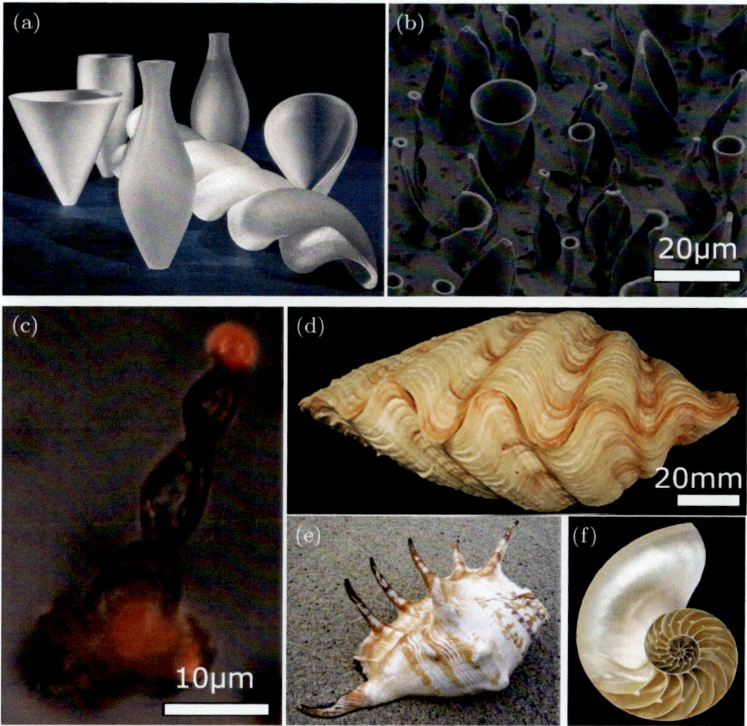

Fig. 1. Self-organized forms. (a) 3D-printed structures from theory [16]. (b) Experimental patterns [11, 16]. (c) Optical waveguiding helix [16]. (d)–(f) Mollusk shells.

From Mineralization to Complex Sensing in Soft Matter: Outlook to Future Developments

Crystallization of complex forms

The main technical challenges with continuum modeling of mineralization of composites and soft multi-phase matter lie in numerical modeling of coupled dynamics at multiple scales, and the mere numerical solutions disallowing closed-form scaling laws, thereby also straightforward experimental testing. Furthermore, continuum formulations of systems ranging from bio-materials to biological suspensions often employ phenomenological control parameters or single-phase descriptions of a multi-phase system [25]. Generalized hydrodynamical and geometrical models concerning all phases of these systems must unify all aspects of relevant reaction and transport processes with well-defined microscopic limits.

In the long term, how can we develop rigorous quantitative formulations for controlled scalable self-assembly based on bio-minerals? Natural shell (Figs. 1(d)–1(f)) morphogenesis requires a general mathematical approach that couples the growth of arbitrarily complex ornamented morphologies to the elasticity of the growth template. To that end, our geometrical theory is sufficiently generic to be adapted to the biomorph morphogenesis. This can yield a unified mechanistic view in natural shell development that may inform processes underlying carbonate-based bio-mineralization, e.g., shape regulation in mammalian bone growth with potential medical applications.

Understanding the formation of bio-mineralized shells greatly benefits from theoretical and computational advances toward the synthesis of their brittle abiotic analogs, such as biomorphs. In order to develop a controlled scalable self-assembly system based on biomorphs, rigorous boundary conditions describing the diffusion-limited growth, the precipitation reactions, and pH-dependent curving at the growth front must be identified [11, 12]. The design of optical materials can be enhanced by quantifying the geometry-dependent optics of the biomorphs, as was done for twisted crystallites [26]. Additionally, microscopic theories of the mechanics of the composite structures will illuminate the structure–form relationship across scales in this system.

Soft matter

Quantifying how multi-phase soft biological and synthetic materials interact with external signals will advance engineered platforms where complex signal processing is critical, ranging from tissue engineering to drug delivery to soft robotics. To this end, our dynamic poroelastic theory enhanced by chemical reaction terms has predicted the core non-equilibrium mechanisms accessible to common hydrogels [17, 18]. These mechanisms are also valid for calcium ion, a ubiquitous signal mediator in biology, as used for locomotion by social amoeba *D. discoideum* [27] and by the biological springs of *Vorticella* ciliates [28]. The way these bio-chemical reaction–transport pathways power the motion of some single-celled organisms may inspire electronics-free micron-scale soft actuators. Future studies shall focus on complexation–transport–deformation dynamics in gels to model settings that can produce mechanical responses with tunable length and time scales. These model settings could be smooth arbitrary geometries of free-standing gels or multi-layered systems of divalent-ion-responsive gels and viscous fluid phases. The challenge is to predict via rigorous continuum models the controllable locomotion and shape change associated with non-uniform geometry and multi-phase composition of gels. This further requires the development of optimal control theories for poroelastic multi-phase systems. Altogether, a unified framework, which couples geometry, mechanics, and control, will address the design of electronics-free soft platforms that integrate how the chemical environment is changing over time, as necessary for soft robotic navigation in complex chemical gradients or specialized sensing and release in medical or agricultural settings.

Conclusion

In nature, reaction and transport phenomena determine both the shape of organisms and the dynamics of how they operate. To that end, augmenting mathematical tools in a continuum setting that have proven successful in characterizing inanimate systems will allow us to quantify dynamics of living systems. Addressing the technical and intellectual theoretical challenges that will couple the laws of thermodynamics, mass, and momentum conservation with reaction diffusion will further enable us to develop novel synthetic routes to advanced materials and devices, guided by the basic principles of biological structures and the economy with which biology solves complex problems of growth, form, and motion.

References

[1] J. von Neumann, The general and logical theory of automata (Lecture given in 1948). In *Cerebral Mechanisms in Behavior — The Hixon Symposium*, L.A. Jeffress (ed.) (John Wiley, New York, 1951).

[2] F.J. Dyson, *Origins of Life* (Cambridge University Press, Cambridge, UK, 1992).

[3] A.M. Turing, *Philos. Trans. Royal Soc.: Biol. Sci.* **237**, 37 (1952).

[4] T.-S. Liew, A.C.M. Kok. M. Schilthuizen, and S. Urdy, *PeerJ* **2**, e383 (2014).

[5] R. Chirat, D.E. Moulton, and A. Goriely, *Proc. Natl. Acad. Sci. USA* **110**, 6015 (2013).

[6] K. Simkiss and K.M. Wilbur, *Biomineralization: Cell Biology and Mineral Deposition* (Academic, San Diego, 1989).

[7] G. Dera, G.J. Eble, P. Neige, and B. David, *Paleobiology* **34**, 301 (2008).

[8] D.M. Raup and A. Michelson, *Science* **147**, 1294 (1965).

[9] Ø. Hammer, *J. Moll. Stud.* **66**, 383 (2000).

[10] K.R. Phillips, G.T. England, S. Sunny, E. Shirman, T. Shirman *et al.*, *Chem. Soc. Rev.* **45**, 281 (2016).

[11] W.L. Noorduin, A. Grinthal, L. Mahadevan, and J. Aizenberg, *Science* **340**, 832 (2013).

[12] J.M. García-Ruiz, E. Melero-García, and S. T. Hyde, *Science* **323**, 362 (2009).

[13] T. Pusztai, G. Bortel, and L. Gránásy, *Europhys. Lett.* **71**, 131 (2005).

[14] R. Kobayashi and J.A. Warren, *TMS Lett.* **1**, 1 (2005).

[15] L. Gránásy, L. Rátkai, A. Szállás, B. Korbuly, G.I. Toth *et al.*, *Metall. Mat. Trans. A* **45**, 1694 (2014).

[16] C.N. Kaplan, W.L. Noorduin, L. Li, R. Sadza, L. Folkerstma *et al.*, *Science* **355**, 1395 (2017).

[17] P.A. Korevaar, C.N. Kaplan, A. Grinthal, R.M. Rust, and J. Aizenberg, in revision (2019).

[18] C.N. Kaplan, P.A. Korevaar, and J. Aizenberg, in preparation (2019).

[19] A. Sidorenko, T. Krupenkin, A. Taylor, P. Fratzl, and J. Aizenberg, *Science* **315**, 487 (2007).

[20] A. Shastri, L.M. McGregor, Y. Liu, V. Harris, H. Nan *et al.*, *J. Nat. Chem.* **7**, 447 (2015).

[21] X. He, R.S. Friedlander, L.D. Zarzar, and J. Aizenberg, *J. Chem. Mater.* **25**, 521 (2013).

[22] A. Sidorenko, T. Krupenkin, and J. Aizenberg, *J. Mater. Chem.* **18**, 3841 (2008).

[23] X. He, M. Aizenberg, O. Kuksenok, L.D. Zarzar, A. Shastri *et al.*, *Nature* **487**, 214 (2012).

[24] Y. Liu, A. Bhattacharya, O. Kuksenok, X. He, M. Aizenberg *et al.*, *Soft Matter* **12**, 1374 (2016).

[25] M.C. Marchetti, J.F. Joanny, S. Ramaswamy, T.B. Liverpool, J. Prost *et al.*, *Rev. Mod. Phys.* **85**, 1143 (2013).

[26] X. Cui, A. G. Shtukenberg, J. Freudenthal, S. Nichols, and B. Kahr, *J. Am. Chem. Soc.* **136**, 5481 (2014).

[27] L. Fets, R. Kay, and F. Velazquez, *Curr. Biol.* **20**, R1008 (2010).

[28] L. Mahadevan and P. Matsudaira, *Science* **288**, 95 (2000).

UNDERSTANDING MOLECULAR TRANSPORT MECHANISMS AT WATER–MEMBRANE INTERFACES

SINAN KETEN

Northwestern University, Department of Civil and Environmental Engineering, Department of Mechanical Engineering, Evanston, IL 60208, USA

My View of the Present State of Research on Dynamics of Water at Membrane Interfaces

Water is a catalyst for human progress. Throughout history, urbanization has taken place most rapidly at the interface between land and water bodies. Despite the fact that the majority of Earth's surface is covered with water, global access to potable water has remained a growing challenge that is central to the water–food–energy nexus and the climate crisis. As we run out of clean freshwater resources, membrane-based water purification technologies such as reverse osmosis and nanofiltration grow increasingly critical. Industrial purification remains energy intensive because water must be driven at high pressure through a polymeric membrane with nanoscopic pores, which must prohibit (as much as possible) larger molecules or charged species from passing through. Balancing selectivity and flux while avoiding fouling requires rigorous nanoscale analysis of water, solute, and polymer interactions at complex membrane interfaces.

Efficient selective transport of water through cost-effective membranes is a challenge that could be addressed by advanced computational methods such as molecular dynamics, as well as learning lessons from biology. Specifically, transmembrane proteins such as aquaporins and ion channels attain very high transport rates and extreme selectivity. Their 3D folded structures exhibit dynamic gating and transport mechanisms for ions and water that we cannot currently achieve with polymeric membranes. Empirical continuum models and experimental trial and error have driven membrane R&D in industry to this point, but fundamentally different designs may be possible through understanding molecular mechanisms underpinning water dynamics at membrane interfaces. Molecular dynamics (MD) simulations uncovered many mechanisms relevant for biological, bio-inspired, and polymeric nanoporous membranes, but challenges remain to translate these computational methods into tools that guide materials design. Here, the focus will be on summarizing recent contributions to this field from molecular modelling, and future prospects for studying water dynamics at solid–liquid interfaces in membranes.

My Recent Research Contributions to Dynamics of Water at Membrane Interfaces

My research group establishes molecular and multi-scale models to understand and design polymeric and bio-molecular materials. Some of our recent work focused on the physics of water and solute dynamics in polymeric and bio-inspired separation membrane interfaces. Our first series of investigations examined block copolymer membranes with aligned sub-nanometer channels consisting of polymer-conjugated cyclic peptide nanotubes (CPNs), synthesized by our collaborators [1]. These cyclic peptides typically consist of eight amino acids and self-assemble into stiff nanotubes that are barely large enough to transport water and monovalent cations. They readily exhibit cation selective transport, which can break down the homeostatic balance of microbes in antibacterial applications. Size exclusion and electronegativity from peptide backbone carbonyl groups facilitate large molecule and anion rejection (e.g., Cl^-), which makes CPNs advantageous for serving as nanopores in separation membranes.

Our analyses focused on transport mechanisms within these nanotubes, where the ideal pore size and chemistry must be tuned for the application in mind. For example, water desalination would require a nanotube that not only transports water rapidly but also rejects ions. We found that this could potentially be achieved by introducing a bio-mimetic point mutation in the sequence of the CPs to functionalize the interior of the nanotubes with a methyl group while maintaining self-assembly capability [2, 3]. Even in these well-defined sub-nanometer pores where most solutes transit in a single file, the transport mechanisms are complex, as lumen electronegativity, conformational flexibility of functional groups (dihedrals), and rigidity of the assembled nanotubes all play a role in transport efficiency and selectivity [4]. A benefit of these well-defined, straight pores is that the free energy landscapes of ion transport can be more easily quantified with advanced sampling techniques such as metadynamics, which can be used to corroborate the effects of coordination and solvation dynamics of ions within the lumen [4]. We also showed that the stacking order of binary CP mixtures with different functional groups could be guided through entropic elasticity of polymer grafts, in order to mimic the chemical diversity and precise positioning of functional groups within the lumen of transmembrane proteins [5]. Our work elucidated the mechanics, self-assembly, and transport behaviour of CPNs, accelerating the discovery of separation membranes. Our investigations on ideal membrane systems with rectilinear CPN pores provided timely insights into transport mechanisms, circumventing the modelling challenges associated with the heterogeneity and tortuosity of polymeric membranes (Fig. 1).

Existing polymeric membranes have fairly optimized performance for specific feed solutions, but to improve membrane chemistry for typical feed solutions, the multitude of charged species and organic contaminants present, as well as the potential for coupled transport effects, must be considered. Membrane charge can drive selectivity up, but the presence of multi-valent ions in the feed can cause

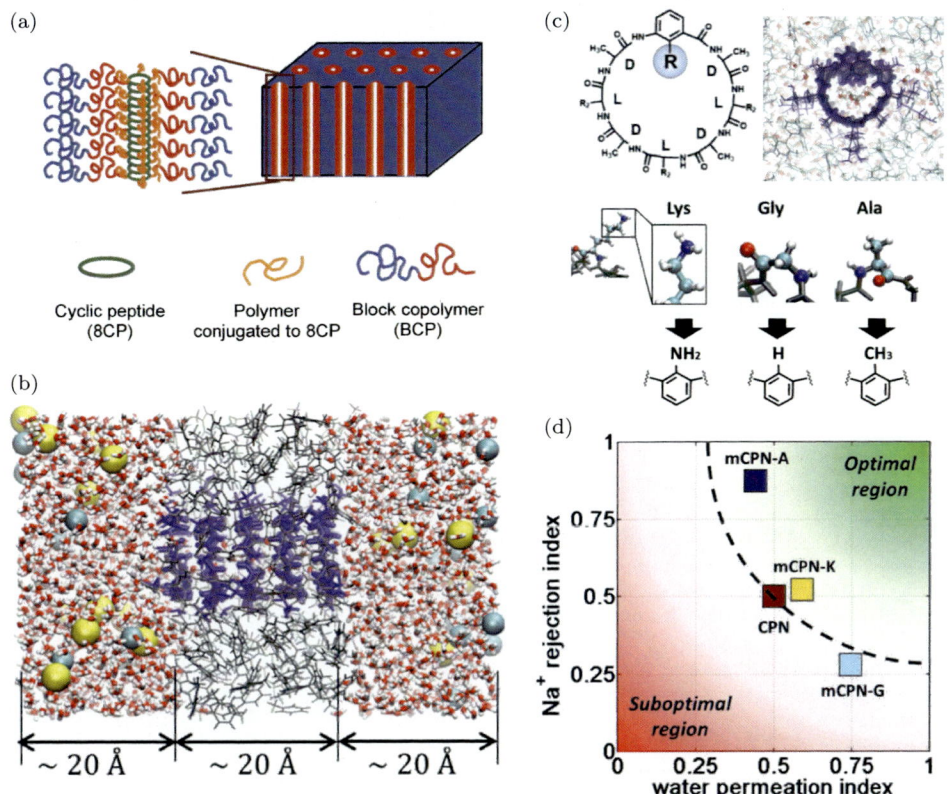

Fig. 1. Cyclic peptide nanotube (CPN) membranes. (a) Peptide rings with polymer grafts can self-assemble in block copolymer membranes to create continuous through pores similar to biological ion and water channels (reprinted (adapted) with permission from Ref. [1], copyright 2011 American Chemical Society). (b) Non-equilibrium MD simulations are carried out to quantify water and ion flux through nanotubes, where pressure or voltage gradients can be applied across the membrane interface to drive flow [2, 3]. (c) The interior of the nanotube can be functionalized to mimic the lumen features of protein channels [6]. (d) Simulations can reveal which functionality yields the best trade-off between ion rejection and water permeation, which in this case is an Alanine-inspired methyl functionality in the CPN that induces a significant desolvation penalty to cations [3]. Panels b,d reprinted (adapted) with permission from Ref. [3], copyright 2015 American Chemical Society. Panel c reproduced from Ref. [2] with permission from The Royal Society of Chemistry.

undesired outcomes in nanofiltration. Our recent research [7, 8] has been focusing on understanding the transport mechanisms of diverse solutes in models of polyamide membranes. Using non-equilibrium molecular dynamics simulations, we examined flux, transport pathways, and solute mean passage time through model membrane systems to understand how solute size and chemistry influence transport metrics. Energy barriers due to the solvation shell size (especially for ions) and hydrogen bond interactions with the membrane influence transport in a way which prevents a simple correlation between decreased shell size and improved transport, as illustrated by urea having a greater rejection rate than molecules of comparable

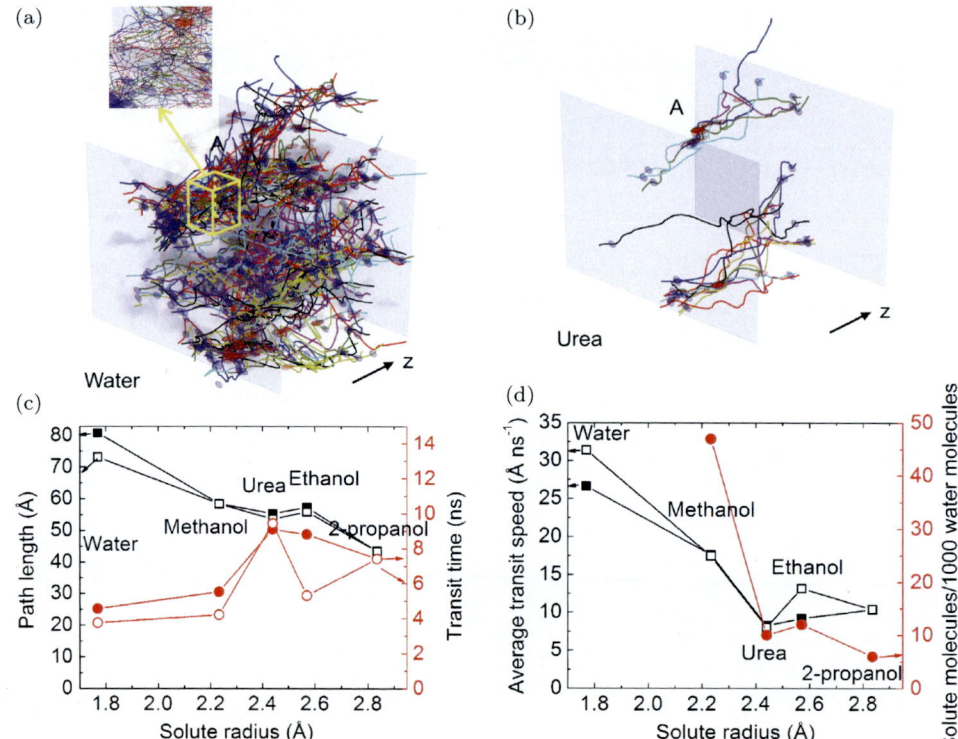

Fig. 2.   Transport mechanisms of water and various solutes through a polyamide RO membrane [7, 8]. (a) Water molecules pass readily through the membrane by meandering through solvent accessible free volume, sampling longer paths owing to their small radius. The planes show the Gibbs dividing surfaces representing the membrane interface with water. A pressure gradient is applied in the +z direction in non-equilibrium MD simulation and passage events are recorded. (b) Some solutes such as urea exhibit straighter paths and fewer passage events, hopping along favorable and accessible sites within the membrane. (c) Path length seems inversely correlated with solute radius; however, transit time depends not only on solute radius but also on other factors including solvation shell dynamics and chemical interactions of the solute with the membrane as well. For example, hydrogen bonding ability of urea with the polyamide membrane results in slower transit. Reprinted from Ref. [7], Copyright 2016, with permission from Elsevier.

size such as methanol and ethanol. We also quantified how swelling of the membrane and porosity control the paths taken by solutes, as well as randomness associated with the synthesis of the membrane. These first steps into studying transport in nanoporous polymeric membranes will now pave the way for future studies that will hopefully accelerate the current understanding of separation membranes for addressing clean water challenges (Fig. 2).

Remarks on the Future of Research on the Dynamics of Water at Membrane Interfaces

Molecular simulations coupled with experiments will address numerous challenges related to understanding the dynamics of water and other molecules near solid–liquid interfaces. In membrane science, the selectivity and flux we see in protein channels are astounding. Reproducing such capabilities in industrial separation membranes requires deeper molecular insight that MD simulations can provide. Bottom-up fabrication of membranes via directed self-assembly can result in better control of pore size and chemistry, for instance, to create pH-responsive lumens [9] with precise pore size definition, while controlling stacking order of these groups through guidance from self-assembly simulations. Computationally, improvements in force-field parameters for water, multi-valent ions, and polymers, better addressing pH effects, and overcoming the spatiotemporal limitations of MD will bring simulations closer to reality. The molecular models will need to account for mixed solute feed streams and membrane charge distribution. Simulation-based design will require high-throughput tools, such as advanced sampling or coarse-graining, while maintaining accurate description of dynamics and transport. Addressing bio-fouling at the membrane interface, particularly by elucidating microbial adhesion mechanisms through simpler molecular models, will be critical. For instance, mechanical models [10] that explain force-enhanced binding lifetimes of catch bonds in bacterial adhesins could allow ways to hinder persistent bio-adhesion. Allosteric dynamics in adhesins can be turned around in membrane frameworks to create nanopores with force-sensitive conformational switching for gating-capable membranes.

Acknowledgments

Sinan Keten acknowledges funding from the U.S. National Science Foundation (NSF Award # 1840816) and a PECASE Award (ONR Award # N000141613175).

References

[1] T. Xu, N.N. Zhao, F. Ren, R. Hourani, M.T. Lee *et al.*, *ACS Nano* **5**(2), 1376 (2011).
[2] L. Ruiz, Y. Wu, and S. Keten, *Nanoscale* **7**(1), 121 (2015).
[3] L. Ruiz, A. Benjamin, M. Sullivan, and S. Keten, *The J. Phys. Chem. Lett.* **6**(9), 1514 (2015).
[4] M.A. Alsina, J.-F. Gaillard, and S. Keten, *PCCP* **18**(46), 31698 (2016).
[5] A. Benjamin and S. Keten, *The J. Phys. Chem. B* **120**(13), 3425 (2016).
[6] R. Hourani, C. Zhang, R. van der Weegen, L. Ruiz, C.Y. Li *et al.*, *JACS* **133**(39), 15296 (2011).
[7] M. Shen, S. Keten, and R.M. Lueptow, *J. Membrane Sci.* **509**, 36 (2016).
[8] M. Shen, S. Keten, and R.M. Lueptow, *J. Memb. Sci.* **506**, 95 (2016).
[9] S.M. Darnall, C. Li, M. Dunbar, M.A. Alsina, S. Keten *et al.*, *JACS* **14**(128), 10953 (2019).
[10] K.C. Dansuk and S. Keten, *Matter* **1**(Oct), 1–15 (2019).

ACTIVE NEMATICS AT INTERFACES AND SURFACES

JULIA M. YEOMANS

The Rudolf Peierls Centre for Theoretical Physics, Clarendon Laboratory,
Parks Road Oxford, OX1 3PU, UK

Background

Active materials take energy from their surroundings on an individual particle level and use it to do work [1]. Examples include motor proteins, cells, microswimmers, and active colloids. Several dense active systems can be described as active nematics [2]. Most current active nematic experimental systems are 2D, and they include thin films of swimming bacteria, such as *E. coli*, or microtubule–motor protein mixtures absorbed at an interface. The hydrodynamic equations of motion of active nematics are very similar to those of passive nematic liquid crystals, but with an additional term in the stress which is proportional to the nematic order parameter tensor [3]. As a result of this contribution, any gradients in the magnitude or direction of the nematic ordering lead to stresses, and hence flows, which have far-reaching consequences.

Most importantly, the nematic state is unstable to fluctuations. It is replaced by active turbulence, which is characterized by chaotic flows, fluid jets, and vortices [4]. The topological defects found in passive nematics still occur: in two dimensions, the defects predominantly have topological charge $+1/2$ or $-1/2$. They annihilate in pairs of opposite sign, just as for passive nematics, but they can also be created in pairs by the activity, leading to a steady-state defect density [5, 6]. Another property peculiar to active systems is that $+1/2$ defects are self-propelled. This is because they correspond to regions where the director is strongly distorted, leading to active stresses. Because $+1/2$ defects have unbalanced head–tail symmetry, the stresses are asymmetric, resulting in a net defect velocity.

The ways in which surfaces and interfaces affect the properties of active materials are just starting to be investigated. Very different behaviour is seen than for the passive case because active systems are out of thermodynamic equilibrium. Even the definition of such familiar concepts as surface tension becomes problematic. We briefly discuss three different geometries: a 2D sheet of active material at an interface or surface; confined active nematics; and the rim of an active colony.

Active Nematics at Interfaces and on Surfaces

An important experimental system that has been key in investigating the properties of active turbulence is a dense mixture of microtubules and two-headed kinesin motor proteins driven by ATP [7]. Bundles of microtubules are propelled by the motors into a cycle of buckling, fragmenting, and reforming, dynamics that can be reinterpreted in terms of the motion of a collection of motile topological defects. In an early experiment, Keber *et al.* showed that a microtubule–motor protein mixture can be confined to a vesicle [8]. For a nematic on the surface of a sphere, the topological charge must be +2, and hence four +1/2 defects are in continual motion, driving active flows on the surface of the vesicle, and causing it to roll if placed in contact with a surface.

If a microtubule–motor protein mixture is stabilized at a water–oil interface, the viscosity of the adjacent oil leads to damping that can control the properties of the active system. For example, increasing the viscosity of the oil leads to an increase in the number of topological defects. If a magnetic field is imposed, the active turbulence can be controlled into laning flows as it is much easier for the microtubules to move parallel to the smectic layers than perpendicular to them [9]. Regular flows will be required to harness power from active materials, and external friction and fields may give ways of controlling active turbulence.

Another 2D example of active nematics is confluent cell layers. Although the shape of cells is on average isotropic, activity allows them to fluctuate into local nematic configurations. Their dynamics can closely resemble active turbulence and motile topological defects have been identified [10]. The behaviour of cell layers is known to be strongly influenced by the properties of the substrate, with cells moving more easily on stiffer surfaces; for example, fibroblasts move readily from a soft to a hard surface, but turn back if they reach the boundary from the soft side. The same phenomenon has now been demonstrated in cell colonies, but such collective durotaxis remains to be fully understood.

Confining Active Systems

When active nematics are confined the hydrodynamics is screened and active turbulence can be replaced by more regular flows. These depend sensitively not only on the fluid parameters and the confinement size but also on the boundary conditions and the strength of intrinsic fluctuations. Simulations of active flow in a 1D-channel show that, as the width of the channel is increased at fixed activity, the evolution in flow configurations is no flow ⇒ laminar flow (shear or unidirectional) ⇒ a 1D-line of flow vortices ⇒ active turbulence [11]. The system starts to flow when the active stresses can overcome the pinning effect of the boundaries, velocity vortices can form once the channel becomes wide enough to accommodate them, and then a further increase in channel width allows relative motion of the vortices, corresponding to active turbulence.

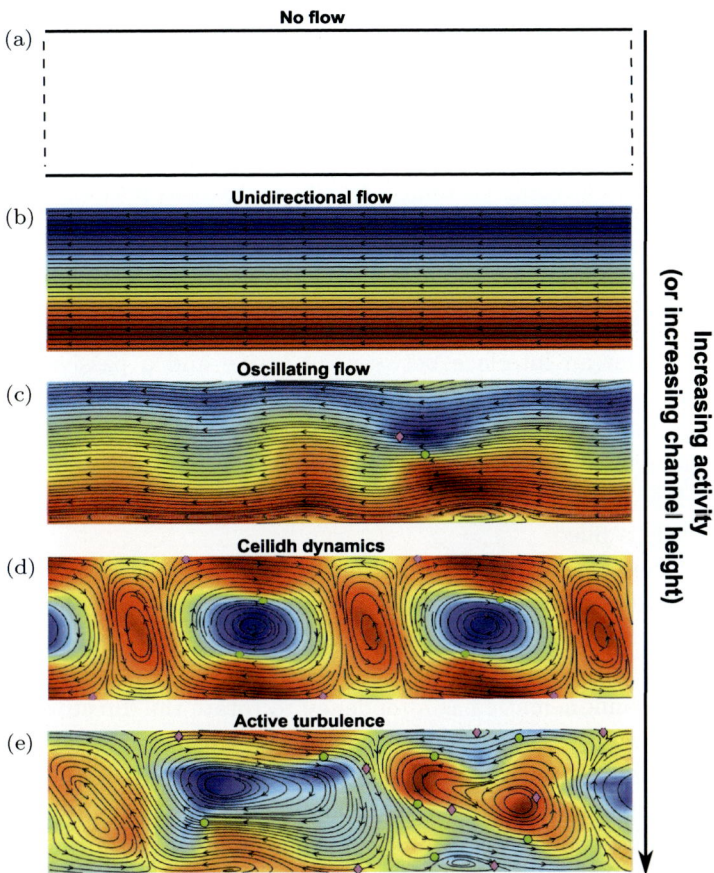

Fig. 1. Flow states of an active nematic confined in a channel. The black lines represent the streamlines of the velocity field and the black arrows indicate the direction of the flow. The colour map represents the vorticity field. Circles (green) and diamonds (magenta) mark +1/2 and −1/2 topological defects, respectively (taken from Ref. [11]).

The motile, active, topological defects add complexity to this sequence. Channel walls are preferential sites for defect formation, particularly if any boundary alignment is weak and if active particles can freely slip along the walls. The stationary −1/2 defects remain close to the walls due to elastic interactions, whereas the self-propelled +1/2 defects move toward the center of the channel. If the flow is laminar, +1/2 defects can traverse the channel to be annihilated by the −1/2 defects at the opposite wall. In the vortex regime, however, they can be captured by the flow vortices and perform a "ceilidh dance", with right- and left-moving defects moving past each other on sinusoidal trajectories in a way reminiscent of the great chain figure in country dancing (Fig. 1).

Recent simulations and experiments [12] which confine microtubule–kinesin mixtures to a channel at an oil–water interface show transitions from shear flow to a

1D-vortex lattice, and then to active turbulence as the channel width is increased. This is a system where defects form easily at the walls because the microtubules have weak, planar anchoring and are able to freely slide along the channel walls. In the shear state, the defects are created in periodic bursts and then move across the channel to annihilate with the −1/2 defects at the opposite wall. In the flow vortex lattice, +1/2 defects become entrained by the vortices to perform a recognizable, albeit noisy, ceilidh dance.

Active Interfaces and Active Anchoring

Finally, we consider an interface between an active and a passive material. An example would be the rim of a colony of bacteria or eukaryotic cells. We have recently been thinking about the concepts of active anchoring, boundary conditions, and surface forces that could act at an active–passive interface [13]. Related physical questions are how a bacterial colony expands, or a wound healing assay: how a hole in a layer of cells is filled.

Recall that gradients in the order parameter of an active nematic lead to stresses. The surface between an active and a passive material corresponds by definition to a gradient in the magnitude of the order parameter. The first effect of this is a stress generating a flow along the interface. The resulting velocity gradient between the interface and the bulk nematic has a tendency to rotate the director (assuming the system is flow-tumbling). In the extensile case, the director is in stable equilibrium when planar to the interface. In the contractile case, however, the flow direction is reversed, rotating the director toward the homeotropic configuration, normal to the interface. Thus, the activity is producing an effective anchoring term, even in a situation with no thermodynamic anchoring energy.

Planar active anchoring can be observed in studies of growing bacterial colonies. In these systems, the division of bacteria along their long axes provides an extensile stress and hence active anchoring may provide an explanation for this behaviour [14].

The second effect of a gradient in the magnitude of the order parameter is a normal force which acts toward the bulk of the active material. Moreover, if the interface is curved, the active anchoring constrains the nematic to splay or bend, also leading to stresses. This pushes convex regions of the surface of the active material outward and concave regions inward. It is interesting to look for examples of active anchoring and active interface forces experimentally, and to assess any relevance to the behaviour of confluent cell layers.

Outlook

We have already touched upon possible research questions. Other directions that have the potential for significant and timely advances in the field of active matter include the following:

- Designing machines that are powered by active nematics. To do this, the active flow will need to be controlled, and confinement or friction is a possible approach.
- Understanding and exploiting the behaviour of active materials in three dimensions.
- Fabricating a wider range of dense active materials that are easy to handle in large quantities.
- Developing theories that can describe the non-equilibrium nature of dense active matter, and assessing the extent to which equilibrium concepts such as surface tension or wettability have applicability in active systems.
- Understanding the extent to which the dynamics of dense cellular arrays, such as bacterial colonies, tissues, and tumours, can be understood in terms of the theories of active matter.
- Studying the role of activity in self-assembly and morphogenesis.

References

[1] M.C. Marchetti, J.F. Joanny, S. Ramaswamy, T.B. Liverpool, J. Prost *et al.*, *Rev. Mod. Phys.* **85**, 1143 (2013).

[2] A. Doostmohammadi, J. Ignes-Mullol, J.M. Yeomans, and F. Sagues, *Nat. Comms.* **9**, 3246 (2018).

[3] R.A. Simha and S. Ramaswamy, *Phys. Rev. Lett.* **89**, 058101 (2002).

[4] H.H. Wensink, J. Dunkel, S. Heidenreich, K. Drescher, R.E. Goldstein *et al.*, *PNAS* **109**, 14308 (2012).

[5] L. Giomi, M. J. Bowick, X. Ma, and M. C. Marchetti, *Phys. Rev. Lett.* **110**, 228101 (2013).

[6] S.P. Thampi, R. Golestanian, and J.M. Yeomans, *Phys. Rev. Lett.* **111**, 118101 (2013).

[7] T. Sanchez, D.T. Chen, S.J. DeCamp, M. Heymann, and Z. Dogic, *Nature* **491**, 431 (2012).

[8] F.C. Keber, E. Loiseau, T. Sanchez, S.J. DeCamp, L. Giomi *et al.*, *Science* **345**, 1135 (2014).

[9] P. Guillamat, J. Ignes-Mullol, and F. Sagues. *PNAS.* **113**, 5498 (2016).

[10] T.B. Saw, A. Doostmohammadi, V. Nier, L. Kocgozlu, S. Thampi *et al.*, *Nature* **544**, 212 (2017).

[11] T.N. Shendruk, A. Doostmohammadi, K. Thijssen, and J.M. Yeomans, *Soft Matter* **13**, 3853 (2017).

[12] J. Hardouin, R. Hughes, A. Doostmohammadi, J. Laurent, T. Lopez-Leon *et al.*, *Comm. Phys.* **2**, 121 (2019).

[13] M.L. Blow, S.P. Thampi, and J.M. Yeomans, *Phys. Rev. Lett.* **113**, 248303 (2014).

[14] A. Doostmohammadi, S.P. Thampi, and J.M. Yeomans, *Phys. Rev. Lett.* **117**, 248102 (2016).

INSIGHT FROM COMPUTATIONAL APPROACHES INTO WATER AT WELL-DEFINED SOLID SUBSTRATES

ANGELOS MICHAELIDES

Thomas Young Centre, London Centre for Nanotechnology,
and Department of Physics and Astronomy, University College London,
Gower Street, London WC1H 0AH, UK
and
Department of Chemistry, University of Cambridge,
Lensfield Road, Cambridge CB2 1EW, UK

My View of the Present State of Research on Solid–Water Interfaces

Under ambient conditions, almost all solid substrates are covered in a thin film of water. As such, water–solid interfaces are of relevance to an enormous range of everyday physical phenomena and technological processes. Indeed, key (connected) global challenges such as climate change, shortages of clean drinking water, and the need for cleaner and renewable forms of energy all call for an improved molecular-level understanding of water at interfaces. For example, improved understanding of how water flows across the surfaces of membrane materials could lead to radical efficiency savings in water purification and desalination. Similarly, deeper insight of ice formation on the surfaces of atmospheric mineral dust could lead to improved global climate models. Global challenges such as these, as well as the simple fact that water–solid interfaces are incredibly interesting scientifically, have led to sustained interest in water–solid interfaces since the time of the first Solvay Conference and before.

Water–solid interfaces are challenging to understand in detail for both experiment and simulation alike. Experimentally, the challenge is simple to state but extremely difficult to resolve in practice: it is incredibly difficult to get atomic- and molecular-level insight of surfaces when outside the idealized ultra-high vacuum conditions of surface science, from where most atomic-level understanding of surfaces has come. From a simulation perspective, the challenge is that simulations of sufficient accuracy on systems of appropriate complexity cannot yet be performed to explore the dynamical evolution of complex aqueous water–solid interfaces. Nonetheless, simulation approaches are now good enough to help in establishing concepts and in interpreting experimental data at water–solid interfaces. Some of the areas in which we have worked over the last few years are now very briefly discussed.

My Recent Research Contributions to Solid–Water Interfaces

Our work on water–solid interfaces has involved a broad range of computational approaches, from the application of very simple coarse-grained simulation approaches, to all-atom classical molecular dynamics simulations, to Kohn Sham density functional theory, right through to explicitly correlated methods such as coupled cluster and quantum Monte Carlo. The approach employed depends on the specific question we are trying to answer, and more often than not this question is motivated by experiment. Two broad examples of topics and areas we have worked on include the following.

1. **Toward a molecular-level understanding of atmospheric ice formation:** Ice formation in the atmosphere invariably occurs on the surfaces of atmospheric aerosol particles. These particles can be almost anything, ranging from clay and silicate components of desert dust to carbonaceous soot particles. We have worked over the years to understand what it is about a material that makes it good or bad at nucleating ice. This work has generally involved classical molecular dynamics studies of heterogeneous ice nucleation and the aim has been to extract general trends and insights about how the physiochemical properties of a substrate affect its ice-nucleating ability. Some of the key insights obtained include the following: (i) If a substrate has a surface structure resembling that of ice, it is likely to promote ice formation. However, this is not a requirement as good ice-nucleating substrates have been observed in our simulations which do not resemble ice. (ii) There is no optimal interaction strength between water and a substrate that determines its ice-nucleating ability. (iii) Classical nucleation theory can often be applied to understand ice nucleation at surfaces. However, it breaks down (or more precisely requires modification) in circumstances where (i) the substrate promotes the formation of specific ice polymorphs and (ii) the substrate favors the formation of a highly non-hemispherical critical nucleus. Further information on these studies can be found in Refs. [1–4].

2. **Toward improved computational accuracy for water at surfaces:** As noted, water–solid interfaces are challenging to explore computationally. Some of the key challenges include the following.

   (a) *Intermolecular interactions*: Water is a closed-shell molecule and interacts weakly with other water molecules and its environment through a combination of electrostatic, hydrogen bonding, and van der Waals dispersion forces. Accurately describing these interactions is a challenge for classical and ab initio techniques. Over the last few years, we have been working to obtain benchmark values for relatively simple systems involving water adsorption at solid substrates. This has involved the development and application of quantum Monte Carlo techniques as well as other high-level electronic structure methods. From these types of studies, well-defined reference information has been obtained as well as insights on the performance of cheaper computational methods such as density functional theory. For an example of our work in this area, see Ref. [5].

(b) *Nuclear quantum effects*: Most quantum chemistry simulations of atoms, molecules, and materials assume that the atomic nuclei can be described as classical point-like particles. Most of the time, this assumption (approximation) is fine. However, when the masses of the atoms are light, the quantum nature of the nuclei needs to be taken into consideration. Water, with its two hydrogens, is obviously one such system where nuclear quantum effects need to be taken into consideration. Indeed, in our research, we have shown that nuclear quantum effects can alter the strength and length of hydrogen bonds involving water as well as the diffusion of water across surfaces. For more information on these findings, see Ref. [6].

(c) *Structural complexity*: Water forms a rich variety of structures as illustrated by its complex phase diagram with at least 15 distinct phases of ice. It is no surprise therefore that water at surfaces also displays a rich and complex phase behaviour with novel water overlayer structures observed even on well-defined substrates. Examples of novel structures proposed in our studies include an ice-like structure built exclusively of water pentagons (for water on copper [7]) and a mixed contract layer of water and hydroxyl (for water on TiO2 [8]).

Outlook to Future Developments of Research on Solid–Liquid Interfaces

Since its inception, computational chemistry has involved the development and application of approaches that enable more accurate predictions to be made about ever more complex systems and to explore the dynamical evolution of these systems over longer timescales. These enduring challenges are particularly relevant to water–solid interfaces and it will take considerable effort to develop new computational approaches that allow us to make better quality predictions of complex water–solid interfaces. Sophisticated electronic structure approaches, such as the diffusion Monte Carlo approach mentioned above, could play an increasingly important role probably in conjunction with the approaches of machine learning.

Acknowledgments

This paper has briefly summarized some of the work carried out in the ICE research group (www.ch.cam.ac.uk/group/michaelides) over the last few years. The key researchers involved in the ice nucleation work were Steve Cox, Philipp Pedevilla,

Gabrielle Sosso, Martin Fitzner, and Michael Davies. Our work to improve the accuracy of water–solid interfaces also involved many people in the group (Yasmine Al-Hamdani, Andrea Zen, Gerit Brandenburg, Ji Chen, Wei Fang, and Javier Carrasco) and close collaborations with the groups of Dario Alfè, Andrew Hodgson, Xinzheng Li, and Geoff Thornton. Support for our work has come from various sources, including the Royal Society and the European Research Council. Computational support for our work has come from the UKCP consortium (EP/F036884/1) and the UK Materials and Molecular Modelling Hub (EP/P020194/1).

References

[1] P. Pedevilla, M. Fitzner, and A. Michaelides, *Phys. Rev. B* **96**, 115441 (2017).

[2] G. C. Sosso, T. Li, D. Donadio, G. Tribello, and A. Michaelides *J. Phys. Chem. Lett.* **7**, 2350 (2016).

[3] G.C. Sosso, J. Chen, S.J. Cox, M. Fitzner, P. Pedevilla, A. Zen, and A. Michaelides, *Chem. Rev.* **116**, 7078 (2016).

[4] M. Fitzner, G.C. Sosso, F. Pietrucci, S. Pipolo, and A. Michaelides, *Nat. Commun.* **8**, 2257 (2017).

[5] J.G. Brandenburg, A. Zen, M. Fitzner, B. Ramberger, G. Kresse, T. Tsatsoulis, A. Grüneis, A. Michaelides, and D. Alfè, *The J. Phys. Chem. Lett.* **10**, 358 (2019).

[6] W. Fang, J. Chen, Y. Feng, X.-Z. Li, and A. Michaelides, *Int. Rev. Phys. Chem.* **38**, 35 (2019).

[7] J. Carrasco, A. Michaelides, M. Forster, S. Haq, R. Raval, and A. Hodgson, *Nat. Mater.* **8**, 427 (2009).

[8] H. Hussain, G. Tocci, T. Woolcot, X. Torrelles, C.L. Pang, D. Humphrey, C. Yim, D. Grinter, G. Cabailh, O. Bikondoa, R. Lindsay, J. Zegenhagen, A. Michaelides, and G. Thornton, *Nat. Mater.* **16**, 461 (2016).

WATER AT INTERFACES: WHERE THEORY NEEDS TO MEET EXPERIMENT

MISCHA BONN and YUKI NAGATA

Department of Molecular Spectroscopy, Max Planck Institute for Polymer Research, Ackermannweg 10, 55128 Mainz, Germany

The Motivation and Challenge of Understanding Interfacial Water

Water, H_2O, has a treacherously simple molecular structure, yet an ensemble of water molecules has unique properties that are difficult to predict from its molecular structure. This is due to the very strong interaction (hydrogen bonding) between water molecules. Hydrogen bonds are strong intermolecular interactions, between the hydrogen atom on one water molecule and the oxygen atom of another. Hydrogen bonding gives rise to the many anomalous properties of water: Water has an anomalously high heat capacity, thermal conductivity, and latent heat of evaporation; it is precisely these properties that allow mammals to regulate their body temperature. Water is a unique solvent due to its small size, high polarity, and large dielectric constant. Water displays a density maximum at $4°C$, ensuring that ice floats on water. In naturally occurring large bodies of water, this means that freezing occurs from the top down, reflecting sunlight and insulating the water from further freezing. Water auto-ionizes and facilitates proton exchange between molecules, of key importance in biological membranes. Owing to the hydrogen atoms contained in water, nuclear quantum effects, such as the zero-point energy, are important in water.

Much of the complexity of water originates from complex and collective interactions between the molecules; the presence of a single hydrogen bond affects the structure and dynamics of water molecules beyond the two water molecules that are directly interacting. The complexity of water's hydrogen-bonded network is connected to its anomalous properties. At the surface of ice and water, and at the interface with other molecules or materials, this network is interrupted, giving rise to yet other fascinating properties, such as the very high surface tension of water and the surface pre-melting, one bilayer at a time, of water molecules at the surface of ice, which is in part responsible for ice being slippery.

The interface of water with a different material — or air, for that matter — is ubiquitous. The following are a few examples: in cells, the cytosolic or extracellular aqueous electrolyte solutions interface with the cell membrane; in nature, water is in contact with minerals; and in electrochemical processes, water is in contact with electrodes. The description of water as a dielectric continuum with $\varepsilon \approx 80$ breaks down at these interfaces. A first step toward an improved description of interfacial water has been to quantify the modification of the dielectric function near the interface. For water confined between graphene and boron nitride, ε is reduced down to ~ 2 [1]. A similar reduction of the dielectric function down to ~ 6 had been concluded for water at the interface with an electrode some 55 years earlier [2]. The large dielectric function of bulk water is primarily due to dipolar reorientation, and the reduction of the dielectric function at these interfaces is associated with the reduction in the rotational mobility of water molecules. The low value for ε observed at interfaces corresponds to the polarization of the electronic shells and nuclei of water, free from contributions due to the orientation polarization associated with the permanent dipole moment.

It is clear, however, that such continuum description of water breaks down on molecular length scales, and a molecular-level description of the structure and dynamics of water is essential for true understanding of interfacial water. Specifically, any quasi-dielectric continuum description of interfacial water does not reflect the geometrical structure in which water is "frozen" at the interface, nor the timescale on which molecular exchange with the bulk and molecular reorientation occur.

Here, we provide a brief overview of one important experimental approach that can provide access to molecular properties of interfacial water, and we highlight the importance of theoretical modelling, not only for understanding but also designing experiments. In this note, we limit ourselves to the most elementary water interface: the water–air interface.

Experiments to Probe Interfacial Water Structure and Dynamics

It is challenging to characterize the molecular structure of the outermost monolayer of water at an interface. Comparing the sub-nanometer thin interfacial layer, it is evident that the interfacial layer is several orders of magnitude thinner than a typically "thin" water film sample with a thickness of $1\,\mu$m. As a result, for most spectroscopies and scattering methods, the contribution from the bulk will be substantially larger than that from the interface. This problem can be circumvented using even-order nonlinear optical spectroscopies. As with any spectroscopy, these spectroscopies have selection rules, and for even-order spectroscopies, inversion symmetry must be broken in order for signal to be generated. This symmetry breaking, per definition, occurs at the interface, where water molecules are located in an asymmetric environment. Combining even-order nonlinear spectroscopy with vibrational spectroscopy, e.g., second-order nonlinear vibrational spectroscopy, one can obtain the vibrational response of specifically interfacial water molecules. In this approach called sum-frequency generation (SFG) spectroscopy, an infrared and a visible laser pulse are combined at the surface, and the sum frequency of the two beams is generated from the interfacial region. If the infrared is resonant with a vibrational transition of interfacial molecules, the sum-frequency signal will be resonantly enhanced, providing the vibrational response of specifically interfacial molecules. The central frequency and lineshape of the O–H stretch vibration of H_2O molecules are sensitive to the hydrogen bond strength and topology. Specifically, an O–H$\cdots$O hydrogen bond weakens the covalent O–H bond of the hydrogen bond donor, leading to a lowering of the O–H stretch vibrational frequency. Therefore, one can directly correlate the hydrogen bond-induced shift in the O–H stretch frequency to the hydrogen bond energy.

To illustrate the surface sensitivity of SFG spectroscopy, Fig. 1 shows the SFG response from the water–air interface. The spectrum reveals, in addition to the response from hydrogen-bonded OH groups in the 3100–3400-cm^{-1} region, a distinct response at 3700 cm^{-1}. While the hydrogen-bonded response also shows up in bulk spectra, the 3700 cm^{-1} peak is absent in bulk vibrational spectroscopy: it is associated with dangling or free OH groups sticking out from the bulk into the gas phase. The absence of hydrogen bonding causes the OH stretch frequency to have a very high frequency, similar to that of gas-phase water. Being a technique that uses pulsed lasers, SFG can be readily extended to perform time-resolved measurements, as illustrated in Fig. 1(b). Such measurements can elucidate not only rates and mechanisms of vibrational energy flow, specifically at the interface, but also the reorientational dynamics of interfacial OH groups [3, 4].

The SFG spectra provide ensemble-averaged *spectroscopic* information for interfacial OH groups, which provides a view on, but is insufficient to generate a comprehensive picture of, the structure and dynamics of the interfacial water. Molecular simulation in combination with the experimental spectra is a powerful approach to answer questions such as the following: What is a typical geometry of interfacial water molecules that possess a free OH group? What defines the "interfacial regions" and what is the probing depth of SFG? How can we resolve and

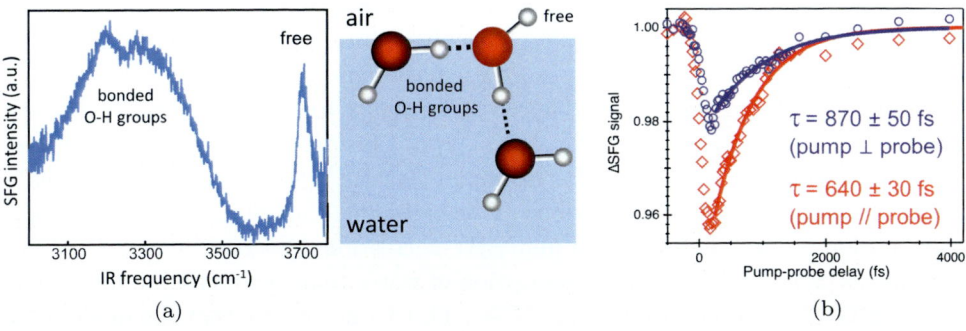

(a) (b)

Fig. 1. Static and time-resolved SFG spectroscopy of the water–air interface: (a)SFG spectrum of the water–air interface, illustrating the surface sensitivity of the technique: about 20% of water molecules at the water–air interface have a free OH sticking out from the water phase into the air, as illustrated in the center panel. These OH groups appear in the SFG spectrum with a frequency of 3700 cm^{-1}. (b) Ultrafast, polarization-resolved SFG spectroscopy of the free OH groups reflects ultrafast reorientation and vibrational relaxation.

assign complex vibrational features to molecular vibrations and local environment? By reproducing interfacial water properties such as the structure and dynamics of the free OH group [5–8], we can assess the accuracy of the structures and dynamics of interfacial water predicted through the molecular simulation technique, allowing us to gain a comprehensive view of the structure of interfacial water.

Theory to Probe Interfacial Water Structure and Dynamics

The molecular trajectory of the water–air interface can be generated through molecular dynamics (MD) simulation using the slab model geometry. To connect the experimental observables for the free OH group with the MD trajectory, one needs to identify the free OH group, among the OH groups in the simulation cell. The free OH group can be identified through the geometrical relation of an OH group of a water molecule and other water molecules [8]. With this definition, one can compute the properties of free OH groups accurately from MD simulation. Interestingly, force fields have been typically optimized to reproduce bulk, rather than interfacial properties of water. As such, it is not *a priori* evident which type of MD simulation is most suitable when studying interfaces.

Among several types of MD simulations, we focus on the *ab initio* MD simulations, because *ab initio* MD simulation can describe the heterogeneous interactions of water at surfaces without complicated force field modelling and thus has a large potential for investigating complex phenomena such as adsorption and dissociation of water on metal oxides, chemical reaction occurring in an atmospheric chemistry, and wetting properties of materials. *Ab initio* MD simulation has been carried out both in the bulk and at the interface, while the accuracy of the description of the interfacial water has not been verified, in contrast to the number of studies on bulk water media.

We evaluated the accuracy of the DFT methods, by computing the deviations of the free OH properties at the various calculation levels (exchange-correlation (XC) functional/van der Waals (vdW) correction) from the experimental data. The averaged deviations (score κ) for various calculation methods are displayed in Fig. 2. In this figure, a large pie indicates better performance for reproducing experimental data. From this pie chart, one can learn several lessons. First, one can see that, within the same XC functional (for example, revPBE), the score varies drastically with various van der Waals corrections. This highlights the importance of the proper combination of the XC functional and vdW corrections. Second, hybrid-GGA XC functionals can provide a better description of water than the GGA XC functionals (revPBE GGA vs. revPBE0 hybrid-GGA). Third, one can see that the modern XC functional such as M06 shows relatively poor performance. This is surprising, given its excellent prediction of gas-phase energetics. This indicates that a critical check should be made for not only in the gas phase but also in the condensed phase. In this respect, the water–air interface is an ideal system, as a water molecule experiences both gas phase and condensed phase at the same time.

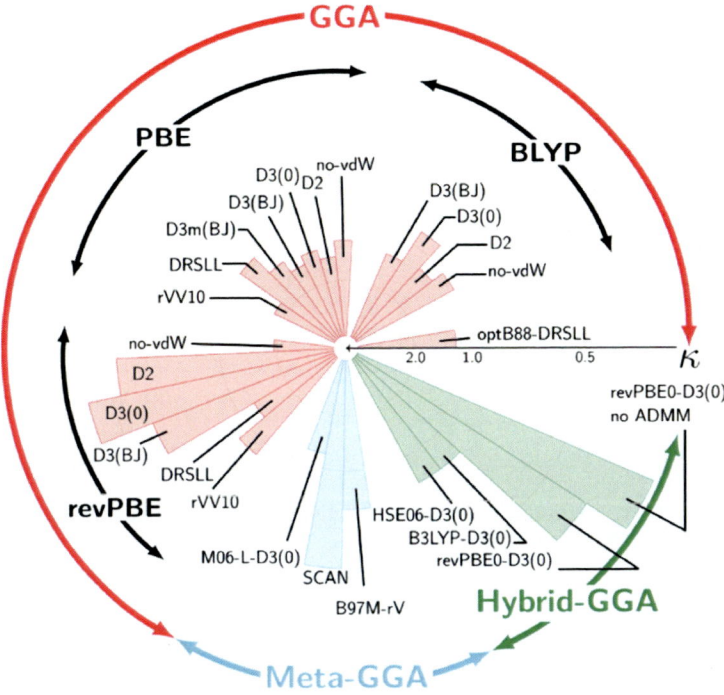

Fig. 2.   Direct comparison of the ability of different functionals to accurately predict water properties. The smaller (larger) score κ corresponds to better (worse) predictive power of the functional [9].

Future Challenges

There are several major water-related challenges that require a molecular-level understanding and description of water in complex environments and relate that understanding to macroscopic phenomena in aqueous systems. These challenges spread across different disciplines, including atmospheric science, geochemistry, energy science, membrane technology required for water purification and desalination, and biophysics.

This list is far from complete and only serves to illustrate the breadth and importance of hitherto unanswered questions relating to water in key scientific disciplines. The common denominator connecting these challenges in these very different systems is that water is either in confinement or at an interface and that non-equilibrium phenomena play an important role. A true understanding in these different systems requires Angstrom/nanometer-level insights into water organization in complex systems on ultrashort (femto- to picosecond) timescales and connecting these insights to macroscopic, long-time characteristics. Obtaining such understanding will require the development of new experimental and theoretical tools, along with an intimate interplay between theory and experiment.

Acknowledgments

We are grateful for financial support from the MaxWater project of the Max Planck Society.

References

[1] L. Fumagalli, A. Esfandiar, R. Fabregas, S. Hu1, P. Ares *et al.*, *Science* **360**, 1339 (2018).
[2] J.O.M. Bockris, M.A.V. Devanathan, K. Müller, and J.A.V. Butler, *Proc. Royal Soc. London. Series A. Math. Phys. Sci.* **274**, 55 (1963).
[3] C.-S. Hsieh, R.K. Campen, A.C. Vila Verde, P. Bolhuis, H.-K. Nienhuys *et al.*, *Phys. Rev. Lett.* **107**, 116102 (2011).
[4] M. Bonn, Y. Nagata, and E.H.G. Backus, *Angew. Chem.-Int. Ed.* **54**, 5560 (2015).
[5] Q. Du, R. Superfine, E. Freysz, and Y.R. Shen, *Phys. Rev. Lett.* **70**, 2313 (1993)
[6] S. Sun, F. Tang, S. Imoto, D.R. Moberg, T. Ohto *et al.*, *Phys. Rev. Lett.* **121**, 246101 (2018).
[7] C.-S. Hsieh, R.K. Campen, M. Okuno, E.H.G. Backus, Y. Nagata *et al.*, *Proc. Natl. Acad. Sci.* **110**, 18780 (2013).
[8] F. Tang, T. Ohto, T. Hasegawa, W. J. Xie, L. Xu *et al.*, *J. Chem. Theory Comput.* **14**, 1357 (2018).
[9] T. Ohto, M. Dodia, J. Xu, S. Imoto, F. Tang *et al.*, *J. Phys. Chem. Lett.* **10**, 4914 (2019).

THE QUEST FOR WATER REPELLENCIES

TIMOTHÉE MOUTERDE, PIERRE LECOINTRE and DAVID QUÉRÉ

Pmmh, Espci, PSL Research University, Paris, France

Water Repellencies

It is significant that the modern research on water-repellent materials started about 20 years ago with papers by engineers [1], on the one hand, and by botanists [2], on the other hand — that is, an opportune combination of biomimetics and potential applications. Since then, studies of all kinds flourished, in particular in fields as diverse as material chemistry (how to make such solids), physics (existence and stability of non-wetting drops), and fluid mechanics (dynamical behaviour of water on repellent materials), among many others.

In our opinion, the most promising studies today still rely on the dual foundations of this field, being pushed by applications often inspired by natural systems. The main question deals with the concept of *functions*: instead of discussing water repellency at large (for instance, through the measurements of contact angles), the aim is rather to define specific anti-water properties and to achieve materials that have clear-cut performances for a given function — such as anti-rain, efficiency to evacuate water, additional oleophobicity, anti-fogging, underwater aerophilicity (and potential dynamical applications or anti-fouling properties), anti-icing, buoyancy and locomotion on large water surfaces, etc.

Hence the temptation to define as many kinds of water repellencies as we have functions, and more ambitiously to interpret the endless variety of texture observed at the surface of hydrophobic leaves or insects as the result of the adequacy of a texture with a function — a correspondence that mostly remains to be established.

The Particular Case of Anti-fogging

The theme of anti-fogging is a good example of a recent development in the field: a material that can expel dew is of obvious practical interest, especially for insects living in highly humid environments — placing this question, there again, at the crossroads between bio-inspired and application-driven studies. It is also challenging: repellency is provided by microtexture at hydrophobic surfaces, so that drops condensing at the scale of the texture will naturally be trapped, a mechanism that destroys hydrophobicity (and can even generate a hydrophilic-like behaviour, the material being infused by water) [3].

In two independent, stimulating papers, Jiang and Chen suggested that the special texture found in natural systems, namely, on the eyes of mosquitoes and on the wings of cicadae, presents anti-fogging properties [4, 5]. In both cases, the features responsible for repellency are at a much smaller scale than usual (around 100 nanometers), which first provides transparency and anti-reflection; yet, they differ by their shapes, being cylindrical and conical, respectively. These preliminary observations raise beautiful questions: (1) Why would a reduction in the texture size repel condensing droplets? (2) What are the relative efficiencies of size reduction and shape effect in anti-fogging? (3) More generally, how can we quantify the abilities of a solid to expel dew?

We developed two tests that enable us to classify the anti-fogging performances of materials [6]: (1) ability of a water drop to slip on tilted solids in the presence of fog, as water bridges nucleate and grow between the drop and its (possibly infused) substrate; (2) ability of condensing droplets to depart the surface as they merge, which can only happen if water adhesion is low enough. We successively describe these two tests.

The first test consists of growing a hot drop on a colder, tilted surface. Adhesion is deduced from the volume of water as it departs, owing to gravity. Hence, condensation-induced adhesion can be quantified, and the intensity of the "fog" precisely tuned by playing on the temperature difference between the liquid and the solid. Using homothetic texture made of cylindrical pillars (and kindly provided by Antonio Checco, Brookhaven, and Gaëlle Lehoucq, Thalès-Palaiseau), we observe that adhesion naturally increases with the intensity of the fog, a consequence of the increase of water nuclei "gluing" the drops to their substrate. Less obviously, the texture size is found to deeply impact adhesion [6]: reducing the pillar size by a factor ~10 (from ~1 μm to ~100 nm) divides the adhesion by ~100, a strong effect reminiscent of that postulated for mosquito eyes. This can be understood by considering the miniaturization of the "glue points" provided by the water nuclei in the texture. Assuming that the density of nucleation of water does not depend on the texture size (not an obvious statement), the surface area occupied by the nuclei is divided by 100 when deceasing the pillar sizes by 10 — and so does the adhesion energy. While the texture size generally has a negligible effect in water repellency (contact angles for instance are the same on these different samples), it is found here to deeply affect the resistance to wetting.

Yet, the most spectacular behaviour occurs when changing, for a given texture size (~100 nm), the shape of the texture, from cylindrical to conical. This change induces a drastic reduction of adhesion, down to the point where we could not measure any effect of condensation [6]. At the macro scale of our experiment, it seems that condensing water does not bridge anymore the drops to their substrate, whatever the intensity of the fog (i.e., whatever the temperature of the deposited drop, up to the boiling point of water). We interpret this result as a consequence of the asymmetric landscape provided by the cones: (1) For an ideal array of nanocones, water should not conform to the tips at the bottom of the cavities, due to surface tension and long-range forces that favor its dewetting from these singular regions [7]. (2) A water nucleus located in the texture between adjacent cones should be driven upward by surface tension — since it reduces its surface area as it becomes less confined. Hence, nuclei should be expelled from the texture, keeping it dry even in these challenging conditions. Our scenario for spontaneous drying is not proved (even if it explains the absence of condensation-induced adhesion), which would require an ultra-fast and highly-resolved technique of observation, but direct observations of drops at a micro scale on hydrophobic nanocones show that non-wetting states are preserved even at this scale [8], which can be seen as an indirect proof of the absence of infused water in the texture.

Our second test is not only closer to applications, but it also directly probes the anti-fogging function — that is, the ability of a material to expel (spontaneously) water condensing from the atmosphere. We just record the so-called breath figure of a material brought and maintained by a cooling stage to a temperature of typically 5°C, so that water from the atmosphere continuously condenses, grows, and coalesces as time goes on [6]. If drops are mobile enough, they can take advantage of coalescence for departing, since this process injects surface energy possibly transferred in kinetic energy, allowing the merged droplet to take off — a mechanism that eventually, at long time, leaves the substrate only with tiny droplets, that which did not yet coalesce with their neighbours.

The number of events is enormous in this experiment (typically 6000 coalescences per square millimeter, per hour), so that a statistical analysis can be performed. This allows us to provide an average anti-fogging performance, which we define as the probability p of depart after coalescence as a function of time (it is constant, apart from early time), and as a function of the droplet radius R — or radii when coalescence is asymmetric. Results are impressively contrasted when comparing the samples: while the probability of takeoff is 0% on nanocylinders, it jumps to a value as high as 99% for symmetric coalescences on nanocones [6, 8]. This probability is independent of the radius R, down to a value $R^* \sim 1$ μm where it abruptly falls to zero — which eventually yields two metrics, namely, p and R^*, for characterizing the performance of the material.

The two tests are correlated. The absence of water bridges reported as water evaporates and recondenses on nanocones preserves the mobility of droplets (even at a small scale), explaining their ability to take off (even at a small scale). More generally, they suggest a hierarchy between size effect and shape effect: nanocones outperform nanocylinders, which strongly suggests that geometry (as often in wetting phenomena) is a major feature for tuning repellency and expanding it to cases where phase changes are present. This could be confirmed by considering microcones [9], for which our arguments predict quite similar performances while anti-fogging fully disappears on microcylinders.

Beyond

We chose here to select our work on anti-fogging as an example of current research in the field. We did it for three reasons.

(1) Superhydrophobicity (in term of contact angle) is a situation easy to generate, but it often has some natural limits. There is today a particular effort to broaden the field of applications to situations where we would first expect a failure of repellency. As we saw, progress can be made on the anti-dew side, but the question remains open whether anti-icing can be promoted by special textures. Stability (as a function of time and speed) of air films on repellent surfaces immersed in water is another question where progress should be made — there is also, in this case, no consensus on the possibility to delay transitions to turbulence, and to affect the drag coefficient as such solids move in water.

(2) We have been flooded for the last 10 years by papers dealing with water-repellent materials (a tsunami to which we contributed). These papers are often redundant, in particular because they are not quantitative enough. We think it is necessary today to address a specific repellent function with relevant tests characterizing the efficiency of a given texture. This allows us to hierarchize different textured materials, including natural ones, that need to be incorporated in such lists: after all, is the lotus leaf the "best" repellent material, as it is often claimed?

(3) The research on water repellency is driven by application. However, there are still very few products on the market, mainly due to the aging characteristics of repellent materials. There are cases (and corresponding products) where a temporary repellency is a satisfactory solution — applying on a solid a colloidal suspension of hydrophobic particles can provide a reinforced hydrophobicity for a mirror or for a windshield. But, in most cases, we rather wish a durable solution and designing a robust texture (from both physical and chemical viewpoints) are a necessity. Fibrous materials, for instance, can be an interesting solution (owing to the flexibility of the fibers), and it is often adopted by animals. Pyramidal texture (such as the cones we considered) can be another option, in particular in the limit of small size for which a mechanical aggression might be less problematic than for a slender, hard microtexture. But, the challenge here remains, and we are curious of what will happen in the coming years.

References

[1] T. Onda, S. Shibuichi, N. Satoh, and K. Tsujii, *Langmuir* **12**, 2125 (1996).
[2] C. Neinhuis and W. Barthlott, *Ann. Bot.* **79**, 667 (1997).
[3] Y. Liu, X. Chen, and J.H. Xin, *J. Mater. Chem.* **19**, 5602 (2009).
[4] X. Gao, X. Yan, X. Yao, L. Xu, K. Zhang *et al.*, *Adv. Mater.* **19**, 2213 (2007).
[5] J. Boreyko and C.H. Chen, *Phys. Rev. Lett.* **103**, 184501 (2009).

[6] T. Mouterde, G. Lehoucq, S. Xavier, A. Checco, C.T. Black *et al.*, *Nat. Mat.* **16**, 658 (2017).

[7] W. Xu, Z. Lan, B.L. Peng, R.F. Wen, and X.H. Ma, *RSC Adv.* **6**, 7923 (2016).

[8] P. Lecointre, École Polytechnique, PhD Thesis, (2019).

[9] W. Ding, M. Fernandino, and C.A. Dorao, *Appl. Phys. Lett.* **115**, 053703 (2019).

SESSION 3: MODELS AND EXPERIMENTAL DATA ON WATER DYNAMIC COMPLEXITY OF SOLID/LIQUID INTERFACES

CHAIR: J. AIZENBERG

AUDITORS: B. CHAMPAGNE[1], A.M. JONAS[2]

[1] *Laboratory of Theoretical Chemistry (LCT), Université de Namur,*
61 rue de Bruxelles, 5000 Namur, Belgium

[2] *Institute of Condensed Matter and Nanosciences, Bio- and Soft Matter (IMCN/BSMA),*
Croix du Sud, 1 box L7.04.02, 1348 Louvain-la-Neuve, Belgium

Discussion among the panel members

<u>Joanna Aizenberg</u>: Well, let us first discuss among ourselves to see whether there is anything that we can learn in addition to discussions that we already had in the presentations; and in some way I honestly want to say that my own take on many projects that I do is trying really hard, in pretty much every case, to develop analytical models. And with that I want to address the first question to David Quéré. If we talk about analytical models, regarding the phenomenon that you described. How much of that can be actually done?

<u>David Quéré</u>: Thank you for asking this question which somehow corrects a little bit my presentation, which I chose to be extremely qualitative, based on images and movies. So you might have the feeling that all of that was just a description of things, which in many cases is enough actually. We are very happy to see phenomena and to try to understand them qualitatively, but of course we need also to develop models. I could have a general answer and a particular one here. The general answer is that in soft matter at large — and I am very flattered to be among you because I am more physicist —, in the physics of soft matter where we have to face complexity, we like very much to use scaling laws. Which is a very powerful tool for describing things in a quantitative way. This is something which allows us to face questions that we could not solve really analytically, or at least it is a first step. In all what I showed, there is a way to solve things by scaling laws. So, they are laws, they are models, which are describing all these facts. Modelling could be also using computers, which personally I do not, but we had many evidences in this conference that this is a fabulous approach, in particular when you go to the limit of microscopic systems. Because we tend to reach smaller and smaller scales when for instance I discussed textures and solids. And the question of reducing the size was obviously something very interesting because it produces new facts. Here, of course we converge with computing because the size of the systems that we are

dealing with is such that molecular dynamics, for instance, provides quantitative answers which very nicely complement what we can do. So, modelling we love it and we need it!

Julia Yeomans: I think we can go a long way with the continuum models, they really seem to work very well down to length scales, maybe even to ten nanometers, maybe a bit less. And we know what the equations are, and so the hard thing is that sometimes they are a bit tricky to solve in a complicated geometry but I think at that level the equations are right. It is an interesting question just where you need to cross over to molecular dynamics.

Mischa Bonn: I completely agree and I think continuum models are extremely powerful and, if we talk about water at interfaces for instance, there are many very straightforward intuitive physically extensions of the Poisson-Boltzmann theory to describe water interfaces, to account for molecular level effects and to take those into account in a more or less phenomenological way. On the other hand, I think it also depends which approach you use, it depends on the question you want to answer. Because if you want to talk about energetics of interactions, as Angelos pointed out, continuum models simply don't give you anything that resembles an answer. So you will need to go to molecular level descriptions.

Eugene Shakhnovich: I am a big fan of continuum models as well. On that, aside, besides dynamics, there is of course thermodynamics. There are a lot of things that we can learn from usual continuum model thermodynamics studying chemical potentials or water effects on solutes, and all of that. It works to a great extent and even for smaller systems. But there are of course cases when this kind of molecular nature of water becomes very important. One aspect of that is related again to biological systems where water plays a structural role. There, continuum models naturally break down and we have to consider explicit energetics. Because in term of just energetics even on the interfaces in some cases you can use simple thermodynamical analyses to estimate or to get these numbers, these parameters from continuum models. We do not always need energetics on the interfaces which are microscopic, molecularly resolved. There are cases when water is molecularly resolved by Nature, in a sense, by its structure. In this case of course we have to treat water obviously as a kind of a single molecule rather than a continuum.

Joanna Aizenberg: I want to continue on that and I was trying to present this as a multi-scale problem when there are solutions and ways we can address a phenomenon on one scale but not on the other. Some macroscopic functions depend on molecular structure and potentials and others do not. So, if we think about challenges associated with scales, what are the best practices of multiscale or reduced order models. Do you want to answer?

Sinan Keten: I think there are usually benefits of continuum models — I'm trained as a mechanician —, so from our perspective you always start with the simpler analytical model that gives you some physical intuition into the problem. But oftentimes you realize the exceptions that come from heterogeneities, chemical features that the small scales are difficult to explain with the simple models. The exceptions are often the most interesting things you see. From things like transport problems *etc.* Then, the question becomes, as you incorporate these details from the molecular scale to higher scale models, how do you pass the information across the scales? In the context of molecular dynamics, we have atomistic models, we have *ab-initio* models, we have coarse-grain models, which are even one scale above. We have many challenges on how to describe the dynamics at these different scales. In the case of transport: how do I coarse-grain, for instance, a membrane? How do I capture the dynamic features that I think are important in those types of systems? This is a big challenge and I think one has to think about ways to enrich in a systematic way to capture the chemical details. There is a lot of work on systematic coarse-grain techniques like that but few have been really applied to things like transport and there is a lot to be done in that field I think.

Joanna Aizenberg: I am an experimentalist most of the time, so if I will be provocative then can you tell me: in addition to a general description of water in particular in this case, are simulations useful? If so, which ones? And if they are useful, how do you link results of your simulations to the scale of experiments? Is there a crossover of scales? Just convince me as an experimentalist that you can do something that I can then take and expend on that experimentally.

Angelos Michaelides: I think that these systems are very challenging, we have heard that a lot already, and you need a very well-defined molecular level understanding to describe a lot of macroscopic phenomena. Mischa showed a nice example on silica, of how specific functional groups can alter the macroscopic properties. We have worked on it. Another example is the wetting of a copper substrate, where the experimentalists saw that under specific conditions the substrate wetted, it was hydrophilic, but they could not understand why. It came down to us doing simulations at the molecular level, understanding that continuum models broke down, in this case, because the water is dissociated and there were specific hydroxyl groups that were acting as nucleation sites. It was only through getting this well-defined information at the molecular scale that we could describe the larger scale macroscopic phenomena. I think the value of theory and simulations, in this area, is providing insights and understanding that can help and interpret the larger scale experimental data.

Julia Yeomans: I think that you need both because simulations can give ideas and then you do the experiments and find out what is wrong with the simulations. In my view, trying to do simulations which look exactly like experiments is a bit pointless because you can do the experiments. But doing simulations which help to understand the basic ideas is often helpful to suggest new experiments.

Joanna Aizenberg: I cannot agree more because honestly the way I try to address many of my questions is first to see whether the simulations can repeat my results but this is not that interesting. But then actually to find interesting features that we have never observed experimentally and be guided by these simulations and knowledge that can be drawn from theory. Can you give me a couple of examples where things that you have done with simulation of water was actually helpful in highlighting new behaviour and then shown experimentally?

Julia Yeomans: An example from the droplet work is that we looked at drops bouncing on a surface which was a superhydrophobic surface but with large holes and these drops spread out like a pancake and then jumped without retracting. It took us ages to work out why. It turns out it was very simple. The thing acts like a trampoline because the water goes in into the holes and back out again. The only reason we have got it in the end (maybe if we had been clever we would have figured it out sooner) was to do the simulations to see what was happening. The simulations did not really match the experiments but the physics behind it matched and then we could do the theory right.

Angelos Michaelides: There was a movie that I was tempted to show but I did not have enough time. There have been nice experiments with STM looking at the diffusion of water clusters across surfaces. Interestingly, when the rates of water monomers diffusion were compared to rates of water dimer diffusion, the STM experiments showed the dimers moved more rapidly than monomers. This was an interesting observation and given the relative mass it was unexpected. We did a lot of simulations to try and understand this effect and we realized that the key to this was the fact the hydrogen bond in the water dimer was exchanging, and it was exchanging through the tunnelling process. We could then predict that if you go off and do the measurements with deuterated water, you would not see more rapid diffusion. Also, we went further and we screened a wide variety of substrates and we made predictions: this substrate tunnelling is possible and we would get fast diffusion or on this other substrate tunnelling is not possible and we will not get fast diffusion. Those predictions were followed up with experiments and verified.

Mischa Bonn: I think there are two useful interplays of simulations and experiments. One is where the simulations explain the experiments. An example that I showed you is that water interacting with the soft charge interfaces behaves differently than water interacting with hard charge interfaces, like mineral charge interfaces. This is simply because you need a molecular understanding of the deformation that can occur at high charge densities for the soft interfaces which is not possible for the hard interfaces. I have also had similar experience to Angelos, where in fact the simulations were predictive. Theory collaborators came to me and said "you know I have been looking at the water/air interface for a long time and if you partially deuterate your water — so you look at mixtures of H_2O and D_2O and HDO —, my quantum simulation predicts that you will have preferentially OH groups sticking out the surface into the gas phase. It is going to be a non-statistical distribution of free OH groups or dangling OH groups" and I have "this is the quantum nuclear effect for sure. We are never going be able to see that." He did the calculations, quantified it and we could see it. It was actually quite significant. For me, it was a very nice example of where simulations predicted an effect that I would have never tried to do the experiments, ever. I would never have believed it is possible. But it worked.

Eugene Shakhnovich: Another example where simulations were quite important was like about six or seven years ago, we were trying to design an inhibitor believed to be implicated in Alzheimer's disease. No luck, nothing worked but then at some point one of the post-docs did simulations very carefully (all-atoms simulations with explicit water), and he predicted that there is one structural water which was not seen crystallographically which slightly changed the conformation of the loop and the close form by, I would say, 1.5 Å probably. And then the design was done against this predicted conformation and this design worked. And then later on there was a high-resolution structure which showed the structural water in the structure. Basically sometimes of course it is a super high-resolution approach that continuum methods cannot reproduce by definition, but this is an example where atomistic water in structural simulation played a really important role.

Joanna Aizenberg: One or two minutes left and I do want to ask another question that nobody touched. I like crystallization, as you probably understood from my presentation. A subject that nobody talked about, not only water at interfaces, about minerals. If you deposit minerals, a majority of things is driven by hydrated ions and their ability to release the hydrated shell. What is done in term of simulations of how hydrated ions, when incorporated into crystalline phases, release their hydrated shell and how this controls crystallographic features and types of minerals that can form. A particularly interesting system is of course carbonates, and carbonates are of interest to Solvay all together. Nobody has ever looked at that.

Angelos Michaelides: We have been interested in understanding the wetting properties of clays. In collaboration with Rahul Nair's group in Manchester, contact angles of water droplets on clay substrates have been measured. They were able to show that the contact angle was a function of the counter-ion that was present in the clay. So, we went about trying to understand what was going on. We could relate the hydration of the counter-ions, comparing different counter-ions to correlate what was observed with the experimental observation.

Joanna Aizenberg: I want to finish this — coffee is waiting — by challenging you, by looking at hydration of ions, somehow it is not done enough and there is a lot of interesting things especially in biological systems, thinking about magnesium with an hydrated shell, calcium with a shell and what is happening due to the fact that water is so tightly bound to some ions but not to others and how that affects the interactions and the ability to control all ranges of phenomena. We know this experimentally but there is a lack of theoretical and simulation data on this subject. And with that, I would like to finish our discussion, have a little bit of coffee, and try to take difficult questions from the rest of the audience.

General discussion

Joanna Aizenberg: This session is open for questions.

Bernd Hartke to Angelos Michaelides: Concerning the nuclear quantum effects in water, would you expect that it is possible to incorporate that into a coarse-grain model that does not do the nuclear quantum motion explicitly? Like, for example, in force fields, I would say force fields coarse-grain the electrons away and this gives you a five-order of magnitude in performance enhancement. Would you expect it to be possible to incorporate nuclear quantum effects into an empirical water model so that you do not have to treat them explicitly?

Angelos Michaelides: In many empirical water models, if they are trained against experiments, then they are implicitly accounting for nuclear quantum effects. If we want to include them explicitly in empirical potentials, then the most widely used approach is path integral techniques. Traditionally, path integrals have been very expensive but there has been a lot of work using clever thermostats that actually make the cost of a full path integral simulation only one or two times more expensive than a traditional classical molecular dynamic simulation. So, those are my answers. They are implicitly included if we are fitting to experiment and if we want to account for them explicitly they were very expensive, but this is not as expensive as it was.

Bernd Hartke: The obvious follow-up question is, is there any need to ever?

Angelos Michaelides: It depends on what you are interested in. There has been a big debate about the properties of liquid water and how nuclear quantum effects alter their radial distribution function, density, diffusion coefficient. It is all on the level of a few percents. That does not necessarily excite me. Also the lattice constants of ice do not change very much. But where nuclear quantum effects are very important and need to be accounted for is if the water is dissociating and we have free protons. That can happen in the bulk, and that can happen at the interface. There, nuclear quantum effects can have a very significant impact and need to be accounted for.

Alán Aspuru-Guzik: I actually have a paper on that topic that I think is relevant. We developed a theoretical force field factor theory which is the equivalent to density functional theory for force fields. It shows the following: there is always a unique classical potential that reproduces the quantum distribution at a fixed temperature and pressure. This basically means that, for each point in the phase diagram, you can have a classical potential that would reproduce the quantum distribution with classical dynamics. This is my opinion how you can do a force field that includes the quantum effects. The theorem is there. We have shown basic examples for van der Waals type potentials but this can be extended to a full water potential. There is theory for that and it is called force field factor theory.

Joanna Aizenberg to Alán Aspuru-Guzik: But the temperatures and pressures that you are developing with, are they relevant to actually reasonable conditions?

Alán Aspuru-Guzik: The way to do it is simple: to develop the force fields you would do a path integral simulation at a particular temperature and pressure. Again, the theorem shows that the force field is not transferable to other temperatures formally. But once you have run the simulation, you can find that force field and the theorem says it would be exact for that particular path integral distribution.

Bert Weckhuysen to Mischa Bonn: I enjoyed very much your talk and I have a first question on your silica and your water. Did you ever have considered to do an experiment where you do silica with ^{18}O and the water with ^{16}O, or the other way around, to try to distinguish, with your method, what is happening so you can really distinguish what you showed in one of your first slides?

Mischa Bonn: That is a very good question. Indeed, one cannot do the isotopic exchange for the deuterium or the hydrogen because that exchanges too rapidly (the surface hydroxyls with the water) to distinguish that. Those two make it impossible. But indeed, that is a very good idea and it is an idea that we implemented after thinking much longer than you thought about it. We did the experiment for $H_2^{18}O$ to show indeed that there is an isotopic shift in the OH that can be assigned to the water and not to the silica.

Bert Weckhuysen: My second question is about silica. Silica is used a lot as a catalyst support and one silica is not the other. You can make silicas in different ways, and I assume your zero point of charge is around two. Do you have ever considered to put heteroatoms in it like aluminium and then try to increase or vary that? And then, what is the effect on it?

Mischa Bonn: That is also an excellent question. Indeed, depending on the source of silica, the silica surface will have different types of surface hydroxyls each with different pKa's. Indeed, the point of zero charge will vary depending on the source of the silica and the way it is made and the way it is purified. We have seen sort of batch to batch variations in the quantitative details. The qualitative trend is always the same. We have not thought about using that chemistry to tune the surface properties to a desired goal but that is a very nice idea.

Joanna Aizenberg: I want to add to this comment and maybe suggest that, in addition to silica with other oxides attached to it, you may try to look at the biological form of silica where different species would have a very nice range of hydroxyls on the surface. All these silicas would have different levels of hydration, not all the OH bonds are actually fully condensed in these systems. You will have a really broad range of surfaces that Nature created from glass surfaces, where OH distributions are quite well described by now. Just to see how that may affect the interaction with water as a function of dangling OH bonds.

Bert Weckhuysen: My last question is about silica. In the introductory comment, it was already mentioned about electrocatalysis. It is very important to know that in electrocatalysis what is really happening at the interface and what is happening with the charges. I was wondering if you can device this as an operando system, where you could look at this first real monolayer under these conditions. What would then happen? Do you believe that your method could establish that?

Mischa Bonn: Yes, that is also again a very good idea that we are working on. The challenge is to actually have the optical beams reach the interface. The infrared beam, when it goes through water, is absorbed over a very short length scale, microns roughly. The trick that we are now developing, and we have some initial work that has just been published, is to use graphene as an electrode. It is optically thin and allows us to access the interface. So, great idea.

Wilhelm Huck: I was wondering, the rearrangement of you interface near lipid bilayers, what do you think the impact is of that on the adhesion of proteins to lipid bilayers?

Mischa Bonn: That is an excellent question. Thank you. Honestly, I don't know. I would certainly expect that there must be an effect because this reorganisation occurs at physiological charge densities. I expect that the effect would depend very much on a protein to protein basis. I think it is very difficult to predict *a priori* if there is a general rule but I honestly do not know. That is a very good question.

Joanna Aizenberg: Just to comment a little bit from the experimental side of things. There is a duality in this effect: when proteins are attaching to biologically-formed silica, you see reorganisation in both. The structure of a protein is affected by the structure of silica to which it attaches. Then, you see changes in both, these proteins affecting hydration of the silica. But then silica changes periodicity, or distances between its charges in correspondence with those in the protein. There is a really interesting dynamical reorganisation when proteins interact with silica, at least in Nature.

Eva Nogales to Julia Yeomans: Could you tell us more about your modelling of these nematic systems applied to the biology? I do not remember your particular example. Was it the microtubules motors or the epithelial layer? Assuming that you are not entering the molecular details of how these things are moving one with respect to the other into your models, what goes into them so that you are able to ultimately represent that kind of motion?

Julia Yeomans: I am talking as a physicist here so I am looking at the generic properties of these active systems. The sort of model I am writing down should apply to all the systems but of course condense all the questions about them. It is like using the Navier-Stokes equation, it works for all fluids (or simple fluids) but it does not tell you about the molecular things that we have been talking about. So, what goes in, is symmetry of the system, as a continuum model. We are doing more coarse-grained models where we are putting more details. We are sort of starting from the large end and other people have done models starting from the small end.

Eva Nogales: The second part of my question. This has to do with the case of the epithelial layer. I found it very interesting that you said that defects correspond with apoptotic cells but you put them as the defects that generate tension, which then make the cells go into apoptosis. But I would have changed the cause and the effect. Because apoptotic cells round up and they lose contact, that would be more what would coincide with that moving defect, rather than what comes first, in a way. I don't know what you think.

Julia Yeomans: Yes, that is what I thought to start with, but what I have said is based on experiments, experiments done in Benoît Ladoux's group where they actually looked at the chemistry behind it. The experiments showed that — they are bio-experiments so there are lots of error bars —, but the experiments showed that what happen is: stress moves YAP (Yes-associated protein), and YAP signals cell death. The stress on the cell moves the YAP from the nucleus to the cytoplasm. This is a known signal for cell death.

Kurt Wüthrich: I think this is a follow-up to Eva's question. Do I understand correctly that all the processes we have been dealing with in this session are stochastic? And what are the frequency ranges covered? I am sure that surface hydration is a stochastic process. I am not so sure whether this is true for turbulence-like motions that you described or for the topological defects.

Julia Yeomans: I think this is a lengthscale story again. So, I think anything molecular is going to be a stochastic process because of fluctuations. The sort of scale

I was looking at is less so because we are looking at larger things and therefore fluctuations are less important. So, the topological defects, no.

Kurt Wüthrich: What are the frequencies?

Julia Yeomans: I can tell you about the velocities but I do not understand frequency in this context. These things are moving at microns per minute.

Todd Martinez to Angelos Michaelides: There was this question that came up about whether or not theory had predicted anything. Before that Angelos showed us the attraction curve for water to graphene with a bunch of different density functionals and showed that they got different results. Clearly whenever you predict anything you have to validate it first. First you have to figure out what density functional should you use, which one is right. What I would like to ask is an open question: how do you know, when you do predict something, whether you have actually already baked the answer in for the validation process? In other words, how do you assess how difficult a particular prediction was, after the facts?

Angelos Michaelides: This is a very difficult and very important question. I think, in the case of density functional theory, as you would know, as many people would know, there are sort of two strategies to developing new functionals. One is about trying to satisfy constraints and the other is about fitting to large data sets, either from experiments of from higher-levels theories. If you take the latter route, then you will implicitly be including some physical effects that you are probably subsequently trying to capture. The route that we have been taking to get more well defined, more reliable *ab-initio* data is not through density functional theory. In these particular systems, the one slide I skipped quickly through was quantum Monte-Carlo. So, I think that quantum Monte-Carlo is a much more promising approach for the types of problems we are interested in, where we have weak interactions when we are talking about wetting on substrates. There are no empirical parameters in that and there is no real scope for including empiricism in that approach. So, that is what I think is a more useful *ab-initio* way for it. Like I have said, in DFT, there is a level of empiricism that is hard to avoid.

Laura Gagliardi to Angelos Michaelides: We have discussed about maybe DFT or quantum Monte-Carlo and then force fields and Alán suggested a high way of doing more advanced force fields. Probably what has not been discussed extensively are *ab-initio* molecular dynamics methods and Car-Parrinello techniques. I would like to have your opinion and also discussion about where do you think these methods are going? Are they useful for these kinds of approaches? Or the time-scale of the simulations is too short? Or the fact that the quantum in DFT is still too much of a simplification?

Angelos Michaelides: There is certainly a big role for *ab-initio* molecular dynamics (MD). We have used it a lot. In particular, where bonds are being made and bonds are being broken. For a lot of these wet surfaces, you have a mixture of intact molecular state and a dissociated water state. If the barrier between these states is close to zero, then you could sample this with any equilibrium *ab-initio* MD simulation. There are lots of scenarios where you might want to look at a chemical reaction. We looked at, for example, the dissolution of salt crystals. We did it with *ab-initio* MD but we enhanced the dynamics with meta-dynamics. Thinking longer into the future, it could be that machine learning comes along and squeezes out *ab-initio* MD from existence. That is a completely foreseeable future. We are working on disposable machine learning potentials where you run an *ab-initio* MD trajectory and you then train a potential based on that trajectory. It does not enable you to go outside the parameter space in which you have trained but to run a longer trajectory and a bigger system.

Sinan Keten: I think it entirely depends on the problem you look at, and in the particular case of reactions certainly there are benefits of a more quantum level, a higher-level treatment. In the case of pressure-driven flow, it might be relevant when you are looking at things like proton transfer *etc.* but I think for the transport of water molecules through a membrane we can get away perhaps with classical treatments. I think the impact on the time-scaling that you have might introduce other errors such as having to run much larger pressures or having to run much smaller size systems. I suspect in those cases the advantages you get from chemical accuracy would be counterbalanced with the errors you introduce from trying to scale the problem to smaller scales and then trying to extrapolate to actual experiments.

Veronique Van Speybroeck: I have a question regarding also the confinement in nanoporous materials for water. We have done now quite some water simulations on various types of nanoporous materials. It was discussed that of course they can interact with an OH group, so that is interesting. But how about hydrophobic materials, where we do not have it and where they actually, sort of, "rip out of the wall" and how do they organize? We also have seen that they then start to make ice-like structures sometimes. It depends also on the confinement and the typical topology of your material. In extension to that, if you, for example, go to metal organic framework materials, there you have the mixture of hydrophobic and hydrophilic sites. What is your opinion about the structural organization of water in these types of materials? How should we tackle it? Maybe both from experimental and theoretical points of view? To get more insight in that, because I have the feeling that it is not completely well understood in my opinion. But I might be wrong.

Sinan Keten: I think it is a really great question. Polarity is not necessarily a necessity for order in confined systems. People have studied carbon nanotubes and you do get all the structures and all kind of complexity, I am sure everyone is aware. The challenge also there is that you change the dimensions of your pore a little bit and then the structure changes and you get complexities that come with that. I think that is a really rich question. There are other aspects that come from not just polarity but even how you restrain a pore. Whether it is dynamics or restrains, or if it is a stiffer system, and that changes again the fluctuations of water with changes of structure, I do not have a good answer for all the things, but I think it is a very worthwhile direction of investigation.

Veronique Van Speybroeck: My question is basically because I also think it is a very broad and open question. So, how should we tackle it also to obtain structural information in these materials and so on? It would be also interesting to know how, also from an experimental point of view, we can learn from that from a theoretical point of view, how we can simulate that? Maybe we can look at symmetry functions also to see what kind of structures are formed. And then, compare to more water-like structures that we know in other environments. Maybe we have to do it somewhat like that — we have some ideas. But then the question also, how do we compare to experimental insights? So, for me, I do not know, is there any suggestion on how to proceed there? It will be very useful.

Mischa Bonn: I am a molecular spectroscopist. I think that the answer to your question is to do molecular spectroscopy. This is of course not completely straight-forward because it means that we have to translate the structures that we obtained from simulations, be they MD or DFT or even higher-level, to somehow translate them to get the molecular response function that we measure in the spectroscopy from those structures. In my experience, that works remarkably well. This allows to translate structures into spectroscopic observables, it allows a direct connection to experiment in precisely the type of systems that you mentioned where you really do not know much and we do not know what is good theory. It is also difficult to do good experiments, of course.

Miquel Solà: Just a methodological question. In the same way that QM/MM are used to analyse proteins or something like that, is it possible to mix QM/MM and these physical methods that you are using, these nematic models, and this kind of things? Is it something that has been done, or possible? Maybe you can analyse better the interactions of some molecules with the surface or something like that with this possible method? I don't know whether this exists.

Julia Yeomans: I think this is a question which has been hanging around for about 40 years and people have tried to do multiscale methods and, up to now, I think it is probably true to say that they do not work. What seems to work better is when you identify the question you are asking and then choose the right lengthscale model to use it. The way the molecular details go into the continuum models is in coefficients like viscosity or parameters for the free energy. You have to realize how limited and useless that is for some sort of questions but maybe right for others. Coarse-graining is really hard when you get to do dynamics.

Ben Feringa: In the last ten years, in the synthetic community, there was a lot of debate and discussion about on-water chemistry. So, the acceleration of reactions on water. Barry Sharpless introduced this concept and nowadays there are a lot of people that are very intrigued by this reactivity at interfaces with water or even in water *etc*. What do we know about it? Can you simulate what is going on? Or why you see such tremendous acceleration effect in some reactions, like for example the Diels-Alder reaction? Or is it just hydrophobic effects and whatever the orientation of the water? I am intrigued by that because we are so much puzzled and people get very exciting results without having explanations, as far as I know, but can somebody comment on that?

Wilhelm Huck: Actually, we have looked at these on-water effects. What we have found is that most people did not study the kinetics, they only studied the yield after certain time. And so, if you look at the kinetics and the changes between on-water and in-water, the differences are often really small. I think, for the Diels-Alder it is big, for maybe a few others it is reasonable to assume that there are some hydrogen bonds to the transition state. For, I think, the vast majority of the reactions, there is hardly any difference. In our experience, at least.

Joanna Aizenberg: I am actually quite surprised because, generally speaking, when we think about what can happen in monolayers, things on water see two environments: there is the hydrophobic environment of air and the hydrophilic environment of water. In principle, your reactions can reach different states that are not necessarily accessible in water. At least, things can happen; again I do not want to bring it again as a crystallization thing, but crystallization on water is always better than crystallization in water. I would say this duality of the interface, having these two media, contributes in many cases, probably not in all reactions, as an accelerating feature.

Wilhelm Huck: One comment on that. The tricky bit is that they are typically not done on air/water interfaces but on oil/water interfaces, so organic solvents. In a lot of cases it is unclear where the reaction actually reaction takes place because these molecules might be soluble in either phase as well. So, it is very difficult to distinguish exactly what is going on in these systems.

Ben Feringa: May I add to that? I agree with that because we do not know absolutely for sure how much organic material is at the interfaces. In some of these reactions, there might be some solvent *etc*. I will tell you why we are so puzzled. Because recently we have found, and not only we but also Italian groups and some groups in England, that when you have butyl lithium or metal lithium, avoid any molecule of water, because it reacts spontaneously immediately with water. Now, we do reactions on water with butyl lithium with very high yields and other groups are also working on this. It is amazing. There must be something special. It could be that there is a monolayer of organic material prevented from reaching the water or whatever, but there is something special there with the hydrophobic interfaces of water or the air/water interface. And I am puzzled.

Eva Nogales to Angelos Michaelides: It concerns your STM studies of the structural organization of water on these atomically-flat metals. I was a little puzzled by the fact that you were not telling us anything about the details of the surface, or lattice parameters and whether that had any effect. I wanted to understand: in the first layer of water, every molecule of water is only interacting with metal, not with itself, is that correct? In the second layer, do they do both, interact with the metal and with other water molecules or just with water? Why is the lattice of the metal not relevant? Or, is it?

Angelos Michaelides: It is relevant. The point of showing that was to give an overview. Explaining every one of those images, each one would have taken ten minutes. It was just to make the point that these are flat substrates and you get this rich variety of behaviour. The variety and the structures that emerge, they entirely do come down to the balance of the lattice structure, of the reactivity of the substrate, so the strength of the bond with the substrate. In all of them, water molecules are hydrogen-bonded to each other. In the contact layer, these are fully hydrogen-bonded structures. In the second layer, the interaction to the substrate is essentially zero. It is screened out so that you have hydrogen-bonding within the contact layer bonding to the substrate; in the second layer it is bonding within the layer and to the layer below. But we spent a lot of time trying to understand why do specific structures form and when, and we have a set of rules. We have a set of two-dimensional rules to predict these structures. But that is not something that I had time to go into.

Tomas Cech: Those of us who work in the molecular macromolecule biological area choose our systems from Nature. But what I saw today was a variety of systems to look at the interface with water, such as these atomically flat metals and silica. It seems to me that there are perhaps a million or several million different surfaces that one could look at. What we have seen is that you probably will get a different answer with each system. You spent, many of your students spent, many years looking at one particular system. Had you looked at a different system, they would have spent a decade looking at that system. How do you choose the systems to look at? What is the rational?

Angelos Michaelides: The metal systems that I showed, there are not any single crystal metals floating about in the atmosphere. One area of our interest is understanding ice formation on atmospheric airborne particles. But we do not know the composition of those atmospheric particles. So, our work on metals is motivated by a desire to understand: are there any general principles that we can extract from these model well-defined systems where people can do very good measurements, getting molecular scale information and we could do good quality calculations? So, it is to extract insights, it is to validate and benchmark the calculations and it is a route to these more complex, more interesting systems.

Joanna Aizenberg: My problem with that unfortunately is that, especially if you talk about ice crystallization on small particles where curvature is extremely important, there would be no connection to crystallization of ice on an absolutely flat surface. Curvatures introduce problems to crystallinity. If it is an extremely curved environment, you can even form amorphous ice. There are some many things, so that I actually agree with this question. Is there relevance? If you tell me that these metal surfaces are important for some industrial application, it would be one thing. But if this is about understanding and trying to address the question of ice nucleation on small particles in a completely different composition of the environment, I have a problem even with the approach *a priori*.

Eugene Shakhnovich: Another system that I think may be interesting to biological applications, which probably you are sort of alluding to, is water in narrow capillaries with hydrophobic surfaces. There has been a lot of work, including our work in the 90's and several work from David Chandler. It is for sure that there are very peculiar behaviours applied in this type of narrow kind of slits/capillaries where you can have even a kind of evaporation under certain conditions at normal temperature, *etc*. The water will escape the capillary and evaporate and there is a definite — again in this story about dry molten globule — connection to what people see experimentally and in simulations, and what is predicted from this type of water. My point is whatever the question you are asking, what systems you are analysing, the surface should correspond to exactly the system you are curious about and I hope this is an example of that.

David Quéré: The question is so general that it should have as many answers as people here on the panel. It is a little bit like considering abstract painting, for instance. You have something which is *a priori* a huge field of freedom. I think that if you are an abstract painter, the first thing you have to do is to frame the things, to add constraints. Because you cannot face this infinity of possibilities, and so we heard already different answers. Another way to face it, which I think we share with Joanna, for instance, is to consider real systems. These systems are existing so we are not facing something which is infinite. But these natural systems, if we think about the surfaces of bugs and leaves that we showed, are amazingly different however. Instead of being infinite, it becomes at least one hundred types of surfaces. Then, there is a very open question there, which is: why do we have this variety? Does it correspond to different ways of facing water, for instance? It should be in some cases. Just the fact that we ask the question opens a very interesting box. In a very few cases we have answers, but I think it is a way to answer your question which is to say: well we have to select and we select in the real world.

Mischa Bonn: I completely agree and I think this also goes for some of the molecular level studies that I have showed. For lipid/water interfaces, of course it is very easy to exchange lipids and look at the effects of different types of lipids. The different lipids behave more similarly than they behave differently. So, there are certainly rules that one can infer there. The same is true for the minerals. Of course, if I talk about oxides, many of the oxides behave similarly. Yes, there are differences but, again, there are more similarities than there are differences. Even the same is true for protein/water interfaces. We study anti-freeze proteins and ice nucleating proteins. Both of these proteins have the same structural surface to both inhibit and trigger ice nucleation. There are all different sorts of anti-freeze proteins but they all have the same structural properties that allow you to extrapolate by studying one. You have to study a few to show that there is this generality, but once you have shown that, it is clear.

Joanna Aizenberg: So, in other words, one should ideally choose a group of representative surfaces that make sense, that are interesting and relevant, and hoping that the entire group would have similar features that can be captured by the model or by the experiments.

Mark Ellisman: Judging from the time, I think this may be the last discussion, but again I will come back to something Tom started, and Eva also communicated on that. The biological systems that you choose could be more evolved. When you use the example of nano-capillaries, I think that there are examples in biology, in the nervous system for example, dendritic spines and the necks, and instead of thinking of these things as hydrophobic because the molecules there are extremely acidic, actin, they are charged, they would organize water. One of the properties that I was waiting to hear about and now I will shut up and try to stimulate you to tell me about. Are there examples that you are promoting your students to work on where you look at organization and disorganization of water within microdomains? Because that is what looks like biology does within the nucleus, with acidic chromatin, within the cytoplasm. You have local domains where water is highly organized but putting it into the Brownian energetic pool does work. Biology figures out probably how to make that work vectoral. There are lots of systems of organizing and disorganizing water with moving cations, which are usually controlled. There is no free calcium in the cell. You may not realize that, that is a misinterpretation of what dyes tell you. There are thieves, they just look at calcium when it is being handed from one molecule to another. When cations enter, they kick water out of bound surfaces into the free pool. What does that do, if you did it locally or vectorally?

Joanna Aizenberg: Who is brave enough to take this question because actually I had asked similar questions at the end of our discussion. I think that this release of water from hydrated states, and what the role of ions is in that is extremely interesting to me, but any answers?

Mischa Bonn: I think that is an extremely excellent point. While we have done work on cells at different levels of hydration to look at water properties, these are still ensemble averages measurements so they contain some information and they show heterogeneity. But, of course, the dream is to be able to develop techniques that allow you, without labels, to look at water as it moves through a cell at the molecular level on the time-scale that it moves. I am afraid that that is a movie that I have yet to produce. It is certainly the holy grail and the dream.

Joanna Aizenberg: That is the last question that we take but I hope that the way to answer this question is while we do not know now but maybe 50 years from now there would be another Solvay conference on water where we will be able to try to address that in more details. Maybe we will hear more about biological systems, including water and other things, in this session that is still to come.

Tomas Cech: Please let us know what happens in 50 years.

Joanna Aizenberg: We will! Don't worry! So, I have to close this session, I am very sorry because picture taking is extremely important and apparently everybody is extremely hungry. Thank you so much!

Session 4

Computational Modeling in High-Resolution Imaging

BEYOND THE PHYSICAL CONCEPTS: COMPUTATIONAL MODELLING IN FLUORESCENCE NANOSCOPY

STEFAN W. HELL

Department of NanoBiophotonics, Max Planck Institute for Biophysical Chemistry,
Am Fassberg 11, 37077 Göttingen, Germany
Department of Optical Nanoscopy, Max Planck Institute for Medical Research,
Jahnstr 29, 69120 Heidelberg, Germany

From Discerning by Focusing to Discerning by Molecular States

Throughout the 20th century, it was widely accepted that a light microscope relying on conventional lenses (far-field optics) cannot discern details that are finer than about half the wavelength of light (>200 nm). However, in the 1990s, it was discovered [1] that overcoming the diffraction barrier is realistic and that fluorescent samples can be resolved virtually down to the size scale of individual fluorescent molecules. Simple yet powerful principles allow neutralizing the resolution-limiting role of diffraction. They are the core of STED microscopy and, more generally, have become the basis of far-field optical "nanoscopy" as a field [2, 3]. In brief, fluorophores residing closer than the diffraction barrier are prepared in different molecular states so that they become distinguishable for a brief detection period. Since the process of focusing is no longer utilized for discerning adjacent features, the limiting role of diffraction vanishes, and the interior of (living) cells and tissues can be imaged with diffraction-unlimited resolution using focused light.

An in-depth description of these basic principles has spawned new powerful concepts such as MINFIELD [4] and DyMIN STED microscopy [5] and, in particular, MINFLUX nanoscopy [6]. Although they differ in some aspects, these concepts harness a local intensity minimum (of a doughnut or a standing wave) for determining the coordinate of the fluorophore to be registered. Most strikingly, by using an excitation intensity minimum to establish the fluorophore position, MINFLUX nanoscopy has obtained the ultimate (super) resolution: the scale of a single fluorescent molecule [6].

MINFLUX, A Photon-efficient Localization Concept Providing Molecular Size (∼1 nm) Resolution in Fluorescence Microscopy

With my groups in Göttingen and Heidelberg, I continue to push the performance of nanooptical molecular imaging in (living) cells and tissues. While, as the Nobel foundation had put it on their widely distributed posters, "[the 2014 Chemistry laureates] had crossed the [resolution] threshold", the actual holy grail of the super-resolution field had remained unreachable: resolution at the size scale of the fluorophore itself (~1 nm). The potential for realizing true molecule size resolution in fluorescence microscopy was recognized already in the 1990s [1]. After substantial development since then, the STED and PALM/STORM concepts have typically afforded resolution of at best 20–30 nm, largely limited by the photostability of the fluorophores. While a small number of studies with severe limitations or special requirements such as using solid-state emitters or cryogenic temperatures had come close, a resolution at the ~1-nm scale remained impossible to achieve. As demonstrated in 2016, this goal has now been reached owing to a major conceptual advance over the STED and PALM/STORM families of concepts, namely, MINFLUX [6].

In MINFLUX, the fluorophores are discerned individually by sequential on- and off-switching like in PALM/STORM. However, whereas in PALM/STORM the localization of a molecule is based on maximizing the number of detected fluorescence photons on a camera, which is inevitably limited by bleaching, in MINFLUX the molecule is localized by probing it with the intensity-zero of a doughnut-shaped excitation beam. The excitation beam is scanned in the proximity of the fluorophore, and the fluorescence is recorded as in a confocal microscope. The position of the molecule is ultimately identical to the position of the doughnut at which fluorescence emission is minimal (see Fig. 1). Importantly, approaching a fluorophore with a position-probing excitation minimum shifts the burden of requiring many photons for localization from the feeble fluorescence to the inherently bright beam of molecular excitation. This gives MINFLUX a fundamental edge over popular camera-based localization in terms of photon detection requirements, and hence speed and precision. Consequently, my key efforts are now devoted to pursuing the implications of the basic idea behind the MINFLUX concept.

By fundamentally reducing the number of detected photons required for nanometre-precise localization, MINFLUX has opened the door to low-light level optical analysis of tiny objects at true molecular-scale resolution (1–5 nm). With MINFLUX, lens-based fluorescence microscopy has thus reached the ultimate resolution limit: the size of the fluorescent molecule itself. Moreover, the resolution is attained at relatively high speed, typically at least ten times faster than in PALM/STORM. The already demonstrated tracking of fluorophores with substantially sub-millisecond position sampling [6] is only the beginning in the quest for the highest spatiotemporal capabilities.

My group and I showed that reducing the required number of fluorescence photons enables MINFLUX to detect molecular movements of a few nanometres, at temporal sampling speeds of hundreds of microseconds, while maintaining ~2-nm precision in standard deviation [7]. The theoretical limits have not yet been attained (Fig. 2(a)). Such performance is out of reach for popular camera-based localization

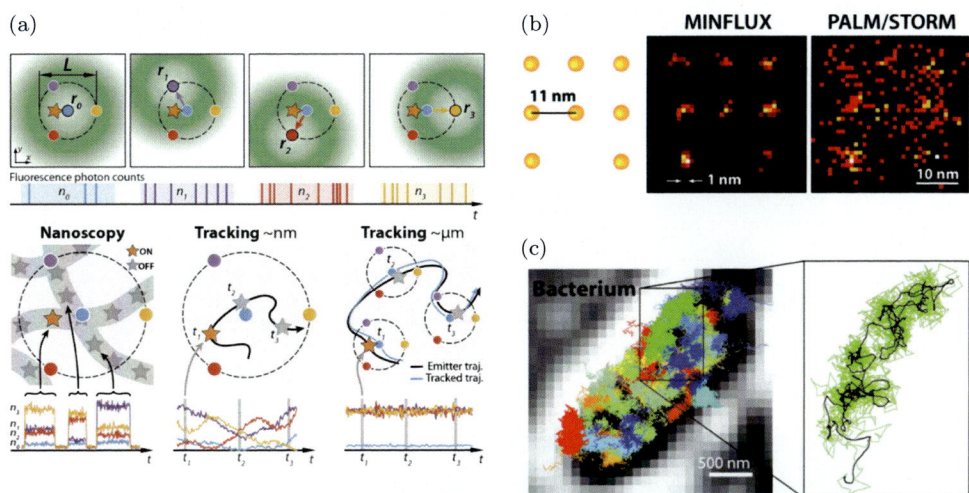

Fig. 1. MINFLUX, a new concept for localizing photon emitters. (a) Implementation of MIN-FLUX in 2D fluorescence imaging and tracking. (Top) Diagrams of the positions of the doughnut in the focal plane and resulting fluorescence photon counts. (Bottom) Basic application modalities of MINFLUX. (Left) Nanoscopy: A nanoscale object features molecules whose fluorescence can be switched on and off, such that only one of the molecules is on within the detection range. They are distinguished by abrupt changes in the ratios between the different $n_{0,1,2,3}$ or by intermissions in emission. (Middle) Nanometre-scale (short-range) tracking: The same procedure can be applied to a single emitter that moves within the localization region of size L. As the emitter moves, different fluorescence ratios are observed that allow the localization. (Right) Micron-scale (long-range) tracking: If the emitter leaves the initial L-sized field of view, the triangular set of positions of the doughnut zeros is (iteratively) displaced to the last estimated position of the molecule. By keeping it around r_0 by means of a feedback loop, photon emission is expected to be minimal for n_0 and balanced between n_1, n_2, and n_3, as shown. (b) With MINFLUX nanoscopy, one can, for the first time, separate molecules optically which are only a few nanometres apart from each other. On the left, a schematic of the fluorescing molecules is presented. Whereas the ultra-high resolution PALM/STORM microscopy at the same molecular brightness (right) delivers a diffuse image of the molecules (here in a simulation under ideal technical conditions), the position of the individual molecules can be easily discerned with the practically realized MINFLUX (middle). (c) Many much faster movements can be followed than is possible with STED or PALM/STORM microscopy. Left: Movement pattern of 30S ribosomes (coloured) in an *E. coli* bacterium (grey scale). Right: Movement pattern of a single 30S ribosome (green) shown enlarged. (a) and (c) adapted from Ref. [6].

by centroid calculation of emission diffraction patterns. Assuming that adequate labelling can be achieved, the full-resolution performance carries over also for cellular imaging, as shown by the first experiments on nuclear pore proteins in mammalian cells at about 2–4-nm resolution (compare Fig. 2(b)) [8]. This molecular-scale resolution now sets the standard, and it is a strong starting point for examining novel measurement opportunities for structure and dynamics at yet unchartered length and time scales.

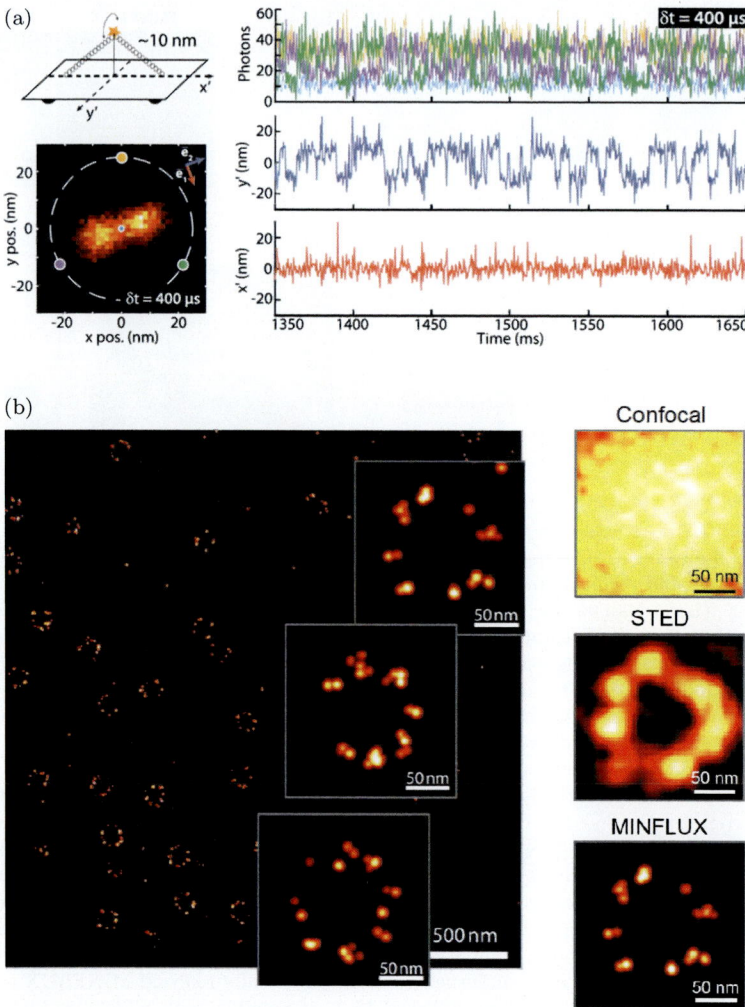

Fig. 2.   MINFLUX offers unprecedented spatio-temporal capabilities and enables cellular nanoscopy at resolution of about one or a few nanometres. (a) MINFLUX tracking of rapid movements of a custom-designed DNA origami. (Top left) Diagram of the DNA origami construct with a single ATTO 647N fluorophore attached at the centre of the bridge (10 nm from the origami base). By design, the emitter can move on a half-circle above the origami and is thus ideally restricted to a 1D movement. (Bottom left) Histogram of 6,118 localizations of the sample with $\delta t = 400\,\mu$s time resolution and a 1.5 × 1.5 nm binning. The predominant motion is along a single direction. (Right, Upper) A 300-ms excerpt of the photon count trace (time resolution $\delta t = 400\,\mu$s per localization). (Right, Lower) Mean-subtracted trajectory (rotated coordinate system). (b) Cellular MINFLUX imaging at 2–4-nm resolution across a large field of view: the nuclear pore protein NUP96, visualized as NUP96-SNAP conjugated to Alexa 647 fluorophores in U2OS cells. The frequent appearance as doublets in the eight-fold symmetric pattern stems from the organization of NUP96 in two rings at axial separation, which fall within the same depth of field of this 2D MINFLUX recording. Comparisons of the best confocal and STED imaging of NUP96 are shown to the right. (a) Adapted from Ref. [7], data shown in (b) are adapted from Ref. [8].

Outlook on Future Developments

While a large number of computational tools have been explored (compare a selection of them in Fig. 3), it is evident that it was the paradigm shift in the physical concepts — and not computational tools — which led to nanometre spatial resolution. Throughout, separation of molecules is achieved by the transfer between discernible states, to date usually between signalling "on" and non-signalling "off" fluorophore states. Without having to rely on *a priori* knowledge, the fluorescence nanoscopy concepts, notably the recently invented MINFLUX, now render molecule distributions down to the ~1-nm scale. Nonetheless, computational modelling and processing may augment the implementation of these concepts. Computational methods most certainly strengthen data interpretation. In fact, a lot of unchartered territory lies before us. So, where are the limits? How much can computation bring in addition to (or, together with) the recent conceptual ideas? The session, with six world-leading contributors, will give some fresh insights. It will very likely provide an excellent survey of the types of routes that have already been taken, both in optical imaging, and at the interface with electron-microscopy analysis and the computational tools established there.

Coming back to my original reasoning from the early 1990s about fluorescent emitters residing in the diffraction zone of the microscope, I would predict that, for large numbers of molecules in the diffraction zone, inference of their exact distribution, i.e., their resolution, is likely to remain impossible unless some key distinguishing features such as a pair of or multiple states ("on"/"off", "immobile"/ "mobile", as in PAINT, different absorption, and emission) is employed for their

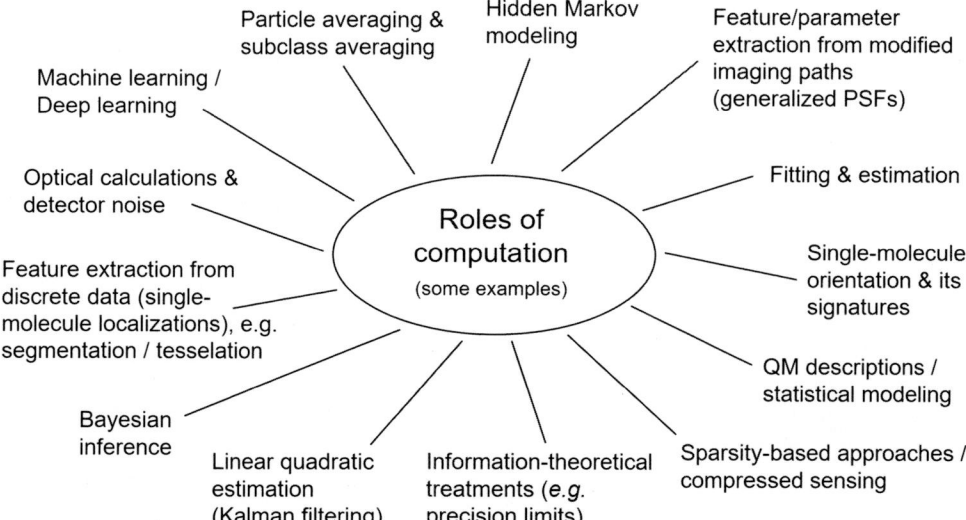

Fig. 3. Some examples of the roles of computation in advanced high-resolution imaging methods and their data evaluation processes.

separation. However, with resolution now zeroing in on the molecular size scale itself, the number of molecules within the resolved spatial domain (area or volume) becomes small. This sparsity of signals should be exploitable, and there is a clear role for computational analysis here. The examination of photon emission statistics from small collections of fluorophores to extract molecule numbers (e.g., Ref. [9]) is a first example. A complete computational accounting for all biophysical and biochemical variables, such as reporter-fluorophore linker lengths, geometries, flexibilities, steric effects, and possibly even protein structural/shape data — i.e., a sophisticated molecular dynamics (MD) calculation informed and constrained by the experimental fluorescence imaging data at molecular resolution — should ultimately provide the full picture.

Acknowledgments

I am grateful to Steffen J. Sahl for helpful discussions and suggestions in my preparation for this Solvay Conference.

References

[1] S.W. Hell and J. Wichmann, *Opt. Lett.* **19**, 780 (1994).
[2] S.W. Hell, *Science* **316**, 1153 (2007).
[3] S.W. Hell, *Nat. Meth.* **6**, 24 (2009).
[4] F. Göttfert, T. Pleiner, J. Heine, V. Westphal, D. Görlich *et al.*, *Proc. Natl. Acad. Sci. USA* **114**, 2125 (2017).
[5] J. Heine, M. Reuss, B. Harke, E. D'Este, S.J. Sahl *et al.*, *Proc. Natl. Acad. Sci. USA* **114**, 9797 (2017).
[6] F. Balzarotti, Y. Eilers, K.C. Gwosch, A.H. Gynnå, V. Westphal *et al.*, *Science* **355**, 606 (2017).
[7] Y. Eilers, H. Ta, K.C. Gwosch, F. Balzarotti, S.W. Hell, *Proc. Natl. Acad. Sci. USA* **115**, 6117 (2018).
[8] K.C. Gwosch, J.K. Pape, F. Balzarotti, P. Hoess, J. Ellenberg *et al.* bioRxiv, doi: https://doi.org/10.1101/734251 (2019).
[9] H. Ta, J. Keller, M. Haltmeier, S.K. Saka, J. Schmied *et al.*, *Nat. Commun.* **6**, 7977 (2015).

PARAMETER ESTIMATION AND INFORMATION THEORY IN SINGLE-MOLECULE AND SUPER-RESOLUTION MICROSCOPY

RAIMUND J. OBER*,† and MILAD VAHID‡

*Cancer Immunology Centre, Faculty of Medicine,
University of Southampton, UK
†Molecular and Cellular Medicine, Texas A&M University,
College Station, TX, USA
‡Department of Biomedical Engineering, Texas A&M University,
College Station, TX, USA

Introduction

Single-molecule-based super-resolution microscopy is special amongst imaging approaches as the acquired imaging data only convey meaning after significant processing of the data using non-trivial algorithms. The validity and the quality of the results, therefore, depend crucially on the analysis techniques that are being used.

The fundamental processing step in these experiments is the determination of the location of a single molecule in the 2D focus plane. More advanced experiments require the determination of other quantities, e.g., the 3D positional coordinates, the number of detected photons, diffusion coefficients, or other parameters related to dynamics properties in the case of the imaging of dynamic samples. These are all questions of parameter estimation that therefore show that these image analysis problems are problems in statistical inference and statistical image analysis.

The question then immediately arises of which analysis approaches provide the "best" results. The criterion for quality is typically given as an analysis approach that, on average, provides the correct results (i.e., is unbiased), and has as low a variance or standard deviation as possible. Related to this is the question of what is the lowest possible standard deviation that can be achieved.

An important follow-on question is that of experimental design, i.e., the question of how an experiment should be conducted so that the final results have suitable statistical properties. Information theoretic approaches have been shown to be very powerful with respect to these questions. For example, the Cramer Rao lower bound [1], given by the inverse of the Fisher information matrix, can be used to characterize the lowest possible standard deviation that can be achieved in the estimation of a specific parameter. Here, the maximum likelihood estimation approach has been shown to be a method that provides an optimal variance/standard deviation as it has been shown to closely match the Cramer Rao lower bound [1–3]. The Fisher information has also proved to be a powerful tool for experimental design.

For example, by analysing the corresponding Fisher information matrices, it was shown that multi-focal plane microscopy can be used to overcome the notorious depth discrimination problem in the determination of the three special coordinates of a single molecule [4]. Novel optical microscopy designs (point spread functions) by optimizing the Fisher information matrix related to the spatial estimation problem have been obtained [5]. Exploiting temporal and spatial dependencies of the Fisher information, it could be shown that the position of a single molecule can be estimated by acquiring a very low number of photons [6].

Results from Our Research Group

In this section, we provide a summary of our prior results related to the problem at hand.

Cramer Rao lower bound and Fisher information for single-molecule microscopy

In early work, we laid careful theoretical foundations to be able to systematically analyse microscopy data using information theoretic tools. A basic result in statistics, the Cramer Rao lower bound states that the variance of the (unbiased) estimate of any parameter that is measured cannot be lower than the inverse of the Fisher information. Specifically, it says that

$$\text{var}(\hat{\theta}) \geq I^{-1}(\theta),$$

where

$$I(\theta) = E\left[\left(\frac{\partial \mathcal{L}\ (\theta|z_1,\ldots,z_K)}{\partial \theta}\right)\left(\frac{\partial \mathcal{L}\ (\theta|Z_1,\ldots,Z_K)}{\partial \theta}\right)^T\right].$$

This is a general expression that is, however, not helpful unless it can be determined for specific estimation problems. We developed extensive tools to be able to compute the Fisher information in many practical circumstances relevant to microscopy image analysis. For example, we showed [1, 2] that a spatial coordinate such as the x-coordinate of an in-focus point source, e.g., as a single molecule, cannot be estimated with a standard deviation as follows:

$$\delta_x = \frac{\lambda}{2\pi n_a \sqrt{N}},$$

where λ is the wavelength of the detected light, n_a is the numerical aperture of the optical detection system, and N is the (expected) number of detected photons. This shows that, however sophisticated the analysis approach, the x-coordinate of a single molecule cannot be estimated with a lower standard deviation than the quantity given by this expression, assuming the experiment is carried out with a classical microscope.

Maximum likelihood algorithm

We introduced and investigated the maximum likelihood algorithm for the analysis of single-molecule data [1–3]. This required a careful determination of the probabilistic description of imaging data in the context of the noise sources that are introduced by the different camera models that are in use. We also investigated the performance of the maximum likelihood algorithm in the presence of model mismatches that may be expected and compared it to the often used least squares algorithm.

A stochastic Rayleigh resolution criterion

Rayleigh's classical resolution criterion has a major deficiency in that it does not take into account the amount of data that is captured. Intuitively speaking, it would be expected that a very low amount of detected photons would make the resolution of two objects very difficult, even if they are very widely spaced apart. Conversely, it is not clear why, if a large amount of data is available, a capable algorithm should not be able to distinguish even very closely spaced point sources. Building on our work on the Fisher information for imaging experiments, we proposed and experimentally verified the following *stochastic Rayleigh criterion* that provides a lower bound on the standard deviation with which the distance between two point objects, such as single molecules, can be estimated [7]

$$\delta_d := \frac{1}{\sqrt{4\pi N \Gamma_0(d)}} \cdot \frac{\lambda}{n_a}.$$

Importantly, this expression shows that the standard constant pre-factor 0.61 of Rayleigh's criterion is replaced by a quantity that provides a trade-off between distance and the amount of available data. More specifically, it shows, for example, that even very small distances can be resolved with an arbitrary low standard deviation provided a sufficient number of photons are captured. It should be pointed out that this result addresses the problem when two point sources are simultaneously imaged at the same wavelength, which is significantly more challenging than the problem when two point sources are sequentially imaged.

Multi-focal plane microscopy and experimental design

We introduced multi-focal plane microscopy [8] for the imaging of fast subcellular dynamics. We subsequently investigated its use for the determination of the 3D position of a single molecule, by proposing an algorithm (MUMLA) which we have shown to achieve the Cramer Rao lower bound [9]. Results that we would like to mention here relate to designing a multi-focal plane microscope. Here, it is critical to determine how many focal planes should be used, what their distances should be, and which objective should be used. Using Fisher information calculations, we developed design tools that allow for different configurations to be analytically assessed and the most suitable to be determined [4].

Algorithmic resolution limit

Determining the location of a single molecule, given an isolated image of a single molecule, is the central step in the analysis of single-molecule localization based super-resolution experiments. However, this step is not the only image analysis step that needs to be performed. Others include the detection of a single molecule in the imaging data or technically speaking the segmentation of the image into regions of interest that contain a single image of a single molecule. A confounding problem here is that two or more single molecules might be too closely spaced to be easily separated for analysis. This is avoided to some extent by stochastic excitation but can never be fully overcome.

Developing techniques based on spatial statistics, we have proposed an *algorithmic resolution limit,* which assesses the resolution capability of a complete image analysis chain [10]. We determined the algorithmic resolution limit for several existing super-resolution algorithms and determined that, despite claiming to be able to deal with multi-emitter problems, they cannot reliably analyse point sources that are simultaneously excited and have distances less than around 500 nm. This shows that significant work remains to be done to address the multi-emitter problem.

Estimation of parameters related to stochastic trajectories

Of immense interest in cell biology and biophysics is the analysis of trajectories of individual molecules, receptors, or larger structures such as vesicles or organelles. From an abstract point of view, this problem is no different from the problems that were mentioned earlier which relate to the extraction of parameters from stationary objects. However, the probabilistic modelling of the data acquisition is significantly more intricate as it is not only the stochasticity of the data acquisition problem that needs to be modelled but also the stochasticity of the underlying trajectories. This has led several authors to either only consider special cases or to introduce several approximations into the underlying problem formulation and derivations. We have recently embarked on a program to consider the problem in the same general framework that was successfully used in the stationary case [10]. The results are computationally very complex, but they show that there is the potential to extend the methodologies that were successfully employed in the stationary case to the analysis of dynamic processes.

Future Directions

Being able to obtain results for the analysis of dynamic processes that match those that were achieved in the stationary case would be a major achievement. The need from the point of view of the applications is clearly present. However, significant efforts would be required to develop the necessary analytical tools. Although there are many similarities to problems that arise in financial mathematics and other areas where the modelling with stochastic differential equations is carried out, the problems in our application area pose unique and very challenging problems due to the specific optical configurations that are required. For example, the optical point spread functions that we use give rise to complex computational problems that pose major challenges.

A problem beyond the technical one addressed above is related to the dissemination of the results that are often in the form of software. A typical approach by many authors is that software is placed somewhere on the web as plugins to various software platforms. The attractive feature is that such software solutions are typically free, but quality control is often absent and support is typically limited if not non-existent. While this may not be a problem for a research laboratory that has extensive image analysis expertise, the laboratories that need to use the tools without having deep in-house experience face more serious problems. In many other scientific fields, commercial companies have entered the field to help to provide such software tools (for full disclosure, the first author is a founder of a software start-up company). Other solutions are also conceivable such as those provided by community-based efforts that are able to provide stable and long-term support. There appear to be many reasons why this problem is not yet solved. Some of them have to do with the very inconsistent funding for software projects by the classical funders of scientific research.

Acknowledgments

Our work that is summarized here has been supported in the past by several grants by the National Institutes of Health.

References

[1] R. J. Ober, S. Ram, and E.S. Ward, *Biophys. J.* **86**, 1185 (2004).
[2] S. Ram, E.S. Ward, and R.J. Ober, *Multidimen. Syst. Sig. Process.* **17**, 27 (2006).
[3] A.V. Abraham, S. Ram, J. Chao, E.S. Ward, and R.J. Ober, *Opt. Exp.* **17**, 23352 (2009).
[4] A. Tahmasbi, S. Ram, J. Chao, A.V. Abraham, F.W. Tang *et al.*, *Opt. Exp.* **22**, 16706 (2014).
[5] Y. Shechtman, L.E. Weiss, A.S. Backer, S.J. Sahl, and W.E. Moerner, *Nano Lett.* **15**, 4194 (2015).
[6] F. Balzarotti, Y. Eilers, K.C. Gwosch, A.H. Gynnå, V. Westphal *et al.*, *Science* **355**, 606 (2017).
[7] S. Ram, E.S. Ward, and R.J. Ober, *Proc. Natl. Acad. Sci.* **103**, 4457 (2006).
[8] P. Prabhat, S. Ram, E.S. Ward, and R.J. Ober, *IEEE Trans. NanoBiosci.* **3**, 237 (2004).
[9] S. Ram, P. Prabhat, J. Chao, E.S. Ward, and R.J. Ober, *Biophys. J.* **95**, 6025 (2008).
[10] E.A.K. Cohen, A.V. Abraham, S. Ramakrishnan, and R.J. Ober, *Nat. Commun.* **10**, 793 (2019).
[11] M.R. Vahid, B. Hanzon, and R.J. Ober, arXiv:1808.02195 (2018).

COMPUTATIONAL MODELLING ENABLES ROBUST MULTIDIMENSIONAL NANOSCOPY

MATTHEW D. LEW

Department of Electrical and Systems Engineering,
Washington University in St. Louis, MO 63130, USA

Present State of Computational Modelling in Fluorescence Nanoscopy

Computation is inescapable in modern fluorescence nanoscopy, ranging from simple tasks such as counting photons (and simple manipulations thereof, e.g., in RESOLFT [1] and MINFLUX [2]) to fusing together multiple images taken under different conditions (e.g., in SIM [3] and ISM [4]) to repeatedly localizing single fluorescent molecules [5, 6] using a model PSF (as in single-molecule localization microscopy [SMLM], (F)PALM, STORM, and PAINT). However, the power of computational imaging lies in its ability to transcend the centuries-old paradigm of point-to-point mapping from object to camera [7] common in traditional microscopy; explicitly integrating computation and physical optics together enables new capabilities that cannot be realized by conventional techniques alone. Further, computational modelling also allows experimenters to optimize and rigorously validate any proposed technology before setting foot in the lab.

Classical Fisher information theory has been used to characterize [8–13] and design [14–17] multidimensional nanoscopes since they were first demonstrated. The power of computing Fisher information lies in its ability to bound the best-possible variance of any unbiased estimator, called the Cramér Rao bound (CRB) [18]. Therefore, independent of the computational algorithm used to generate super-resolved images, Fisher information can be used to compare the performance of any variant of fluorescence nanoscopy. Another advantage is the quantitative specificity of this metric; unlike many other (loose) statistical bounds [19], the CRB can be optimized computationally to design the best-possible optical system for a certain imaging task. Such analyses show that the tetrapod family [16, 20] of point spread functions (PSFs) achieves higher localization precision in 3D SMLM than other approaches [17].

The classical model-based approach for designing optical nanoscopes involves choosing (1) the appropriate forward model for a given imaging task (e.g., localizing an isolated fluorescent emitter in 3D space), (2) the desired performance metric to evaluate a proposed design (e.g., CRB with Poisson shot noise), and (3) the methodology to achieve the best-possible performance (e.g., a gradient decent algorithm with certain constraints). However, it is difficult to maximize both emitter detectability, which requires a PSF to be compact on the camera, and measurement precision, which requires the PSF to change significantly with the parameter of interest, e.g., z-position, dipole orientation, or emission wavelength. However, recent deep learning methods promise to overcome these challenges. Neural networks have been proposed to replace traditional optimization algorithms for axial localization and colour identification in standard microscopes without colour filters [21]. Further, neural networks can be used to design optical nanoscopes that maximize colour classification accuracy, again without the use of traditional filters [22].

Recent Contributions to Computational Modelling in Fluorescence Nanoscopy

My lab is developing new technologies to augment standard SMLM with new capabilities, such as long-term imaging of amyloid fibrils [26]. Another example is the Tri-spot PSF [23, 24], which measures all degrees of freedom related to the orientational dynamics of SMs without angular degeneracy. A key insight of our analysis is that the orientational dynamics of any dipole emitter may be parameterized in terms of six orientational second moments, $\mathbf{M} = [\langle \mu_x^2 \rangle, \langle \mu_y^2 \rangle, \langle \mu_z^2 \rangle, \langle \mu_x \mu_y \rangle, \langle \mu_x \mu_z \rangle, \langle \mu_y \mu_z \rangle]$, which are a function of a dipole's orientation $\mu = [\mu_x, \mu_y, \mu_z]$ averaged over a single measurement (i.e., a camera frame). (This orientation vector μ may also be expressed in spherical coordinates using a polar angle θ and azimuthal angle φ.) Therefore, to measure the brightness and orientational dynamics of an SM, a PSF must have at minimum six degrees of freedom to measure these parameters. It follows that designing a PSF to contain six discrete spots would enable scientists to measure all possible orientational dynamics without degeneracy, while also maximizing detectability, even for weak SMs. Thus, we designed a three-sector linear phase mask to create the Tri-spot PSF within a modified polarized fluorescence microscope. The Tri-spot PSF reveals depolarization within fluorescent beads that is difficult to detect using other methods, and it can also be used to observe rotational dynamics of fluorophores within polymer thin films that are not observable by conventional SMLM (Figs. 1(a)–1(d)). It achieves an orientation measurement precision of 5° with 3000 photons detected from an SM.

Our imaging models have revealed a surprising fundamental limit [25] for measuring accurately the rotational dynamics of SMs. We expect an unconstrained, uniformly rotating dipole emitter to absorb varying input polarizations of light uniformly, to emit uniformly across all possible detection polarizations, and to emit an isotropic angular (energy) spectrum. However, any practical measurement will capture a finite number of photons from the dipole emitter, yielding a finite signal to noise ratio (SNR), and therefore, this expected symmetry is routinely broken. Detailed modelling of the imaging process, including photon shot noise, shows that

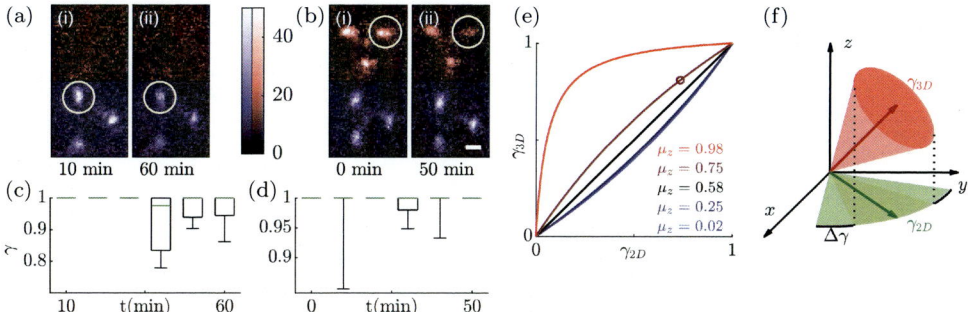

Fig. 1. (colour) Measuring orientation dynamics of single molecules. (a) and (b) Tri-spot images of Atto 647N molecules (a) 1 and (b) 2 at the (i) beginning and (ii) end of time-lapse imaging. Both molecules are embedded in a thin polymer film under continuous exposure to humid air. Circles highlight changes in spot brightness. Scale bar: $1\,\mu$m. Colour bar: detected photons/pixel. (c) and (d) Effective rotational constraint of molecules (c) 1 and (d) 2 measured over 50 min. Green: median; Box: first and third quartile; Error bars: minima and maxima. (e) The relation between in-plane γ_{2D} and 3D γ_{3D} rotational constraint varies with an SM's average orientation along the z axis μ_z. (f) A dipole emitter with $\mu_z = 0.75$ and $\gamma_{3D} = 0.81$ (cone half-angle of 30°) appears to be more rotationally free in the xy plane ($\gamma_{2D} = 0.73$, wedge half-angle of 38°). Reprinted figures with permission from Refs. [23–25]. Copyright (2018) by AIP Publishing and (2019) by the American Physical Society.

the expected rotational constraint γ measured in 2D for an isotropic emitter is given by $\mathrm{E}(\gamma_{2D}) = (\pi)^{1/2}/\mathrm{SNR}$, where $\gamma = 0$ represents an isotropic emitter and $\gamma = 1$ represents a rotationally fixed dipole. Therefore, measurements will be biased for all but the highest SNRs. To provide physical intuition, the expected bias in measuring rotational constraint is 0.16 for 1000 signal photons and 30 background photons/pixel, corresponding to a cone angle measurement of 77° (instead of the true value of 90°) for uniform rotational diffusion within a hard-edged cone. Therefore, the molecule always appears to be *more constrained* than it actually is, similar to how a non-moving molecule appears to have a non-zero translational diffusion coefficient because of shot noise.

Further modelling also shows that 2D orientation measurements, i.e., those that capture the in (xy)-plane dipole orientation, and 3D orientation measurements actually perceive identical 3D orientational motions differently (Fig. 1(e) and Ref. [25]). Because 2D methods are blind to the out-of-plane component μ_z, one must have prior knowledge of this quantity in order to compute an equivalent 3D rotational constraint. For small μ_z, the difference between 2D and 3D measurements of motion is small, but for a large $\mu_z = 0.98$ (polar angle $= 11°$), a highly constrained molecule in 3D ($\gamma_{3D} = 0.80$ or a cone half-angle of 30°) appears to be almost completely rotationally free in two dimensions ($\gamma_{2D} = 0.20$ or a wedge half-angle of 75°, Fig. 1(f)). Therefore, one must exercise caution when using 2D methods to infer rotations in 3D space.

My group also develops algorithms for robustly analysing SMLM datasets that contain images of overlapping molecules [27, 28]. Our analyses have found that

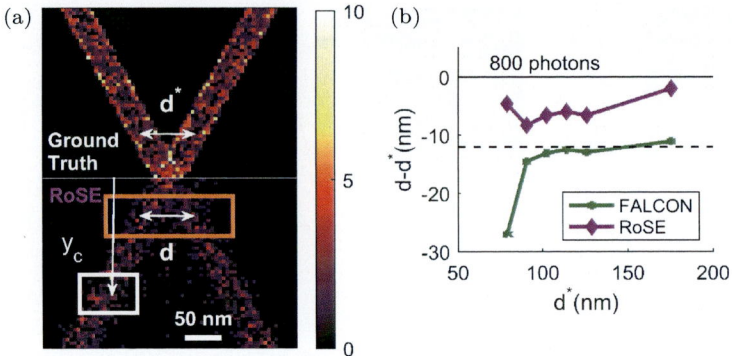

Fig. 2. (colour). Structural bias of two crossing microtubules (MTs) recovered by RoSE [27] (purple) and FALCON. [30] (green). (a) (Top) Simulated ground-truth structure and (bottom) structure obtained by RoSE. (b) Mean separation error between centres of the MTs along the length of the structure for 800 photons detected. Dashed line represents two times the best-possible localization precision given by $(\text{CRB})^{1/2}$. Scale bar: 50 nm. Colour bar: number of localizations per $5 \times 5\,\text{nm}^2$. Adapted with permission from Ref. [27].

localization artefacts tend to not be simply random but instead are structured and correlated with the sample of interest. For example, localizations from overlapping molecules tend to be biased towards their collective centre of mass, making separated microtubules appear closer than they really are and causing circular clusters to appear elliptical. Therefore, scalar error metrics commonly used to evaluate SMLM algorithms, like root mean square error, fail to quantify how the *structure* of the sample itself and the *structure* of the PSF induce systematic *vectorial* artefacts in super-resolved images. Further, these errors are difficult to detect using simple image-based quantities, like the apparent width of the localized PSF or the brightness of the localized molecule. Towards addressing these challenges, we have recently built a methodology, called Wasserstein-induced flux (WIF) [29], to compute the measurement accuracy of any SMLM image without ground truth knowledge of the sample. By measuring computationally the statistical *stability* of each localization within an SMLM reconstruction, WIF can quantify the degree of mismatch between experimental data and a computational model of the imaging system, as well as enhance the accuracy and resolution of SMLM reconstructions. While shot noise makes detecting minor model mismatches difficult, WIF has excellent sensitivity for detecting overlapping molecules and dipole-like emission patterns at SNRs typical of SMLM.

Outlook on Computational Modelling in Fluorescence Nanoscopy

Given the popularity of modern data-driven deep learning and image analysis techniques, one may question the role of traditional physics-based imaging models and statistical models of noise. More fundamentally, why design and build a complicated imaging system if deep learning can shoulder the burden of creating multi-colour, 3D nanoscope images? First, one must note that photon detection is nonlinear with respect to the photon wave function; cameras are only sensitive to the intensity of light. This nonlinearity destroys some information contained within this wave function, and once this information is lost, no computational algorithm can recover it without using prior knowledge. Therefore, to preserve this information as much as possible, physical optical elements (e.g., lenses, masks, and polarizers) can transform this information into an optimized (intensity) PSF to yield superior measurements. Recent innovations that leverage physical interactions of light with the sample, like MINFLUX [2], DONALD [31], and spectrally resolved SMLM [32–34], show that powerful observations can be made using relatively simple models and imaging algorithms. Second, when executing an imaging task, deep learning algorithms by design impose strong prior knowledge learned during the training process. In most cases, this functionality is useful because it excludes unlikely outcomes, but for scientific imaging, such biases could hinder scientific discovery and lead to erroneous interpretations of an experiment. There exists a need to adopt deep learning architectures whose inner workings are interpretable, so that failure modes for edge cases can be predictable and so that the confidence or trustworthiness of their outputs can be quantified on a *per-experiment* basis. Our work on WIF [29], in which we compute the trustworthiness of each individual localization within a SMLM dataset, is one step towards this goal.

Computational models have the potential to bridge the gap between the chemical and physical processes within living systems and the experimental images produced by fluorescence nanoscopy. Imaging is uniquely suited for revealing the inner workings of these systems because of its ability to correlate dynamic events across space and time by producing rich, high-dimensional datasets. However, end-to-end multiphysics modelling from the target of interest to the measurement (e.g., modelling molecular dynamics within a solvent, interactions of a biomolecule with neighbouring macromolecules, environmental effects on the electronic states of fluorophores, and electromagnetic wave propagation through tissues) is challenging because of the complex physical processes, timescales, and size scales involved. The next frontier is to develop multi-scale techniques that integrate first-principle models and data-driven methods to express how information about an object manifests itself in the measurement domain. Such models will be pivotal for maximizing information transfer from object to measurement and for building imaging systems that directly quantify biochemical dynamics and mechanisms within living systems. These models could bring a paradigm shift in the use of imaging technologies like fluorescence nanoscopy. Instead of considering fluorescent labelling, imaging, and data analysis as separate steps in a protocol, a unified computational model could optimize the entire pipeline simultaneously, thereby maximizing and quantifying statistical certainty on the specific phenomena of interest given the observations in an experiment. Such an intelligent imaging system is yet to come.

Acknowledgments

Hesam Mazidi and Oumeng Zhang made pivotal contributions to the findings discussed in this work with assistance from Jin Lu, Tianben Ding, and Eshan King. Research reported in this publication was supported by the National Science Foundation under grant number ECCS-1653777 and by the National Institute of General Medical Sciences of the National Institutes of Health under grant number R35GM124858.

References

[1] S.W. Hell, *Angew. Chemie Int. Ed.* **54**, 8054 (2015).
[2] F. Balzarotti, Y. Eilers, K.C. Gwosch, A.H. Gynnå, V. Westphal *et al.*, *Science* **355**, 606 (2017).
[3] M.G.L. Gustafsson, *Proc. Natl. Acad. Sci.* **102**, 13081 (2005).
[4] C.B. Müller and J. Enderlein, *Phys. Rev. Lett.* **104**, 198101 (2010).
[5] E. Betzig, *Angew. Chemie Int. Ed.* **54**, 8034 (2015).
[6] W.E. Moerner, *Angew. Chemie Int. Ed.* **54**, 8067 (2015).
[7] J.N. Mait, G.W. Euliss, and R.A. Athale, *Adv. Opt. Photon.* **10**, 409 (2018).
[8] R.J. Ober, S. Ram, and E.S. Ward, *Biophys. J.* **86**, 1185 (2004).
[9] S. Ram, P. Prabhat, J. Chao, E.S. Ward, and R.J. Ober, *Biophys. J.* **95**, 6025 (2008).
[10] M. Badieirostami, M.D. Lew, M.A. Thompson, and W.E. Moerner, *Appl. Phys. Lett.* **97**, 161103 (2010).
[11] M.A. Thompson, M.D. Lew, M. Badieirostami, and W.E. Moerner, *Nano Lett.* **10**, 211 (2010).
[12] H. Deschout, F.C. Zanacchi, M. Mlodzianoski, A. Diaspro, J. Bewersdorf *et al.*, *Nat. Meth.* **11**, 253 (2014).
[13] A. Small and S. Stahlheber, *Nat. Meth.* **11**, 267 (2014).
[14] A. Greengard, Y.Y. Schechner, and R. Piestun, *Opt. Lett.* **31**, 181 (2006).
[15] S.R.P. Pavani, M.A. Thompson, J.S. Biteen, S.J. Lord, N. Liu *et al.*, *Proc. Natl. Acad. Sci.* **106**, 2995 (2009).
[16] Y. Shechtman, S.J. Sahl, A.S. Backer, and W.E. Moerner, *Phys. Rev. Lett.* **113**, 133902 (2014).
[17] A. Von Diezmann, Y. Shechtman, and W.E. Moerner, *Chem. Rev.* **117**, 7244 (2017).
[18] J. Chao, E. Sally Ward, and R.J. Ober, *J. Opt. Soc. Am. A* **33**, B36 (2016).
[19] V.I. Morgenshtern and E.J. Candès, *SIAM J. Imaging Sci.* **9**, 412 (2016).
[20] Y. Shechtman, L.E. Weiss, A.S. Backer, S.J. Sahl, and W.E. Moerner, *Nano Lett.* **15**, 4194 (2015).
[21] T. Kim, S. Moon, and K. Xu, *Nat. Commun.* **10**, 1996 (2019).
[22] E. Hershko, L.E. Weiss, T. Michaeli, and Y. Shechtman, *Opt. Exp.* **27**, 6158 (2019).
[23] O. Zhang, J. Lu, T. Ding, and M.D. Lew, *Appl. Phys. Lett.* **113**, 031103 (2018).
[24] O. Zhang, J. Lu, T. Ding, and M.D. Lew, *Appl. Phys. Lett.* **115**, 069901 (2019).
[25] O. Zhang and M.D. Lew, *Phys. Rev. Lett.* **122**, 198301 (2019).
[26] K. Spehar, T. Ding, Y. Sun, N. Kedia, J. Lu *et al.*, *ChemBioChem* **19**, 1944 (2018).
[27] H. Mazidi, J. Lu, A. Nehorai, and M.D. Lew, *Sci. Rep.* **8**, 13133 (2018).
[28] H. Mazidi, E.S. King, O. Zhang, A. Nehorai, and M.D. Lew, in *2019 IEEE 16th Int. Symp. Biomed. Imaging (ISBI 2019)* (IEEE, 2019), pp. 325–329.
[29] H. Mazidi, T. Ding, A. Nehorai, and M.D. Lew, BioRxiv 721837 (2019).
[30] J. Min, C. Vonesch, H. Kirshner, L. Carlini, N. Olivier *et al.*, *Sci. Rep.* **4**, 4577 (2015).

[31] N. Bourg, C. Mayet, G. Dupuis, T. Barroca, P. Bon *et al.*, *Nat. Photon.* **9**, 587 (2015).

[32] Z. Zhang, S.J. Kenny, M. Hauser, W. Li, and K. Xu, *Nat. Meth.* **12**, 935 (2015).

[33] M.N. Bongiovanni, J. Godet, M.H. Horrocks, L. Tosatto, A.R. Carr *et al.*, *Nat. Commun.* **7**, 13544 (2016).

[34] S. Moon, R. Yan, S. J. Kenny, Y. Shyu, L. Xiang *et al.*, *J. Am. Chem. Soc.* **139**, 10944 (2017).

CONVERTING TEMPORAL INTO SPATIAL INFORMATION

JÖRG ENDERLEIN

3rd *Institute of Physics — Biophysics, Georg August University,*
Göttingen 37077, Germany

My View of the Present State of Research on Computational Modelling in High-resolution Imaging

The last approximately 20 years have seen an explosion of various super-resolution methods and techniques. All these methods and techniques have a common basis which is best summarized by the concept of reversible saturable optical fluorescence transitions (RESOLFT), i.e., all are based on the quantum-mechanical nature of fluorescence that involves transitions between different energetic states in fluorescent molecules. The most commonly used methods such as stimulated emission depletion (STED) microscopy, photo-activated localization microscopy (PALM), or stochastic optical reconstruction microscopy (STORM) use these state transitions in a static or quasi-static way. STED microscopy uses a clever interplay of excitation and de-excitation with differently shaped spatial light distributions. PALM and STORM use these transitions for generating sequential images of sparsified single-molecule distributions. However, a special class of methods such as dynamic saturation optical microscopy (DSOM), super-resolution optical fluctuation imaging (SOFI), or metal-induced energy transfer (MIET) imaging uses the intrinsic temporal dynamics of these transitions, and converts this temporal (spectroscopic) information into spatial super-resolution by applying dedicated image-processing algorithms.

My Recent Research Contributions to Computational Modelling in High-resolution Imaging

In 2005, I proposed an idea on how to evaluate the intensity-dependent temporal dynamics of intersystem crossing in fluorescent molecules for increasing the spatial resolution of confocal microscopy [1]. This concept was named DSOM, and it relied on measuring the temporal kinetics of singlet-to-triplet transitions within the focal region of a confocal microscope. Although there had been some experimental realizations of this idea, see, e.g., Ref. [2], the main drawback was the limited photo-stability of dyes that prevented one from achieving a decent signal-to-noise ratio in most cases. However, it should be mentioned that the principal idea of DSOM was recently revived within the context of STED [3], where it was used to improve resolution and to suppress unwanted background.

(a)   (b)

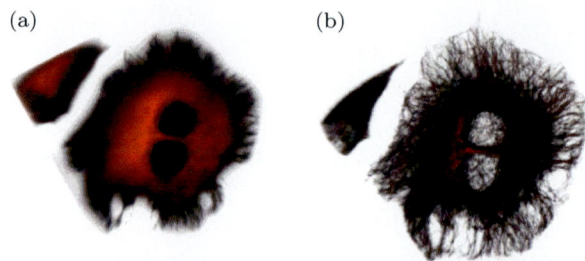

Fig. 1.   Comparison between conventional wide-field and SOFI microscopy. (a) Superimposed z-stack of 32 wide-field images of HEK cells with fluorescently labelled tubulin network. (b) Same sample, but recorded with SOFI. Image size ∼50 × 50 μm. For details, see Ref. [5].

A more successful idea was SOFI, an idea that originated from our long-year experience in fluorescence correlation spectroscopy (FCS). We could show than one can apply an FCS analysis to a movie of microscopy images to increase the spatial resolution of an image beyond the diffraction limit [4], see Fig. 1.

Besides yielding a higher spatial resolution, SOFI also delivers three-dimensional sectioning (with a wide-field imaging microscope) and superior background suppression (all temporally static signals are eliminated from the final image). Meanwhile, SOFI has found manifold applications, and one of its most promising variants is simultaneous three-dimensional multi-plane imaging using a dedicated multi-plane beam-splitter [6].

Finally, the latest development in the direction of merging time-resolved fluorescent spectroscopy with imaging for achieving super-resolution is MIET imaging. It is based on the peculiar effect whereby a fluorescent molecule is increasingly quenched in its emission when approaching a metal. The physical basis of this effect is the near-field coupling of its excited state to electron oscillations in the metal (surface plasmons). We first introduced this idea in 2010 [7], where we applied the technique for profiling the height of cellular membranes with nanometre axial resolution, see Fig. 2.

Although the fluorescence quenching reduces the measurable fluorescence intensity, it also increases dye photo-stability (excited dyes are de-excited more quickly), so that MIET works well even for single molecules [8]. In the latest variant of the technique, we have replaced the metal by a single-sheet graphene layer, which allowed us to measure the thickness of individual lipid bilayers with Angström precision [9].

Outlook on Future Developments of Research on Computational Modelling in High-resolution Imaging

So far, the general idea of exploiting temporal kinetics information for improving spatial resolution was most successfully realized by MIET imaging, where one computationally converts a measured fluorescence lifetime into an axial distance.

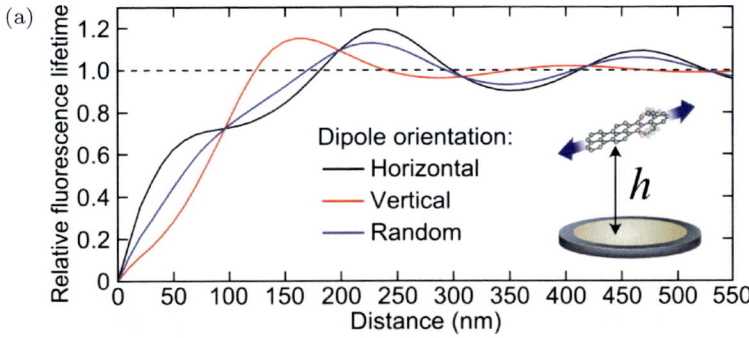

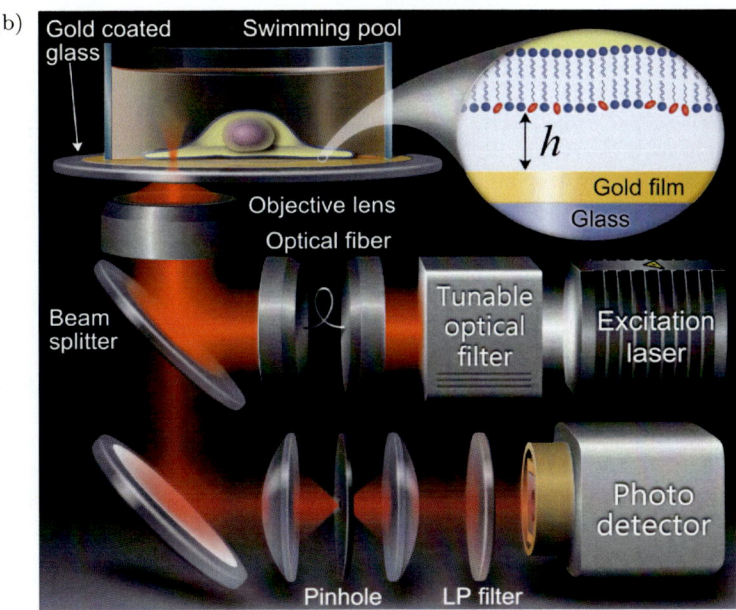

Fig. 2. (a) MIET is based on the distance dependence of fluorescence quenching close to a metal surface, which can be measured as a reduced fluorescence lifetime. This lifetime vs. distance dependence can be used to convert a measured fluorescence lifetime value into a distance from the surface. (b) MIET measurement setup, based on a conventional confocal microscope equipped with a pulsed laser excitation source and a single-photon counting detector with high-speed electronics for time-correlated single photon counting (TCSPC).

What we need for a broader application of this technique is the availability of single-molecule sensitive wide-field fluorescence-lifetime imaging microscopes (FLIMs), which would boost its imaging speed by several orders of magnitude. With the recent developments of new single-photon avalanche diode (SPAD) arrays, there seems to be some hope to see such FLIM systems emerge in the near future.

The biggest bottleneck for a wider application of SOFI is currently the lack of reliable fluorescent labels with well-defined blinking behaviour. Here, new probes based on intramolecular conformational transitions with associated fluorescence intensity changes may offer an interesting perspective. Another computational problem still to be solved is the intrinsic nonlinearity of SOFI: To an nth-order SOFI image, emitters contribute with the nth power of their brightness. New advanced algorithms are required for rectifying this nonlinearity without sacrificing spatial resolution.

Acknowledgments

I thank my fantastic group at Göttingen University, my collaborators, and my former co-workers. In particular, my deep thanks to Thomas Dertinger, Shimon Weiss, Martin Hof, and Narain Karedla.

References

[1] J. Enderlein, *Appl. Phys. Lett.* **87**, 094105 (2005).
[2] J. Humpolíčková, A. Benda, R. Macháň, J. Enderlein, and M. Hof, *Phys. Chem. Chem. Phys.* **12**, 12457 (2010).
[3] L. Lanzano, I.C. Hernández, M. Castello, E. Gratton, A. Diaspro *et al.*, *Nat. Commun.* **6**, 6701 (2015).
[4] T. Dertinger, R. Colyer, G. Iyer, S. Weiss, and J. Enderlein, *PNAS* **106**, 22287 (2009).
[5] T. Dertinger, J. Xu, O.F. Naini, R. Vogel, and S. Weiss, *Opt. Nanosopy* **1**, 2 (2012).
[6] S. Geissbuehler, A. Sharipov, A. Godinat, N.L. Bocchio, P.A. Sandoz *et al.*, *Nat. Commun.* **5**, 5830 (2014).
[7] A.I. Chizhik, J. Rother, I. Gregor, A. Janshoff, and J. Enderlein, *Nat. Photon.* **8**, 124 (2014).
[8] N. Karedla, A.I. Chizhik, I. Gregor, A.M. Chizhik, O. Schulz *et al.*, *ChemPhysChem* **15**, 705 (2014).
[9] A. Ghosh, A. Sharma, A.I. Chizhik, S. Isbaner, D. Ruhlandt *et al.*, *Nat. Photon.* **1**, (2019).

DATA FUSION IN LOCALIZATION MICROSCOPY

HAMIDREZA HEYDARIAN*, MARK BATES†, FLORIAN SCHUEDER‡,

RALF JUNGMANN‡, SJOERD STALLINGA* and BERND RIEGER*

*Technical University Delft, Lorentzweg 1, 2628 CJ Delft, The Netherlands
†Max-Planck-Institute for Biophysical Chemistry, 37077 Göttingen, Germany
‡Ludwig Maximilian University, Munich,
Germany & Max Planck Institute of Biochemistry, Martinsried, Germany

Present State of Research in High-resolution Localization Microscopy

For more than a century, the law of optical imaging has been that no details smaller than the diffraction limit $\lambda/2NA$ $\sim$200 nm, with λ the wavelength and NA the numerical aperture of the microscope objective lens, can be usefully imaged in the far field. Several methods have been proposed to circumvent the diffraction limit in the last 10–15 years. Here, super-resolution fluorescence microscopy uses important contributions from single-molecule biophysics and fluorescent protein labelling technology, and conceptual proposals by Hell [1] that date back from as early as the 1990s. The hallmark publications of PALM and STORM in 2006 [2, 3] provided the basis for super-resolution microscopy that has become common in today's labs. The Nobel Prize in chemistry was awarded in 2014 to Hell, Betzig, and Moerner for these developments.

We have introduced the concept of Fourier ring correlation (FRC) into super-resolution microscopy for taking all resolution factors into account [4]. The FRC quantifies the available image information as a function of spatial frequency, i.e., across all length scales, and finds a resolution limit by a threshold criterion. In general, the best resolution that can be achieved is limited by the total photon count from a single emitter *and* by the density of the emitters labelling the structure [4, 5]. Therefore, we need to have high (effective) labelling density *and* high localization precision in all spatial dimensions at the same time for obtaining images with the highest resolution.

Localization precision

The recent introduction of Minflux [6] is an extension of SMLM where structured illumination in the form of a scanned doughnut-shaped excitation beam is used to incorporate extra prior knowledge into the localization process. The beam is scanned in a region of interest (ROI). This breakthrough concept can in principle give the same localization precision as with a flat illumination but at about 3–10 lower illumination dose. This idea was adopted by several groups in 2019 [7–10]. This is the most promising way forward if one wants to avoid cryogenic imaging which allows one to collect millions of photons from one fluorophore [11–15].

Labelling density

There are a number of emerging developments that address the limitations of labelling technology, i.e., larger linker size and incomplete labelling efficiency. Advances in bio-photochemistry have resulted in the development of labels that make a direct covalent bond to the target molecule, such as click-chemistry [16], effectively giving a small (<1 nm) label size. Also, techniques to achieve high labelling densities [5, 17] have surfaced, but, to date, labelling a structure with sufficient labels is a very hard biochemical problem that often cannot be solved such that a high degree (>50%) of active labels is achieved [18]. The only way to improve the effective labelling density is by combining information, that is, data fusion, a technique we have pioneered in the field of super-resolution light microscopy [19–21].

Our Recent Research Contributions to Data Fusion in Localization Microscopy

Traditional methods for fusing multiple SMLM images of a single underlying structure can improve SNR and resolution, but use templates or are very sensitive to initial registration errors. These approaches typically copy directly algorithms from cryo-TEM data fusion or single particle averaging (SPA), whereas the physical image formation in cryo-EM and in SMLM is completely different. In cryo-EM, the interaction of the electrons with the electrostatic potential of the specimen is imaged on the atomic scale, whereas in SMLM, single fluorescent molecules labelling the structure of interest are localized. The imaged particles in cryo-EM are truly the same in all acquisitions while noise is randomly distributed. The noise is averaged out after alignment, increasing the SNR with each added particle. In contrast, in SMLM, the structure itself is not imaged, but the positions of the fluorophores attached via a linker to the underlying structure are imaged. Furthermore, in practice the labelling is often incomplete where a 30–70% density of labelling (DOL) is typical [18]. In addition, statistical variations in localization uncertainty, false positive localizations [22, 23], and the statistical distribution of repeated localizations of the same fluorophore are additional sources of randomness that make the image data more complex than in cryo-EM. We have developed a method that takes these considerations into account and works on the localization data itself while considering the individual localization uncertainties for each point [21]. We also introduce a 3D particle fusion [24] approach for SMLM which can incorporate, but does not require, any a priori knowledge about the structure. It works directly

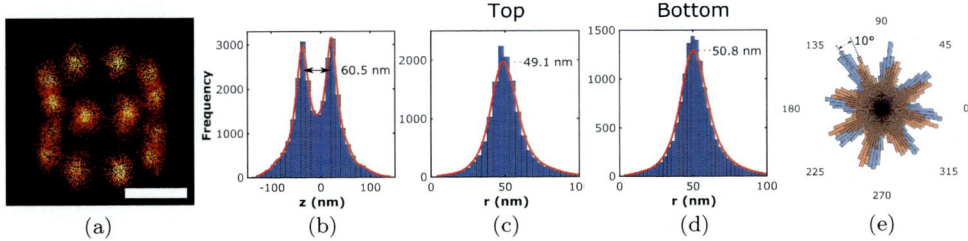

Fig. 1. (a) PAINT 3D astigmatic acquisition of NUP107 and reconstruction of 306 NPCs. (b) Histogram of the z-coordinate of the reconstruction, showing a separation of the two rings by 60 nm. (c) and (d) Radii of the cytoplasmic and nuclear ring localization. (e) Rose plot of the localizations over azimuthal angles 24. Scale bar 50 nm.

on the 3D coordinates of the localizations, considers the anisotropic localization uncertainties, and can perform joint alignment of multi-colour SMLM data. This pipeline is built upon our previous 2D method [21] with specific modifications of each step to handle 3D localizations. In Fig. 1, we show the result on such a 3D data fusion for NPC NUP107 acquired with PAINT in 3D via the use of an astigmatic lens in the emission beam path.

What are the Limits of Data Fusion and Resolution?

The key benefit of resolution in the range of a few nanometres with light microscopy is that it is the *molecular scale*. On this scale, the structure and function of biomolecules organizes life. It is most important to mention that light microscopy then can be combined with cryo-EM in one workflow, allowing both structural characterization and functional identification. In the past, super-resolution imaging could for example reveal protein distribution along neurons [25]. These structures are relatively "large" but required 3D super-resolution with multiple colours. To resolve the composition of macromolecular complexes, another resolution improvement is needed. I would envision that data fusion is the key to enable structural biology with light microscopy, hereby allowing the time to identify structures and its molecular components from currently years to months once the workflow has been established. As a most promising result of correlative super-resolution light microscopy combined with serial block face imaging, I would like to point to the result by the group of Harald Hess [15]. This publication does not apply data fusion, but certainly sets the pace in correlative high-resolution light microscopy and correlative EM.

Acknowledgments

This work was supported by the European Research Council (Nano@cryo, grant no. 648580 to H.H. and B.R. MolMap, grant no. 680241 to R.J. CellStruct).

References

[1] S.W. Hell and J. Wichmann, *Opt. Lett.* **19**, 780 (1994).
[2] E. Betzig, G.H. Patterson, R. Sougrat, O.W. Lindwasser, S. Olenych, J.S. Bonifacino, W.R. M.W. Davidson, J. Lippincott-Schwartz, and H.F. Hess, *Science* **313**, 1643 (2006).
[3] M.J. Rust, M. Bates, and X. Zhuang, *Nat. Meth.* **3**, 793 (2006).
[4] R.P.J. Nieuwenhuizen, K.A. Lidke, M. Bates, D. Leyton Puig, D. Grünwald, S. Stallinga, and B. Rieger, *Nat. Meth.* **10**, 557 (2013).
[5] W.R. Legant, L. Shao, J.B. Grimm, T.A. Brown, D.E. Mildie, B.B. Avants, L.D. Lavis, and E. Betzig, *Nat. Meth.* **13**, 359 (2016).
[6] F. Balzarotti, Y. Eilers, K.C. Gwosch, A.H. Gynnå, V. Westphal, F.D. Stefani, J. Elf, and S.W. Hell, *Science* **355**, 606 (2016).
[7] L. Gu, Y. Li, S. Zhang, Y. Xue, W. Li, D. Li, T. Xu, and W. Ji, *Nat. Meth.* accepted (2019).
[8] J. Cnossen, T. Hindsdale, R. Thorsen, F. Schueder, R. Jungmann, C.S. Smith, B. Rieger, and S. Stallinga, *Nat. Meth.* accepted (2019).
[9] L. Reymond, J. Ziegler, C. Kanpp, F.-C. Wang, T. Huser, V. Ruprecht, and S. Wieder, *Opt. Exp.* **27**, 24578 (2019).
[10] S. Leveque-Fort, unpublished (2019).
[11] C.N. Hulleman, M. Huisman, R. Moerland, D. Grünwald, S. Stallinga, and B. Rieger, *Small Meth.* 1700323 (2018).
[12] C.N. Hulleman, W. Li, I. Gregor, B. Rieger, and J. Enderlein, *ChemPhysChem*, **19**, 1774 (2018).
[13] S. Weisenburger, D. Boening, B. Schomburg, K. Giller, S. Becker, C. Griesinger, and V. Sandoghdar, *Nat. Meth.* **14**, 141 (2017).
[14] W. Li, S.C. Stein, I. Gregor, and J. Enderlein, *Opt. Exp.* **23**, 3770 (2015).
[15] D.P. Hoffman, G. Shtengel, C.S. Xu, K.R. Campbell, M. Freeman, L. Wang, H.A. Milkie, D. Pasolli, N. Iyer, J.A. Bogovic, D.R. Stabley, A. Shirinifard, S. Pang, D. Peale, K. Schaefer, W. Pomp, C.-L. Chang, J. Lippincott-Schwartz, T. Kirchhausen, D.J. Solecki, E. Betzig, and H. Hess, bioRxiv, doi: http://dx.doi.org/10.1101/773986 (2019).
[16] P.J.M. Zessin, K. Finan, and M. Heilemann, *J. Struct. Biol.* **177**, 344 (2012).
[17] S. Strauss, P.C. Nickels, M.T. Strauss, V. Jimenez Sabinina, J. Ellenberg, J.D. Cater, S. Gupta, N. Janjic, and R. Jungmann, *Nat. Meth.* (2018).
[18] A. Burgert, S. Letschert, S. Doose, and M. Sauer. *Histochem. Cell Biol.* **144**, 123 (2015).
[19] J. Broeken, H. Johnson, D.S. Lidke, S. Liu, R.P.J. Nieuwenhuizen, S. Stallinga, K.A. Lidke, and B. Rieger, *Meth. Appl. Fluorescence* **3**, 014003 (2015).
[20] A. Löschberger, S. van de Linde, M.C. Dabauvalle, B. Rieger, M. Heilemann, G. Krohne, and M. Sauer, *J. Cell Sci.* **125**, 570 (2012).
[21] H. Heydarian, F. Schueder, M.T. Strauss, B. v. Werkhoven, M. Fazel, K.A. Lidke, R. Jungmann, S. Stallinga, and B. Rieger, *Nat. Meth.* **15**, 781 (2018).
[22] P. Fox-Roberts, R. Marsh, K. Pfisterer, A. Jayo, M. Parsons, and S. Cox, *Nat. Commun.* **8**, 13558 (2017).
[23] J. Sinko, R. Kakonyi, E. Rees, D. Metcalf, A.E. Knight, C.F. Kaminski, G. Szabo, and M. Erdelyi, *Biomed. Opt. Exp.* **5**, 778 (2014).
[24] H. Heydarian, F. Schueder, B. v. Werkhoven, R. Jungmann, J. Ries, S. Stallinga, M. Bates, and B. Rieger, unpublished (2019).
[25] K. Xu, G. Zhong, and X. Zhuang, *Science* **339**, 452 (2013).

MULTI-COLOUR, DENSE, AND VOLUMETRIC SINGLE-MOLECULE LOCALIZATION MICROSCOPY AND POINT-SPREAD FUNCTION DESIGN BY DEEP LEARNING

YOAV SHECHTMAN

Department of Biomedical Engineering & Lorry I. Lokey Interdisciplinary Center for Life Sciences and Engineering, Technion — Israel Institute of Technology, Haifa 320003, Israel

Introduction: Computational Modelling and Deep Learning in Localization Microscopy

Computational modelling is at the basis of many modern super-resolution techniques. This is closely related to the fact that the resulting image of interest can be very different from the acquired data, in which case solving an inverse problem is required to obtain an image. An extreme example is point-spread-function (PSF) engineering in localization microscopy. In this case, the shape of the PSF of the microscope is deliberately and drastically altered, to encode information about the emitter — most commonly its depth, but possibly also colour and/or molecular orientation [1–6]. This enables, for example, scan-free volumetric imaging, using an additional image-processing step, that can rely on a computational model that maps the shape of the measured PSF back to the 3D position of the emitter [7].

In recent years, there has been growing use of machine learning algorithms in image processing, and those proved to be extremely successful in various tasks such as classification, denoising, segmentation, and de-convolution [8, 9]. These techniques, and specifically deep learning (DL), have also been applied to microscopy with great success [10–15]. The great appeal of DL for microscopic computational imaging lies in that it can alleviate the need for sophisticated image processing algorithms or even computational modelling. Broadly, if you can simulate high fidelity data, namely, data that would look similar to what you would measure experimentally under a microscope, then a neural net can be trained to extract amazing amounts of information.

Perhaps more intriguingly, a net can be trained to *design* the optical system itself. This is a highly promising direction that my lab is currently pursuing.

Deep Learning Enables Efficient Information Extraction and System Design for Challenging Localization Problems

My group has been leading several efforts in the past few years in applying DL to localization microscopy. These efforts can be broadly classified into *analysis* challenges, and *synthesis* challenges. The *analysis* projects include the following:

(1) Tackling the 2D overlapping emitter problem in localization microscopy [13] (Fig. 1(a)): A trained neural net yields super-resolution images from highly dense fields of emitters, orders of magnitudes faster than existing state-of-the art approaches, all with no parameter tuning by the user.

(2) Overcoming the dense 3D localization problem, intensified by large-footprint PSFs: We trained a net to resolve the positions of multiple emitters using a depth-encoding PSF, enabling unprecedented volumetric imaging of dense fields of emitters [14] (Fig. 1(b)).

(3) Algorithmic classification and characterization of single-particle diffusion characteristics: A trained neural net classifies single-particle trajectories to different anomalous diffusion models and estimates their relevant parameters, requiring orders of magnitude less data than current common approaches and achieving significantly higher accuracy, all with no parameter tuning by the user [16].

Examples for *synthesis*, namely, system design, projects include the following:

(1) Using a neural net to design the optimal PSF for distinguishing between differently coloured emitters, on a single optical path, using a grey scale

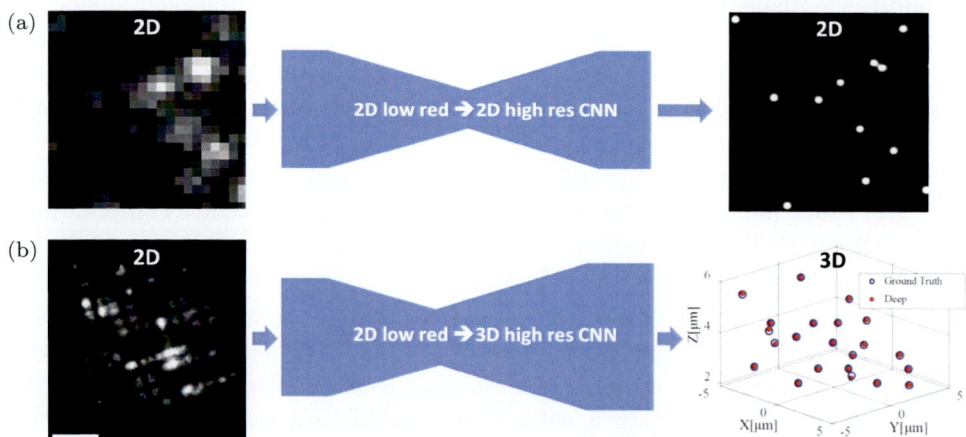

Fig. 1.   DL for super-resolution localization microscopy *analysis*. (a) 2D imaging: A set of diffraction-limited images of stochastically blinking emitters is fed into the convolutional neural network (CNN) to produce reconstructed high-resolution images [13]. (b) A trained neural network receives a 2D low-resolution image of overlapping engineered PSFs and outputs a 3D high-resolution volume which is translated to a list of 3D localizations; scale bar is $2\,\mu m$. Adapted from Ref. [14].

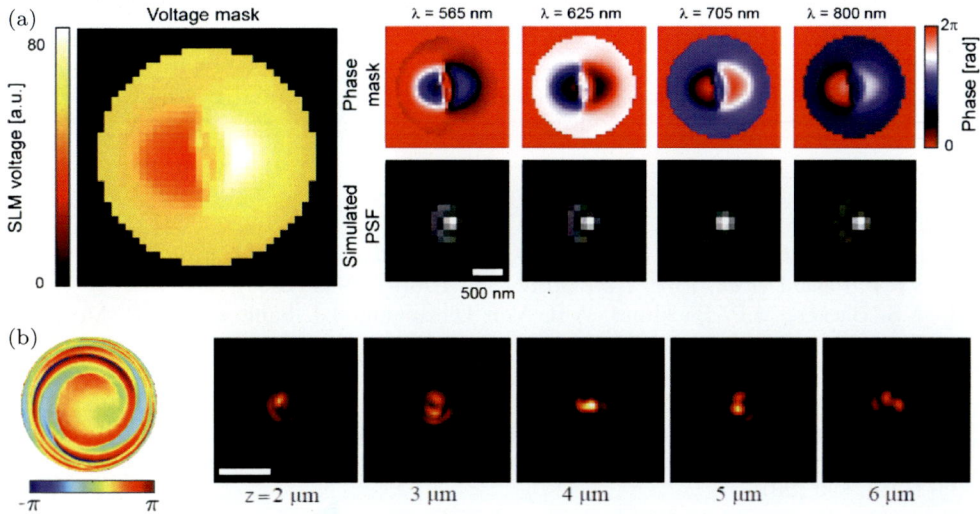

Fig. 2.   DL for microscopic PSF *design*. (a) The optimal SLM voltage mask for distinguishing differently coloured emitters by multicolour PSF encoding is learned by a neural net jointly with the decoding (analysis) part (left: voltage mask, right: PSF vs. wavelength). Adapted from Ref. [17]. (b) Following similar logic, the optimal PSF for encoding the 3D positions of dense emitters with overlapping PSFs is learned by a neural net (left: phase mask, right: PSF vs. z). Adapted from Ref. [14]. Scale bar is $3\,\mu m$.

sensitive camera: The net is optimized at the same time to decode the emitter colours from the imaged PSFs. This translates into efficient simultaneous multicolour localization microscopy with no need for spectral filtering [17] (Fig. 2(a)).

(2) Using a neural net to design an optimal PSF for encoding the positions of dense emitters in 3D: This yields significantly better results than using an "off-the-shelf" PSF that might be optimal in other scenarios, e.g., for single emitters [14] (Fig. 2(b)).

Outlook: End-to-end Optimization

End-to-end neural net optimization of super-resolution microscopy systems is expected to become increasingly useful and enable optimally designed systems. Using this approach, the emphasis is put on increasing the information content of the system (encoding), with less regard for the analysis (decoding) algorithm, since the use of a neural net replaces the need for sophisticated image processing.

There are many open questions and design parameters that can be optimized using such methods, for example, adaptation to coherent imaging, incorporation of temporal information, and task-specific tailoring of the optical system (e.g., Ref. [18]).

Acknowledgments

This work has been funded by the H2020 European Research Council Horizon 2020 (802567), by the Israel Science Foundation (ISF) (450/18), and by the Zuckerman foundation.

References

[1]   B. Huang, W. Wang, M. Bates, and X. Zhuang, *Science* **319**(8), 810 (2008).
[2]   S.R.P. Pavani *et al.*, *Proc. Natl. Acad. Sci.* **106**(9), 2995 (2009).
[3]   A.S. Backer, M.P. Backlund, A.R. Von Diezmann, S.J. Sahl, and W.E. Moerner, *Appl. Phys. Lett.* **104**(19) (2014).
[4]   Y. Shechtman, S.J. Sahl, A.S. Backer, and W.E. Moerner, *Phys. Rev. Lett.* **113**(3), 133902 (2014).
[5]   M.P. Backlund *et al.*, *Proc. Natl. Acad. Sci. USA* **109**(47), 19087 (2012).
[6]   Y. Shechtman, L.E. Weiss, A.S. Backer, M.Y. Lee, and W.E. Moerner, *Nat. Photon.* **10**(9), 590 (2016).
[7]   A. Von Diezmann, Y. Shechtman, and W.E. Moerner, *Chem. Rev.* **117**(11), 7244 2017.
[8]   A. Krizhevsky, I. Sutskever, and H. Geoffrey E., *Adv. Neural Inf. Process. Syst.* **25**, 1–9 (2012).
[9]   N. Nitta *et al.*, *Cell* **175**(1), 266 (2018).
[10] Y. Rivenson, Z. Göröcs, H. Günaydin, Y. Zhang, H. Wang, and A. Ozcan, *Optica* **4**(11), 1437 (2017).
[11] M. Weigert *et al.*, *Nat. Meth.* (2018).
[12] N. Boyd, E. Jonas, H.P. Babcock, and B. Recht, bioRxiv, 267096 (2018).
[13] E. Nehme, L.E. Weiss, T. Michaeli, and Y. Shechtman, *Optica* **5**(4), 458 (2018).
[14] E. Nehme, D. Freedman, R. Gordon, B. Ferdman, L.E. Weiss, O. Alalouf, T. Naor, R. Orange, T. Michaeli, and Y. Shechtman, *Nature Methods* **17**, 734–740 (2020).
[15] C. Belthangady and L.A. Royer, *Nat. Meth.* (2019).
[16] N. Granik *et al.*, *Biophys. J.* (2019).
[17] E. Hershko, L.E. Weiss, T. Michaeli, and Y. Shechtman, *Opt. Exp.* (2019).
[18] R. Horstmeyer, R.Y. Chen, B. Kappes, and B. Judkewitz, arXiv (2017).

RECONSTRUCTING EMBRYONIC DEVELOPMENT FROM HIGH-RESOLUTION LIGHT SHEET MICROSCOPY DATA

JAN HUISKEN

Morgridge Institute for Research, 330 N Orchard St, Madison, WI 53715, USA
Dept. of Integrative Biology, University of Wisconsin,
250 N Mills St, Madison, WI 53706

State of Research and Current Challenges

In an attempt to image, reconstruct, and understand the fundamental principles of a biological system, we are pushing the limits of conventional microscopy to further increase resolution and speed in microscopy. At the same time, we try to expand the applicability of imaging techniques and to span an increasing range of scales, from sub-cellular features to entire living organisms. Importantly, the optical penetration depth is becoming a crucial parameter, which dictates how much of the theoretical resolution is actually achievable as we peer deeper and deeper into scattering tissue. We also begin to realize that in the context of a living organism, the photo toxicity, i.e., the damage that is caused by the imaging, is particularly critical and needs to be kept at a minimum.

Light sheet microscopy (LSFM, selective plane illumination microscopy (SPIM)) has changed the field of fluorescence imaging substantially by offering a versatile technique to obtain optical sectioning in large specimens with high speed and minimal phototoxicity [1]. For the first time, it has become possible to image living embryos of a fly, fish, and mouse *in toto* over the course of several hours and days. The challenge now is to retrieve the relevant information from the enormous amounts of data, ideally in real time during the image acquisition [2]. In addition, we try to merge data from multiple samples for a comprehensive, statistical model of embryogenesis. Data analysis and visualization have now become an integral part of the imaging process. In particular it is now desirable to merge information from different imaging modalities and from *in vivo* and fixed tissue data.

Our Recent Research Contributions to the Field

Light sheet microscopy has been shown to be especially well suited to non-invasively image dynamic events in fragile biological samples at high resolution [1]. *In vivo* applications primarily include time-lapse imaging of developmental processes over the course of several hours or days and high-speed imaging of dynamic and transient

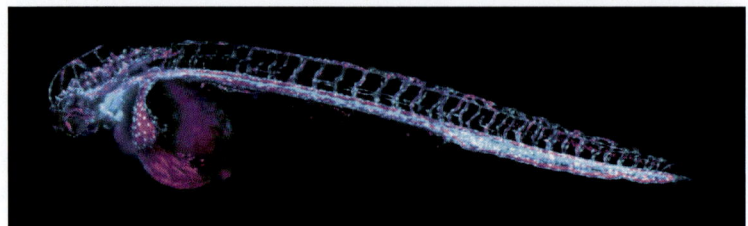

Fig. 1.   Large live specimen reconstructions. Zebrafish vasculature reconstructed from several 4D volumes (size of the final FOV ca. 3.5 mm). Cyan: vasculature, Magenta: blood cells [4].

events in organs such as brain or heart inside intact organisms [3]. We have developed state-of-the-art light sheet instruments and protocols tailored to live embryo and fixed-tissue imaging. Crucial to the success of the techniques is a comprehensive approach in which all aspects of the experiment are optimized, including sample preparation, image acquisition, and analysis [4].

Large data acquisition and segmentation

To quantitatively understand dynamic biological processes that occur over many hours or days, it is desirable to image multiple samples simultaneously and automatically analyse and merge the resulting datasets into one unifying model. We have developed a complete multi-sample preparation, imaging, processing, and analysis workflow to determine the development of the vascular system in zebrafish [4]. Up to five live embryos were mounted and imaged simultaneously over several days using selective plane illumination microscopy (SPIM). The resulting large image dataset of several terabytes was processed in an automated manner on a high-performance computer cluster and segmented with a novel segmentation approach that uses images of red blood cells as training data. This analysis yielded a precise quantification of growth characteristics of the whole vascular network, head vasculature, and tail vasculature over development. We now expand these techniques to address other biological questions, e.g., in zebrafish neurogenesis, in 3D human organoid morphogenesis, and in cortical excitability of Xenopus oocytes.

Smart multi-view image acquisition

Complex and inhomogeneous optical properties of biological samples limit the optical penetration in microscopy, making it challenging to perform *in toto* imaging on large samples. Multi-view imaging is a popular paradigm in many modalities, especially in light sheet microscopy, to improve the overall sample coverage: multiple imaging orientations are acquired and the well-imaged portions of all views are fused into a single image stack with improved sample coverage [5]. However, the number of views and the specific angular views are often empirically selected: multi-terabyte datasets contain less than optimal imaging views. We built a GPU-accelerated image analysis software that estimates the information content distribution in each

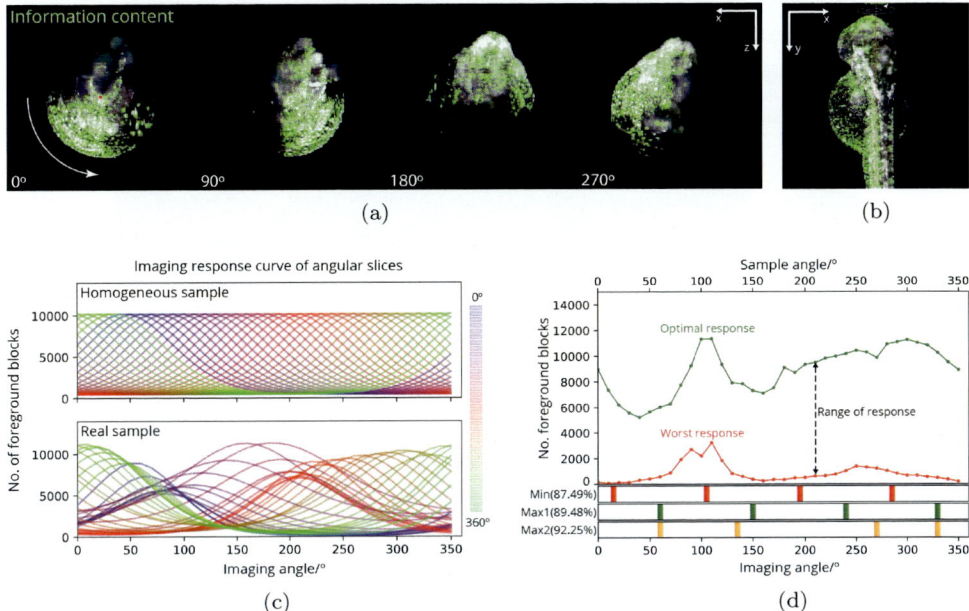

Fig. 2. Sample coverage evaluation and optimization for a zebrafish imaged with 24 views. (a) Anterior view of a 48hpf zebrafish embryo (Tg(*h2afva:h2afva-mCherry*)) at four different rotation angles. Information content (green) overlay with grey raw data. (b) Dorsal view. (c) Comparison of the image response curve between a completely homogeneous sample and a real zebrafish embryo. (d) Variation in imaging response for different angular regions (top) and comparison of the sample coverage for 4 view multi-angle imaging strategies. Minimum coverage by four equally spaced views; Max1: Maximum coverage given by 4 equally spaced views; Max2: Maximum coverage with flexible view spacings.

3D image stack and finds the optimal combination of views that gives the best sample coverage [6].

Optimized fixed tissue preparation

When it comes to even larger biological samples (5–20 mm) such as tumour biopsies and intact organisms, they pose several imaging challenges owing to their size. While it is desirable to image them at high resolution in three dimensions across their entire extent to study system-wide phenomena such as development and disease, imaging depth is generally limited to several hundred microns due to visible light scattering in tissue. Tissue clearing is a promising approach, which leaves the sample intact and uses fixation and refractive-index matching to render tissues transparent [7]. These biological techniques combined with SPIM allow parallelized imaging and reconstruction of centimetre-sized tissues which are sub-micron resolution orders of magnitude faster than other volumetric imaging techniques with comparable resolving power. The imaging of cleared tissues has now become a second, invaluable application of light sheet microscopy in our lab and other labs.

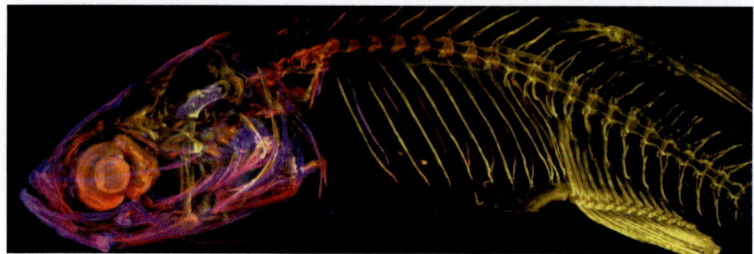

Fig. 3.   Optical clearing. The recently improved optical clearing techniques offer insights into otherwise opaque tissues, such as adult zebrafish. One of the goals now should be to use the additional high-resolution information to improve in vivo data that suffer from the typical limitations such as deep tissue scattering, noise, etc.

We have developed custom SPIM systems to address the wide range of sample sizes and growing collection of clearing techniques to effectively image a wide range of samples using various optimized clearing techniques in collaboration with dozens of labs.

Deep learning to enhance dynamic time-lapse data

During a time-lapse experiment of a fragile biological specimen, image quality is usually affected by several factors: images are acquired with low laser power, long working distance lenses, short exposure times, etc. However, at the end of the time lapse, samples may be sacrificed, treated, and imaged again to obtain a superior quality snapshot dataset. We have developed a deep learning approach capable of enhancing the information content of time-lapse images by acquiring highly resolved images of the sample at the end of the imaging. We optimized an antibody-staining protocol with primary antibody against the fluorophore of interest and conjugated to a fluorescent dye which emits into the near-infrared (NIR) regime. We image the stained sample in both native fluorescence and exogenous NIR channels. Crucially, the NIR image is less affected by scattering and autofluorescence, yielding a higher-resolution image with increased optical penetration. Moreover, chemically designed dyes are typically orders of magnitude brighter than fluorescent proteins, generating a higher signal-to-noise ratio image. We train a deep neural network on the NIR images and use this network to reconstruct NIR-like images for each of the low-resolution time-lapse microscopy images previously acquired.

Outlook on Future Developments

The ability to custom design a SPIM instrument around a sample has empowered many research labs to do experiments that have been impossible with commercial instruments. Unfortunately, only physics and engineering labs have been able to custom design such an instrument to enable demanding biological applications. It is therefore a necessity now to strengthen the dialogue between the developers and the users: discoveries in biology often depend on cutting-edge technologies such

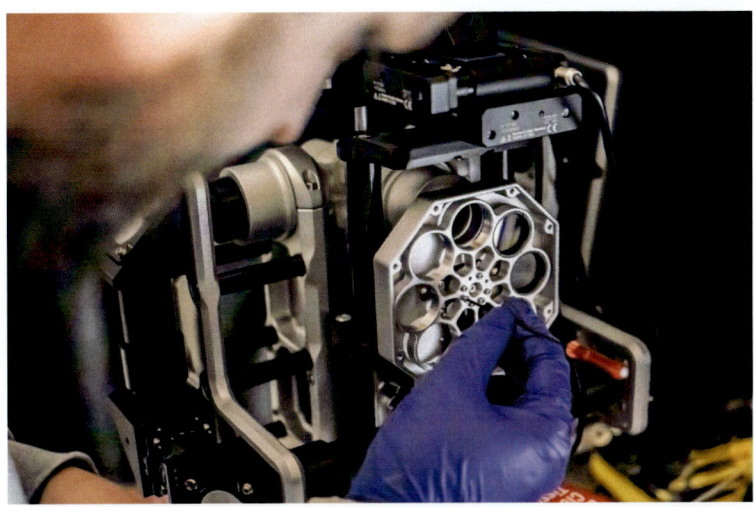

Fig. 4.   Flamingo. Development of a modular, portable, and shareable microscope platform.

as microscopy, but unfortunately, the best and latest microscopes are found in engineering and physics labs, where the dissemination of the technology becomes a challenge. We hope that other labs will support us in our efforts to create an open-hardware platform to facilitate the collaborations between physics and biology. Our recently created Flamingo microscope framework is the beginning of an initiative to pass advanced microscopy hardware back and forth between two or more labs. Modularity and portability of such devices will enhance the possibility that tailor-made instruments can be created that offer unique experimental opportunities for the biologists. Sharing advanced microscopy technology freely will not only democratize the access to such hardware but also offers the chance for better reproducibility in science [8].

Acknowledgments

I would like to acknowledge all present and former members of the Huisken lab and financial support by the Max Planck Society and the Morgridge Institute for Research.

References

[1] R.M. Power and J. Huisken, *Nat. Meth.* **14**, 360 (2017).
[2] N. Scherf and J. Huisken, *Nat. Biotechnol.* **33**, 815 (2015).
[3] M. Mikoleit *et al.*, *Nat. Meth.* **11**, 919 (2014).
[4] S. Daetwyler *et al.*, *Development* **146**, 6 (2019).
[5] S. Preibisch *et al.*, *Nat. Meth.* **11**, 645 (2014).
[6] J. He and J. Huisken, *Nat. Commun.* (in press).
[7] D.S. Richardson and J.W. Lichtman, *Cell* **162**, 246 (2015).
[8] R.M. Power and J. Huisken, *Nat. Meth.* **16**, 1069 (2019).

SESSION 4: COMPUTATIONAL MODELING IN HIGH-RESOLUTION IMAGING

CHAIR: S.W. HELL

AUDITORS: R.G. EFREMOV[1], J.N. HARVEY[2]

[1] *Center for Structural Biology, Vlaams Instituut voor Biotechnologie, and Structural Biology Brussels, Department of Bioengineering Sciences, Vrije Universiteit Brussel, Brussels, Belgium*

[2] *KU Leuven, Department of Chemistry, Celestijnenlaan 200F, 3001 Heverlee, Belgium*

Discussion among the panel members

<u>Stefan Hell</u>: Thank you to all the panelists. This was fascinating and has revealed a number of interesting aspects. I am going to start off by asking maybe one or two questions to each of the panelists and let's see how the discussion goes. If you leave aside the last talk by Jan Huisken which deals with many cells and configurations and large systems, basically, all the others somehow work with the same object. Whereas in the past, the objects were all different, now the objects are all the same: molecules, individual molecules. This is something that really changes the perception of what we do in fluorescence imaging. I remember the days, twenty years back, when people argued about who had the best image. They argued by saying: oh, I can see this and I can see that, and I have that feature in it and so on. This was very tedious. At some point some colleagues suggested to make an image repository where people would download their image and some arbitrary experts would look at this image and then confirm that the quality of the image is there or not, and confirm that the claims are justified. But now, since we are dealing with identical objects, these single molecules, all the objects are in the end the same. So, we have a very fair view of judging of what quality the image actually is. This brings us back to the first talk by Raimund Ober. What are the odds of coming up with very standardized measures of the quality of an image, given that our image consists of molecules and their positions?

<u>Raimund Ober</u>: That is a very interesting and very important question. I think we're still not quite there yet. We have a lot of tools to investigate when, whether or not, an individual molecule is localized well. And I think we can do that to some extent and it relates to some of the things I've been talking about here. Where I think we are still missing a lot, is by then analyzing a full single molecule super resolution image. I was right at the very end of my presentation where I said that I think we were making a first step in that direction where we try to define what

we call the algorithmic resolution of a full algorithm. Which means, the algorithm that takes the images as they're acquired from the microscope and at the end of the day spits out all the final localizations. Not just one molecule at a time, but the full ensemble. When we looked at the various approaches there, we discovered that some of these problems that are published right now, which refer to my multi emitters problem, are still very badly defined. And that doesn't seem a very good way at the moment to analyze those algorithms to determine whether they work well. What we found was that, when you look at an algorithm from that full holistic point of view, many of these algorithms actually don't perform very well at all. So basically, you have algorithms which measure with resolution in the region of 500 nanometers or something, so just the classical Rayleigh criterion. Therefore, I would only trust many of these structures for distances that are in excess of 500 nanometers. So, I think there is still a lot of work to be done there.

Matthew Lew: To speak to that point: The fact that we look at individual molecules, and that we know physics so well, Maxwell's equations and how these imaging systems work, allows us to start working towards that point. That was the topic of the very end of my talk. What is critical now, especially in localization microscopy, is that the model, the computational model of the imaging system and the algorithm are part of what we get — at the end — as the quality of that image. We can ask very fundamental statistical questions, knowing that we are looking at individual molecules and that we assume this model. How stable are the localizations that we get and can we assign a per molecule score and then, if you want to average over the ensemble, you can. I think that going beyond the metric of just simple resolution, allows us to interrogate the question for any algorithm, whether it is a machine-learning based algorithm or a classical deconvolution one. You can now ask questions of the type: is that model actually describing the raw data with high fidelity? I think that is going to move us towards being able to quantify these images in a non-biased, in a systematic quantitative, way.

Bernd Rieger: I agree. Typically, the problem at the moment is, that the image is not only generated by the microscope on the camera, but there is a lot of processing in between. This could disturb the ideal microscope. At the moment, we are not in the way to analyze this. Maybe with the approaches that Matthew Lew mentioned, we should be able to get a bit more of an objective measure on the question: is the image good? We should also be able to compare images between labs a bit better, and between different modalities. As you have seen there are a lot of acronyms out there which means there are a lot of technical differences in the different ways how to acquire an image. Also, that makes a difference in what you, in the end, get back to the screen.

Stefan Hell: But could we agree on the fact that this kind of comparison or standardization is on the horizon? I think, ten, twenty years back, it was unconceivable.

Let's look further down: in ten years from now, maybe there is a way of putting the data, the raw repeats of photons that we get, the single photons that we get, put in a sort of algorithm and you could tell the content of information that is there and the bias of the estimation of the image, the fidelity of the image etc. I think this should be further down the road. But realistic.

Matthew Lew: I think it is a really tantalizing possibility because, unlike traditional lower dimensional scientific instruments, where you maybe get a few curves out, with images now, we have this very high-dimensional data, where we can ask these questions about fidelity, quality and how accurate these datasets are. That's just something that's a unique advantage for the imaging field. Quite possibly an exciting thing to come.

Raimund Ober: I'll add something as a challenge to some chemists in the audience. What we are still lacking to some extent — it is *partially* there — are good prototypes in terms of samples. Primarily because the labeling, usually, is pretty random concerning the underlying protein that we are trying to image or the underlying sample that we are trying to image. And much more reliable chemistries to allow us to come up with prototypes that could then be used as reference standards, would help us very significantly to validate both the instruments and possibly the subsequent algorithms that we are using to come up with final imaging and final results.

Jörg Enderlein: But maybe, Raimund, we have now this DNA origami, which also Stefan used for calibrating MINFLUX, and I think this works really nicely. We basically now have samples where we know with something like nanometer accuracy where the labels sit. Also you, Matthew, showed this with PAINT, coming down to two nanometers and verifying this with origami. Also now there is this initiative from Jonas Ries at the EMBL where they have the nuclear pore complex as a standard, a biological standard, for super resolution microscopy. I think nowadays we have much better reference samples than 10 years ago.

Raimund Ober: I am not denying that. Absolutely not. We've made massive progress in that regard. I think there are still issues with consistent labelling. Questions about how reproducible are these samples, how reproducible is DNA origami over larger structures rather than smaller ones. Issues of that sort. But I would agree: we've come a long way.

Stefan Hell: This brings me to another question: once we have the molecules and we display them on the screen, we put dots in there that are surrounded by a Gaussian blur. But the question is what does that mean, because the molecule is still smaller than the Gaussian that we plot. And what does the brightness of that Gaussian mean? Actually, there is little meaning. Some try to gauge it with their number of occurrences, number of localizations that come up, others do other things. But clearly there is no good representation in my view, at this point, that would reproduce the fact that there is a little point that has the size of a nanometer, and it comes up at a certain occasion or with certain statistics. Are there any ideas here how we can visualize what we actually measure, so the fact that the molecules appear with a certain probability means that in principle, the images that we show shouldn't be static. They should be dynamic or show wiggling molecules, or popping up with certain probability. Of course, computation will allow us to do this in the end, I think. The way we represent these images is still old fashioned, it is still 20th century type. This is not the way it should be done. Any suggestions, ideas?

Bernd Rieger: For the visualization, that's an issue. For the computation luckily, there is no need to do that. You just can take the information that you have, for example from the fitting, and proceed with that, so that's fine. To gain insight, that is of course a problem. Typically, what you do is that the blurring is proportional to the uncertainty of the measurement. That is a reasonable idea. The downside is: the intensity scale is very arbitrary and I have not seen any good ideas on how to tackle that problem. Typically, the way how you measure that is not quantitative. So, you get a number of localizations. But that does not represent the concentration of molecules per se. If you would have that, that would be easy. In order to get that, that's really a pain. A lot of people are working on that, but it is tough.

Stefan Hell: I agree. It is an unsolved problem and maybe computation helps us sorting that out in some way or another. So, we have to take into account the statistics of localization and have to relate it with the concentration and with the photon statistics and I don't know what. But there is something that has to be done in the field given that we have now this spatial resolution at the size of a molecule.

Matthew Lew: I think also one of the key areas that the topic of this conference can help us with, is in terms of modeling, is bridging the biological and chemical questions that you all are interested in with the data that we are actually collecting. In the end, no one in the audience probably cares about how many localizations we get per pixel, you probably don't care very much how we blur it. What you really want to know is: What is the probability of colocalization of this ligand in this receptor or what is the folded state of this receptor, is it open or closed? You can think of direct biological questions and then model bridging our data, the fundamental data that we are able to quantify, with those questions and directly visualizing that. There is one more step after our images and, of course, there is a step before in terms of encoding the information through our fluorophores.

Stefan Hell: I have a question directed to Jörg Enderlein. You came up with this SOFI method. A related question now. What is the highest number of molecules that I can still identify as individual emitters given the photon statistics that we have now from typical fluorophores, say blinking fluorophores? Is there any rule-of-thumb that you can give us or do you think that further research will lead to a higher molecule density that you can cope with in a decent way without producing the localization artifacts that Matt talked about in his presentation?

Jörg Enderlein: I mean, the beauty, in my point of view, of SOFI is that you don't have to care about any overlapping molecules because it is a rigorous method and basically what you really show mathematically by doing the cumulant analysis is that every molecule has a framed point spread function. Of course, as in fluorescent correlation microscopy where it comes from, you have the problem that you then have too many emitters at the same time in your detection focus then your signal to noise goes down. Fortunately, at least in our experiments, you never hit this level where the density of labelling was so high that we could not extract any decent SOFI signal anymore. And I think before this will happen, you will rather be hitting the label density limit than from the *labeling* point of view. For that limit, I mean I think we are talking about something like, per μm^3 roughly 100 fluorophores, and I think this is rarely reached in real sample labeling. From this point of view, it was always a good experience that SOFI never had a problem that we lost the signal contrast because the label density was too high. Of course, what you need to do is the more labels that you have in one area, the longer you have to measure. So that is the signal to noise statistics problem. But theoretically SOFI is super relaxed concerning label density. We don't have any of these problems with overlap like PALM/STORM do, that's the good thing. Of course, you will never be able to reach the resolution of PALM/STORM but on the other side, we don't have all these artifacts which Matthew, for example, was addressing.

Stefan Hell: Coming back to that question: What can we learn, actually? That was my question. What can we learn from the SOFI concept to increase the number of emitters say in a PALM/STORM type of recording or is there nothing that we can do about it?

Jörg Enderlein: There was a nice study by Theo Lasser at EPFL some years ago where they compared PALM/STORM with SOFI and also where is the transition. And they clearly could show when you can tolerate a relatively low density of labels in one image then PALM/STORM, compared with SOFI, is much better concerning the resolution but when you go to a higher density then SOFI becomes superior in terms of speed. I think both methods are somewhat complementary and if you want to go with live cells and fast and relatively relaxed concerning label density, SOFI can be really advantageous. If you want to be superhigh resolution and you have a lot of time, and you can work with fixed samples then, of course, PALM/STORM is the method of choice.

Yoav Shechtman: I would like to follow up on Stefan's questions about density and the relation between SOFI and localization microscopy. What we can learn from

SOFI is that you should use temporal information and extending the neural network approach to include temporal information, over the frame by frame method we use, would make a lot of sense. At that point, I think that there is not going to be a distinction between SOFI and localization microscopy because you input the movie and you output a high-resolution image.

Bernd Rieger: A few years back, Susan Cox made this 3B Approach where they tried this in a Bayesian approach. The same idea to have the whole movie in and then the localizations out. That was computationally so complex that it just did not work. It took too long based on the way how you would approach this from a Bayesian view. Now if you could learn how to do this then, maybe, SOFI and PALM, this localization will be one at the end. I would expect — or hope.

Matthew Lew: So one other possibility here is just leverage the blinking and the blinking statistics as another dimension over which we can get some contrast from our datasets. The fluorophore is going to be intrinsically coupled to its local chemical environment. The lifetime of the triplet state, the lifetime of whatever dark blinking state we see, might be a great way of sensing something about what is going on around our fluorophore and therefore answering a very interesting question about whatever biological or chemical dynamics that our technology might be directed towards. So, there's a huge possibility of just leveraging all these different things that we can measure for very powerful results.

Stefan Hell: I have a question for Jan Huisken. There have been attempts of marrying super resolution with light sheet microscopy. What is your opinion on that? What are the hurdles, what are the benefits and can you do that decently in, perhaps, living cells?

Jan Huisken: I think you raise a valid point. The question would be: where can we meet the middle. I think you shouldn't probably try to image a cm^3 piece of tissue at super resolution. It would just enormously increase the amount of data and would probably not necessarily yield relevant information from a developmental biology standpoint. Where the goal should primarily be to increase the n, in that case, the number of samples. Which, for the biologist, is probably more valuable, at least in the context of a whole tissue, to increase the number of samples. The higher resolution for the fixed tissues can always be achieved by simply cutting the tissue into pieces, and then using techniques that are established or even going down to electron microscopy. On the whole tissue level, in the fixed tissues, it is probably not desirable to achieve even higher resolution than what we already have. In the live tissue case that is more interesting. Maybe not again for in-toto imaging, but if we are imaging in just individual vessels, neurons, individual cells that are relatively superficial, it would definitely be desirable to increase the resolution, as long as the photo-toxicity doesn't increase. Which of course is very important for all the live cell imaging that we are usually concerned with.

Stefan Hell: Further along these lines, where do you see the limits, or where should light sheet microscopy go to? Do you see needs for development in the way the data is handled? Or do you think there is still something that can be done in terms of the physics?

Jan Huisken: Well of course I think there are still opportunities optically to further increase the technique. If we had labels more in the far red or infrared, that would definitely be an opportunity for us, to further increase penetration and ultimately also resolution deep inside living tissues. But, as I mentioned in my talk, currently the main limit we are hitting is on the computational side. Processing all the data. Making a wise decision what exactly we're trying to record in a living embryo. What is the information that we don't need? What is the information that we already have? And what is the phenotype? What exactly should we focus on? And, ideally, just image the differences and not what is already commonly known.

Stefan Hell: So, we are going to have a break now, I have been told by the chairman. We will resume afterwards the discussion with the panel.

General discussion

Stefan Hell: Welcome again to the next hour of discussion. And now of course we invite you for questions.

Mischa Bonn: I'm Mischa Bonn and I am confused. So, all of you gentlemen showed magnificent results in extremely high-resolution imaging down to essentially the molecular level. So, what more can we do? If we know where exactly with the nanometer resolution, the molecules are in 3D in our sample, what modelling needs to be done and what are the remaining challenges?

Stefan Hell: I can answer some aspects of your questions. First of all, keep in mind that the fluorescence microscope image is nothing but fluorescent molecules. But users usually are not interested in the fluorescent molecules. They are interested in proteins or other types of bio-molecules that are labeled with those fluorescent molecules. There is a residual distance for example between the fluorescent molecules and the actual position of the protein or the epitope or whatever you are thinking of. And so there will be a need of course for modelling the residual distance between the fluorescent molecule and the actual position of the protein. That is something that is not really addressable by the physics, it is beyond our capability as microscopy developers. This is part of the labelling procedure and that also requires mathematics. But that is just one small aspect that I am adding here at this point.

Jörg Enderlein: I think one other really fundamental aspect is that the labelling of our structures in cells is always a stochastic process. So, I mean for example I want to see the cytoskeleton in a cell's actin, microtubules, etc. So, I label them stochastically. I cannot physically have every nanometer a fluorophore — and then I have a collection of points in 3D from my measurement and I know I want to reconstruct this network. So, I think there is the big challenge now: to come up with clever, maybe content-aware, computational procedures to convert this number of random spots into, let's say, filament structure, crosslinks, whatever. This, I think, is the next step you have to address.

Raimund Ober: I would say dealing with live cells is the next really big challenge. There are lots of fundamental differences: for example in a fixed cell you can play lots of games that you cannot play with live cells. You can wind up your lasers that would immediately destroy a live cell. So, the modeling will be much more complex in live cells to capture the dynamics that the biologist might be interested in. The underlying mathematical modeling then becomes very very hard. The computational challenges become really very significant. Even the image processing becomes very challenging in live cell environments because you want to then expose the cell to as little light as possible:

number 1: to not get into phototoxic regime,
number 2: to preserve the fluorophores from bleaching.

So, therefore you get into an extraordinarily low signal environment that is extraordinarily challenging to then even analyze and deal with from an image analysis point of view. I see massive challenges there, still in the live cell, in the context of live cells.

Wilhelm Huck: Stefan, I have a question about the MINFLUX system. It struck me that in localizing your molecule you're in a way also tracking its movement or potentially have the ability to track its movement. And I was wondering whether you can elaborate a bit more about options you see for studying dynamic processes maybe even zooming in and out or following multiple dyes at the same time.

Stefan Hell: Yes, you are absolutely right. I didn't prepare a slide for that but as a matter of fact we have done experiments where we showed that the fact that we need only a few or much fewer fluorescent emissions in order to localize the molecule can be converted into imaging speed, so localization speed. And so, you can imagine that we can see translations of fluorescent molecules, dislocations and so on much faster than in PALM/STORM because we need fewer emissions to localize the molecules. To give you an estimate: back of the envelope calculations show you that, if you detect the molecule within a spatial range of 10 nanometers, so with an anticipated spatial range of 10 nanometers, current fluorophores with current emission rates would allow you to localize the molecule with 1 nanometer precision, so with molecular precision, within 10 microseconds. And this is very fast! Meaning that, if something moves within this 10 nanometer range, within

10 microseconds, you can follow its movement with 1 nanometer precision. And in my view, this is a unique strength of MINFLUX. There is no other method that I know of now, that could see such small movements and be totally benign to a cell. There is also a strength of the concept that the molecule is in the center of that minimum of light intensity. Of course, the molecule is not fried. Don't worry that the donut fries something or bleaches something. These are normal excitation rates and so I think there is a tremendous potential for seeing dynamics. In my view, this will be as important as just seeing structures because at some point we will be bored with molecular structures, we want to see molecular dynamics. If I were now 20 or 25 and I wanted to go into some sort of Structural Biology, this is exactly where I would go to as a bio-physicist, because I know that I can see things, movements, interactions that I couldn't see otherwise.

Wilhelm Huck: Over what kind of distances do you think it would be possible to track the molecule?

Stefan Hell: You can track them over long distances, say over 20 microns if it moves, because you can just follow that molecule. If you take advantage of the fact that you know roughly where it is, which in most cases you can, because the MINFLUX has this feature that you are in control of the distance between the intensity zero and the molecule and you don't have to waste time for collecting photons in order to get to a certain localization precision. Then you gain speed because you need much less time. So, anything is possible, 50 nanometers, 100 nanometers. It will always be faster than standard localization in PALM and STORM because you can do away with this need to wait for many photons, thousands of photons.

Dean Astumian: With regard to the last question: Is there any fundamental reason that you couldn't follow two molecules on different channels?

Stefan Hell: I can follow two molecules, if I can distinguish them. In imaging, the mode of distinction has been on/off, which is the most fundamental mode of distinction right now because you can have any number of molecules by looking at them sequentially. That's the on/off, but you can distinguish also by two colors or three colors or five colors or six colors and as long as you can distinguish them you can follow six molecules at the same time, with the same speed that I was mentioning: 10 microseconds, 1 nanometer precision with a 10 nanometer range, six molecules. And then watch how they interact. Of course, this will, in my view, be a rival of FRET. I am not saying it will do away with FRET because there will be a use for FRET in the future as well. But for many things where people thought that you would have to use FRET, they won't use FRET anymore because FRET would be not very powerful in comparison with this.

Andreas Walther: We heard about this deep learning technique to optimize the PALM method. And you said you have to simulate high-resolution and low-resolution images to train the system. This would mean that if we anticipate that, you always have the same kind of problems in these images, that your training algorithm is universal. If you have done this for your structure that would mean that the training should work on any kind of image. Is that correct? Or would you train on different structures first to increase the resolution?

Yoav Shechtman: Let me address the more general question that I think is being asked which is: how robust is the training to the images that we are trying to reconstruct? So, this is a very good point and one that you should be beware of whenever you are doing something like learning based reconstruction is that your network learns whatever you teach it. In localization microscopy, which makes images out of blinking point sources you can, if you do this frame by frame, then you don't really have to include any underlying structure in these training images because you can just have molecules blinking in random positions in each frame. Ok, this is when you do it frame by frame. This is important because, if you simulate the whole movie, then you are going to start having correlations between the position of the blinks. So, this is, in order to avoid such biases, this is exactly what we do. We train on completely random positions of emitters and therefore you are not biased to some sort of shape like filaments or points or something else. On the other hand, if you do know that you are looking at filaments, then you would get better results if you train on filaments. But it would never be your scope. So, now you have to start being really careful that you really know that you are looking at filaments. That is the tradeoff you always have to make.

Bernd Rieger: I would like to add on that: In the literature there are examples where people would train on one type of cell structure maybe in actin network and apply it to something else, and of course, you get out what you trained on. That is a very difficult procedure if you want to use the microscope to discover something new. Or, sometimes people even acquire let's say, face contrast images and then from that they said "OK, we can even acquire different color channels that we did not observe". Now, that is a very difficult, I mean, especially how could you check that this is real because a network will always give you an output. See, you have to be very careful because all these procedures will not fail gracefully and will produce just an image and then you might believe it or not. So, there is still a lot to do for the field.

Mark Ellisman: First, a note of caution about biological systems. The genome of the vertebrate organism produces 20,000 Lego pieces, macromolecules. I am a little worried about the scalability. Even now I think that what you guys have represented is probably one of the biggest breakthroughs in modern biology. It is the ability to use photon-based imaging to study things at higher resolution than we expected, than the resolution barrier. So, how do you think we'll scale to the complexity? First of all, if you're going to localize fluorophores, as proxies for macromolecules.

Every time you add a fluorophore you have the risk of modifying the behavior of the molecule. We usually do that by some genetic method. CRISPR is thankfully available, but most people don't do such things at this point. You're working with overexpression or molecules like Nile red or whatever which may modify this system. So, that's the first thing. I'd like some comments about where you think the limits are. And then, I wanted also to try to force again a connection between your discussion and your session and some of the other sessions. The one that I am sensitive to is the session on water. Where, if you extend thinking about water to biological systems, we understand that water, free water, is really probably not as available as if it had been thought. Now, when you introduce the energy, even though it has been minimized, into the system, which you know, photon loading in a complex system, you may increase the temperature, which is going to change the activity of the water in the vicinity of the macromolecular complex. It sometimes depends on the availability of the water as a currency to talk between subunits. How will you get around some of these risks or calibrate them or test? What molecule would you look at? Or could we discuss that?

Bernd Rieger: Maybe I can take the second part of the question. It depends a lot on if you have a look at fixed cells or at live cells. Maybe if you fix something, maybe that is not a big issue. And also, the powers that you would use typically in these imaging modalities are very high. So, they're, let's say, maybe a thousand-fold above sunlight, what you would expose. Now, these methods that Stefan Hell discussed and maybe also the light sheet that uses light strengths that are comparable with outside light levels, that should be relatively OK with the heating. And the other thing is that you use visible light. So, the absorption by water is not so strong. Maybe other molecules will absorb it. But the water directly? I am not so sure. This is an issue that you really heat the sample.

Stefan Hell: If I may add to Bernd Rieger's reply, I think that at the time Winfried Denk introduced two-photon microscopy, this has been discussed extensively. And I did a back-of-the-envelope calculation back then, as Winfried did. And we found that in the near infrared, which of course, has a much, much higher absorption, 100 times higher absorption than in the visible range, even the much, much higher average power that was used in former microscopy didn't induce much heating. I think that we are fortunate that in the visible range water absorption is low, as Bernd just mentioned, and so I don't think that heating will be a major issue. And I am saying that very consciously. Especially in living cells. Why? Because there is a lot of water around and water is a decent heat conductor. I am not saying a perfect heat conductor. A decent heat conductor. And so, just as an anecdote: we imaged, we tried to image a certain protein within mitochondria. We did fixed cells because: we said "Oh, let's do fixed cells! Because the fixed cells are sort of more forgiving when it comes to that". Later we learned we should have done it with living cells. Because in living cell conditions, the situation is even better! In terms of the heat conduction and the rest of it. Because of the abundance of the water around. So, I wouldn't subscribe to that. And I would also like to add: what happens is, that you want to see the molecules, the individual molecules. And as we mentioned we are now down at the highest spatial resolution. We are still losing out molecules that we bleach, that includes PALM/STORM quite heavily, because 50% or so of the molecules, fluorescent molecules, are not detected because we don't get enough emitted photons, localization photons, to detect them. So you throw these molecules away because you cannot localize them. And MINFLUX also helps in that regard since we need fewer fluorescent emissions or fewer fluorescent detections, so you discard much fewer molecules because you need fewer emissions so you can keep many more of the molecules. What I am saying by this is: there is progress in the field. Four years back I heard colleagues giving talks like: you can either have high resolution or you can have high speed or you can have, I don't know, a live cell, but not at the same time all of it. And this is not true! I am saying human ingenuity should not be underestimated. MINFLUX pushes all these demands together and so you can have now a higher resolution with fewer emitted photons and a better live cell compatibility. So, I really trust that human ingenuity will come up with solutions for these questions and I don't think this story is finished yet.

Matthew Lew: About the first question: I think a paradigm that can change our field dramatically with regard to perturbations is actually thinking about molecules, not solely as light bulbs that you hang off of a bio-macromolecule to figure out where it is, but actually as sensors themselves. Just like biology has leveraged proteins and other sorts of biomolecules to do the same. So, just imagine like in the case that I showed you where Nile red is actually not hanging off of amyloid at all. It is transiently bound with a life time of 10–15 milliseconds. We don't necessarily expect that to be too perturbative although you can argue that continuous bombardment of Nile red may be a bit perturbative. But just imagine a regime of imaging where these fluorophores are actually detecting the target that you want to image and give you a flash of light or a change in life time, or whatever else you can read out with our imaging technology and use that as a proxy for looking at activity of water or whatever. I think that could really change the paradigm that we operate in. So, we are looking for a functional fluorophore, not just a light bulb.

Stefan Hell: If I may add to Mark's question, I think there is a big, say, physical question that is still out there, and which means that I may be wrong when I state that physics cannot contribute anything to the field anymore. The question is: can we do away with the fluorophore at some point? This is a big question out there and this is a Nobel Prize worthy question if that is addressed. Can we do away with the fluorophores? Can we get a contrast modality that doesn't require labeling, yet a specific one. It must be something that gives enough contrast. It's very hard, but that is something that would make a big difference of course.

Eva Nogales: I just wanted to have a conversation about what we define as resolution. Especially now that you're in a regime of accuracy that claims to be smaller than the molecule that you are visualizing. Because, obviously, it has a very different meaning for you guys than it has for a structural biologist that is doing cryo-electron tomography. If you say you have 2 nanometer resolution, and you, for example, have a microtubule, which is an object that you guys like a lot, the repeat of your labeling is 80 ångström, or 50 ångström in the shortest direction. If you have a resolution of 20 ångström, you will be able to define the structure of the microtubule, the grooves of the outside, the striations at an angle in the lumen. Obviously that is not happening. So, I think it is very important, and I only mention this because this is so obvious, right? Because at a certain point, and I think it was Bernd, you used an expression of structural biology and I just think it is just not what you mean, because if you were doing structural biology at 1 nanometer — two nanometer resolution of those volumes, it would be truly spectacular. Even if you were just looking at, say, 1 fluorophore, so one structure, but this is not the case. It is not possible to have a structure at a certain resolution if you are not sampling the molecule with twice that resolution. So, it has a different meaning. And it is fantastic! But it has a different meaning. So I think this is important.

Bernd Rieger: So, I can get the problem here. The point of "what is resolution really?", that is not so easy because in the end we acquire points and from these points we want to infer some information. So now there are maybe two ways out that are used. One is called Fourier ring correlation or Fourier shell correlation, which is also used in electron microscopy. So, you have a look at the decorrelation of the signal. And there is another way out that also has a look at decorrelation of phases across the image. You have a look on what is the information content of the image. And that is typically what is stated. So, that does not tell you exactly where the resolution comes from: it comes from the localization precision or the sampling density. It just has a look at what is the information content in the image. That is exactly how you do it also in electron microscopy. That is, if you compute it for your shell correlation, that is what you do. This at least gives you a number what you can infer from the image.

Eva Nogales: So, when you do that kind of Fourier shell correlation, you do it with the data that basically is a localization that, once you have done it through the center of your Gaussian or whatever, is what? A single pixel? In your system, each one of your fluorophores gets converted to a single pixel or something like that? That is the accuracy with which you decide what is the pixel in the image that it comes from. Is that correct?

Bernd Rieger: No, that is not needed. So, there is no need to pixelize the image. So, once you have the localization, that is, coordinates with given uncertainties, you can compute the resolution. So, there is no need to go back to a pixelized or voxelized representation.

Stefan Hell: I think I can sort out this confusion once and for all. We must not forget, as I said, that a fluorescent microscope images nothing but the fluorescent molecules. If I have 2 fluorescent molecules, 2 nanometers apart and I can tell them apart and say: one is here, the other one is there, I have two-nanometer spatial resolution. There is no discussion about it. What you are alluding to is the fact that people don't show images, say, with this 2-nanometer resolution, of complex objects. Why is that the case? Like the microtubules or something. Because the labeling is not there. But that is not a problem of the microscope. It is a problem of the labeling. So what needs to be tackled now, actually, is the labeling. Getting the fluorescent molecules in the right place. I wish I could have showed you images that are extremely dense with everything. I promise you: I could have shown you those. My system would have been able to reveal it, if I had had the labeling. The labeling is the problem. But this is why it is completely correct to say the microscopy per se can relate this resolution, can get this resolution. But the labeling is still years behind and we have something to do about that.

Bert Weckhuysen: It is fantastic what you have shown about this resolution. Even when I just heard about time and space resolution, 10 milliseconds and 1 nanometer. That is fantastic. And most of the examples of almost all the discussions so far, has been about life science applications. So, I am just curious: material science has, although with a different pace and also with a bit lagging behind life sciences, has tried to use fluorescence microscopy to apply it, and even single molecule has already been applied to a certain extent. If you want now to apply or use these methods, so these approaches you have shown today here. What should it take to have it also applied in materials, functionals materials, functional nanomaterials. And I would like then also to go away from water and away from room temperature conditions. What are then the challenges? I think they are numerous, I believe.

Bernd Rieger: So, people have done that. Maybe Johan Hofkens (KU Leuven) here in Belgium and also Christy Landes (Rice University) have used fluorescent microscopy single molecules, so to tackle that. To have a look...

Bert Weckhuysen: I know that. I know these works. I have worked with Johan Hofkens and we did it also ourselves. So, I am using single molecule fluorescence myself in my group or in collaboration. But that work, if you look very carefully, it's always under what I would call, very mild conditions far away. The layered double hydroxide (Nature paper), it was room temperature, single molecules which are still far away from real life applications in functional nanomaterials. So, I am not criticizing that work. I am trying to give to you another dimension. If I want to do 200°C, the water is gone. I take as a solvent cyclohexane and I take a material which has a refraction index which is also much more cumbersome as what you do have now. What then?

Bernd Rieger: Yes, what then? You would need a fluorophore, a probe, that would be able to go into this on/off stage at that temperature. Because in principle, the microscope, we can make a small chamber. You can heat it up. That's fine. And the light can still travel through as long as the sample is transparent. So, you need a fluorescent marker that would show this on-off blinking kinetics at your ambient temperature. So, that would also mean it should also work at that buffer condition. So, what we did not talk about: most of these molecules don't show this on-off transition just if you throw them in water. We have to add certain chemicals to make that work. Probably all of us here, from a bit or more the physics or data side, we would not know how to make the fluorophores. But maybe you would.

Stefan Hell: I think this highlights very clearly that, in the end, this super-resolution microscopy is all about fluorophore states. And if you don't get the states that you need in order to make this separation, it won't work. So. In the past, light microscopy was all about making the light focus sharp. But now, it is all about the molecular states. And of course, as I said, if you don't get them, if you don't get your molecules on-off as Bernd just mentioned under these conditions, with certain kinetics, with a certain on and off time, you're lost. The power of the microscope is a feature of the fluorophore.

Bert Weckhuysen: Some of the reagents in catalytic conversions are aromatic compounds. Hence you should be able to have, provided that there are good side groups, something that is at least close to or even in the visible region. With the laser you should be able to get it somehow excited. No?

Stefan Hell: You get them excited. But at lower temperatures there are many states populated that don't get populated at room temperature. Like, say, all kinds of hidden dark states. Triplet states are being built up. The triplet states have a very much longer life time. There is little or no oxygen diffusion that decreases the life time of the triplet states and so on. And there is little experience of course with fluorophores in that wavelength regime. And under those same material, say, molecular environments, let's put it that way. And if you want to sort it out, you can. But you have to sort out the states of the fluorophores under your conditions. And once you have sorted it out, you can take the images. I am sure. But no one has done that yet because, if I may say that: it's a highly specialized application. Life science, of course, there are thousands of labs out there. And of course, the need for coming up with fluorophores that meet those conditions has been very high traditionally. And therefore they are there, those fluorophores. There have been GFP's. There have been switchable GFP's, activatable GFP's, all kinds of fluorophores that have been used by biologists for ages. But I am not so much aware of fluorophores that have been optimized for low temperature, I don't know what, ceramics or things like that. But once that is sorted out, you go.

Bert Weckhuysen: To finalize because we can't continue on that. Yes, life science is an enormous field. I agree. You can see it in the budgets of the National Science Foundation, then you know it. But if you would add up batteries, fuel cells, catalysts,... it is also a rather massive investment. And if you think about sustainability where all the countries are now spending money on, you would, and also the batteries and all that, you would expect that at some point also that would deserve the same type of attention. That is all I wanted to say. Thank you.

Mischa Bonn: I completely agree. And I'd like to advertise that there is a fluorescent molecule that we discovered by accident. That doesn't require a specific blinking buffer to operate. It just works! It works in air, it works in liquids, it works in apolar liquids, it works in water, it works under all conditions. So, we haven't checked the temperature dependence which would be relevant for you. But at least there is a marker that we envisioned to work in material sciences.

Bert Weckhuysen: I am very much interested in that molecule!

Mischa Bonn: I will send you the paper.

Mark Ellisman: I wonder, the live imaging, how you get deep, is a grand challenge. And there is some promise from something called photo acoustics. Where you excite a fluorophore with light and you read out the signal with hydrophones as a 3D-localised calculated source by photo acoustics. It's an old phenomenon described by Alexander Graham Bell actually to begin with. Washington University had a big activity there until Caltech stole your prince. So, I wonder if, within the expertise here, the computational expertise, the localization microscopy expertise, if you can think of a way that one can use for photo-acoustics: one, by pulsing the activation, two, by doing some kind of refinement of the localization of the acoustic signal, to make a hybrid technology that would improve resolution in deep imaging.

Matthew Lew: If I may just put one sidenote before we address for the photo-acoustics directly. A very very hot problem in optics right now is inverting scattering. Tissues... if we, let's say subtract the blood movement. The scattering of light through a complex medium is deterministic and that is in principle an invertible problem. And there are many groups working on modulation detection, of characterization of complex media, focusing through egg shells, through chicken tissue, a whole bunch of things. To actually get light in and get light out very, very deep. Photoacoustics has already surpassed the limit of what pure optics can do, but it is not a hopeless cause to get light in and out of tissue. And that is an active problem that people are working on.

Jan Huisken: I would like to add that there's a huge potential in doing multi model imaging. Not just to bridge scales, but also to use one technology to improve the other technology. Ideally using one technique, for example holographic techniques, to learn more about the refractive index of the specimen, and use that information to tailor your light sheet or trim your adaptive optics. I think that is definitely a huge potential for us and the field to increase penetration or increase resolution. And we not just focus on one technique but use many techniques ideally on the very same sample. So we've used optical projection tomography to gain additional information about this specimen and it works really well in combination with light sheet microscopy. I mentioned holographic techniques, which I consider very valuable, especially learning more about the optical properties. Photo-acoustics is probably more on the larger scale/lower resolution end of things. But I think once, maybe before we start clearing the sample that could be a good alternative. I think there is a huge potential in combining different techniques on the very same sample. And when the techniques are fast enough and not invasive, there is usually enough time, at every single time point, to acquire multi-model data.

Yoav Shechtman: Just to continue the discussion about photoacoustics and how it relates to what I have shown. You have seen the point spread function engineering in my talk where we modify the way that the point source appears on the camera to encode the depth of the point source. You can turn this around and you can do this in the illumination side. That would mean that you could engineer the illumination such that you would overcome this scattering in the medium, which is deterministic as Matthew said, such that it will focus inside. So, there is this relation.

Mark Ellisman: Can you work on the detection side in a similar way with the acoustic output?

Yoav Shechtman: Well, I don't do acoustics. But I think, the main difference is, that it is a coherent detection. I have to think about it more. In principle, to some extent you can probably do it. But I think that if you know where your emission is originating from, then that is all you need. Because your localization comes from the illumination. Similar to MINFLUX for example. So as soon as your localization information is from the excitation, you don't really care what the scattering looks like, as long as you collect everything.

Stefan Hell: Exactly like in STED, where the position of the emitter is determined. In fact, there have been attempts. People have attempted to combine sort of STED kind of things with photoacoustic detection. But I haven't seen any decent data so far.

Bernd Rieger: In acoustics, there is sometimes now use of what is called super resolution. That works when they use injected small bubbles or scatterers. Because the speed of sound is so slow and if they use a very fast camera so that they can see the signal arriving at the camera, they get the same sort of sparsity that we get in the localized image, they get it also at the camera and it works exactly the same way. So they achieve this "on/off" because the sound travels so slowly and they have a very fast camera, but the idea is the same.

Eva Nogales: I have a question concerning averaging. So this is for Bernd again. Especially because the nature of the data that is used in electron microscopy and the purpose of the averaging are so completely different. I just think that there must be better ways of designing your own software with having in mind what the issue is. Because in cryo-EM we are just limited by signal to noise ratio. And we can assume that the molecule is always the same, that the signal is the same, the noise is not. And that is how we, as long as we can do accurate alignment, we are going to recover the structure. But in your case that is not the issue. You don't have noise. You are in no way limited by that. But maybe you are limited by the relative position of one label in one molecule with respect to the equivalent label for an equivalent molecule or, as you were you saying, the sampling. So, I think you probably could do much better. Because the cryo-EM will even blur things when you don't need to. And I think you would benefit from a software that is specifically dedicated to your type of data. What do you think?

Bernd Rieger: I would have hoped that you would have been one of the reviewers of one my papers. That would have helped me a lot. That is because we are the only ones who do not use EM algorithms, but we did exactly what you just asked why we did *not* do that. I did not show any technical details how to do it. But I can assure you that we did it very, very differently than how this is done in electron microscopy because, the point is, the software works if the images are relatively ok. But the output result is not as good as you should get and it is also in the end what we showed. If you make a dedicated algorithm that is tuned to the image formation, that is different. There are so many differences. I can tell you how we did it. But I thought it was a bit too tedious to go through the mathematical model how to deal with that.

Eva Nogales: It is because I saw that the EMAN software package for cryo-EM image analysis is specifically in one of your slides.

Bernd Rieger: We used this to show that we could do better than EMAN on this data.

Kurt Wüthrich: In the formal presentations we got descriptions in detail of peak picking, moving target peak picking, deconvolution of peaks. These routines are widely used in all sorts of spectroscopies that chemists and biologists use nowadays. What is unique here, in addition to all that? What do you need in addition to this routine set that is widely used in spectroscopy? It seems to me that deconvoluting two peaks in your datasets is directly related to the resolution of the method. This is not the case in, say, NMR, or in infrared or Raman spectroscopy. What do you need, in addition to these widely known techniques, for the analysis of your data?

Yoav Shechtman: When you think about the problem of localization, of single molecule localization microscopy, it's a parameter estimation problem rather than a deconvolution problem. You have a very strong, very strong, prior information, which is that you're looking at a single point. And not at some unknown structure that you now have to deconvolve with your impulse response. In a sense, it is much easier. It transforms the problem to a bunch of very easy problems, instead of one complicated one. I hope that answered something.

Stefan Hell: I fully agree with Yoav. In a way, the mathematical treatment becomes more powerful because the object is becoming more simple, it is becoming binary. It's basically molecules, that's for sure. If there is an entire new mathematical formalism required, is that specific to our field? I am not so sure about that. I am not saying it is not going to happen. But in a sense, you are right. The vast majority of the mathematical formalisms are there and have been around and applied to other problems. But still, somebody has to be aware of them, take them and transform them to this field, to make them useful and powerful. And we are at the beginning. Think 10 years back, 20 years back. People, especially physicists, said, you have to decompose your object into spatial frequencies. And then you have to check out if the system transfers out all the spatial frequencies. And then you recompose again the spatial frequencies to get your image and see what is lost. This is what has been the mantra throughout the 20th century. This mantra is not, this model is not, this way of perceiving the light microscope is not useful anymore because we are dealing with discrete objects.

Andreas Walther: There is a lot of magic which always goes into sample preparation. You fix cells and so on. And my question would be: in classical confocal microscopy, the cell is considered as fixed. If you really go down to 1 nanometer resolution, how much does actually Brownian motion smear out the resolution which you need? Would that mean that the imaging should be done at say colder temperatures or do you need new fixation methods and so on to also improve resolution?

Yoav Shechtman: If you have a single molecule that is fixed then it emits as a dipole. And then you have some bias. It turns out that you have a biased estimation of its position because of the shape of the radiation. If the molecule is wobbly enough, which is typically the case, then it all averages out and then you can localize it better. Maybe Matthew can say more about that?

Matthew Lew: The typical rotation or correlation time of fluorophores in water is nanoseconds. And our integration times are, at least even in the case of MIN-FLUX, a bit longer, by an order of magnitude. So, we're going to average out the rotational diffusion, which is key for making accurate localizations. On the translational diffusion side, yes: we are going to be averaging over, whatever integration window the position of the fluorophore and all the emission locations in which that fluorophore was, into a final measurement. And so, until Stefan's development of MINFLUX we hadn't quite hit that Brownian motion limit. Now that needs to be an active modeling process just like the linker spacing between the fluorophore and the biomacromolecule. These are all effects that need to be integrated to produce higher fidelity images on the nanoscale.

Thomas Hermans: I am just wondering if you can use circularly polarized fluorescence emission and, for example, have dyes that are right- and left-handed. Or to increase the information density in the separation. If you can imagine two dyes that are very close to one another, but you can separate them due to circularly polarized emission. Would that help you in your case?

Jörg Enderlein: You don't even need circular polarization. You have basically a dipole emitter which is a linear, highly linear polarized emitter. We have worked out, some time ago, the theory whereby you can use polarization for distinguishing molecules. You can, but unfortunately only two. Beyond two, this is not possible any more. I mean, if you have only two emitters in one spot, you can use polarization information to really disentangle the emission of both molecules and also the position of both molecules. That's an option. Unfortunately, light has only two polarizations so it's only two molecules.

Stefan Hell: There have been attempts. There has been a paper in Nature Methods. Basically, they tried to use different orientations as a means to separate, but it didn't go well. Let's put it that way.

Thomas Hermans: Maybe, to make it more complicated, if you would consider Mueller matrix polarimetry. You mix all, not just circularly polarized states, but all polarized states, which have been shown to be, you can quite easily deconvolute that, for example using photoelastic modulators. Would that then resolve your limit of just two states?

Jörg Enderlein: When you look from a group theoretical point of view and usually develop your emission into spherical harmonics and physically the n to the second order, the second principal orbital number. There can never be anything higher than $L = 2$ and that's the limit. When you then do the mathematics, you will see that you can only disentangle in the end two molecules. If the emission would be higher order, let's say it would be non-linear, and this was the paper where Stefan was alluding to, if you use non-linear excitation and then you can basically work with order four, six and so on, then people claim that they can better resolve more molecules on one spot. But as Stefan already told, in practice experimentally that was not really convincing.

Matthew Lew: Because we can take this limitation of two fundamental dipoles, being able to be resolved and even make the imaging system more powerful in the sense that even our integration times are on the order of milliseconds, we can now analyze how much wobble was there of the fluorophore during every individual acquisition and give you a mean orientation and a wobble cone angle. In addition to the fluorophore position, to really understand those dynamics. That's a feature that is still being explored.

Eva Nogales: Can I ask a biology question? This is for Matt. It's concerning your amyloid fibrils. I think I miss what you were imaging. Were you imaging brain sections or something, neurons growing on a substrate? That is not my question. That's part of the information that I need to get. You were talking about looking in time at the sample. You have shown before this sigmoidal behavior in terms of fibril formation. I was wondering whether you were looking at a particular window, where you had triggered something, and you knew in which window to look. How much were you covering? What was the resolution, the temporal resolution and things like that? I would like to have more detail please.

Matthew Lew: Right now, where we are is starting, we are developing a new technique to understand the beta sheet structure of these amyloid fibrils. We started with just taking the peptide, purifying it, monomerising it and allowing it to aggregate in physiological buffer and temperature conditions on glass. The temporal resolution of our technique: we need to integrate all these molecule flashes over a period of a few minutes, but because there is continuous replenishment of new fluorophores from the bath, we essentially have unlimited sampling. We have looked at the activity of EGCG, so called catechin that has been shown to dissolve amyloid fibrils. We looked at that over a period of two days and watched these fibrils actually dissolve. Now, that is the published work. What we are currently working up right now is imaging growth of the fibrils in real time and also looking at laser-induced damage. Those are just again, examples of making sure that we can actually see structural transformations that we can't pick up with other techniques.

Stefan Hell: Since there are no further questions: thank you again for participating in this discussion. Thank you all the panelists. And so, there is now time to have a break and relax.

Session 5

Modeling of Non-Equilibrium Systems and Simulation of Molecular Machines

NON-EQUILIBRIUM SYSTEMS AND MOLECULAR MACHINES

BEN L. FERINGA

Stratingh Institute for Chemistry & Centre for Systems Chemistry,
University of Groningen, Nijenborgh 4, 9747 AG Groningen, The Netherlands

Introduction

The move from molecules to dynamic functional molecular systems that operate out of equilibrium sets the stage for a new era in the chemical sciences. It confronts us with the multifaceted challenge of how to control among others information processing, reaction networks, motility, and adaptive and autonomous behaviours intrinsic to life itself. Molecular motors are key to the functioning of biological systems and offer a great source of inspiration toward the design of artificial molecular machines capable of inducing motility and maintaining far-from-equilibrium capabilities. However, the future design of molecular motors and more general non-equilibrium dynamic systems requires computational methods that go far beyond analyzing simple model systems to the prediction of multicomponent dynamic systems and responsive functions while guiding the synthetic chemist in this largely uncharted territory. Challenges and opportunities for the "molecular motorist" will be briefly discussed.

Taking inspiration from living systems, that use molecular motors in nearly every essential function ranging from protein synthesis, transport, and cell division to muscle movement or the process of vision, the molecular designer is confronted with a quest of how to exploit non-equilibrium behaviour and synthesize systems that undergo controlled motion. Understanding and mimicking the basic principles of the fascinating biological machines [1] and the complex dynamic chemical networks [2] of living systems will not only provide crucial insights to advance chemical biogenesis [3], not hampered by the building blocks selected during evolution of life on the early earth, but perhaps more importantly also provide ample opportunities for future smart materials and novel chemical functionality [4]. Taking the next major step, it is rewarding that impressive breakthroughs have been realized in recent years in molecular recognition and sensing [5], dissipative self-organization [6, 7], molecular switches, motors and machines [7], and biohybrid systems (enzymes, DNA, etc.) [8] to sustain responsive functional molecular systems. However, toward the future design of molecular systems and facing its complexity, a merging of experimental and theoretical approaches is essential and, relying on major advances in computational methods, in an iterative process predicts better models as well as understanding and fully exploring motion and constructing complex dynamic networks.

Present State of Research and Our Contribution

Targeting dynamic molecular systems that ultimately operate far from equilibrium and show autonomous behaviour, one has not only to take into account structure, reactivity, and specific functions but organization, multi-molecular nature, spatial–temporal control, interface phenomena, and hierarchical levels are also key parameters. In order to control spatial–temporal behaviour, programmable reaction networks have been recently designed and the complex kinetics analyzed [9]. Important challenges include the selection of components, the predictive power of mathematical modelling, kinetic networks and circuits, feedback loops, transmission, amplification, and adaptation.

The field of molecular machines and motors has seen remarkable progress since the dedicated Solvay conference on this topic [7] and, besides major advances in design of mechanical interlocked systems (rotaxanes, catananes) and chemical and photochemical powered motors, the focus has shifted to dynamic functions exemplified by responsive materials, adaptive catalysts, pumps, and muscles [7, 10, 12]. Following our discovery of light-powered rotary molecular motors, recent efforts have resulted in control of speed, visible-light-driven motors, and control of motion in liquid crystals, polymers, as well as organization of motors on surfaces and in crystals (MOF's) tackling the fundamental problems of amplification and cooperative action [11, 12]. DFT methods in combination with transient spectroscopy have been shown to be crucial in elucidating isomerization pathways, whereas mapping the excited state potential energy surface is one of the key challenges to facilitate both better control and design of future generation motors [13]. Computational methods have also suggested new-generation motors with improved efficiencies [14]. Dissipative catalysis with rotaxane-based machines and rotatory motion powered by catalysis are based on ratchet mechanisms reminiscent of biological molecular motors. The theory analysis on the key role of microscopic reversibility, substrate binding, and catalytic conversion governing directionality has been reported [15]. Similar schemes apply to metal-redox catalysis driving rotary motion [16] (Fig. 1).

In contrast a power-stroke mechanism is operating in the case of a light-driven unidirectional rotary motor.

Considering responsive systems, the distinct differences between molecules in an equilibrium state, molecular switches allowing access to metastable states, and genuine motors enabling far-from-equilibrium states need to be emphasized. Molecular assembly is another fascinating area where "moving away from equilibrium" has made remarkable progress [7, 10, 12, 17]. Light-driven rotary motors controlling responsive polymers, supramolecular foldamers in out-of equilibrium states, and rotaxane-based molecular pumps illustrate how dynamic functions can be used to drive assembly of materials far from equilibrium. The assembly in confided space, i.e., 2D- and 3D-like metal organic frameworks (motor MOF in Fig. 2), provides a means to address pertinent issues like cooperativity and synchronization of motion [18]. Also, dynamic systems involving self-assembly–disassembly driven by catalysis

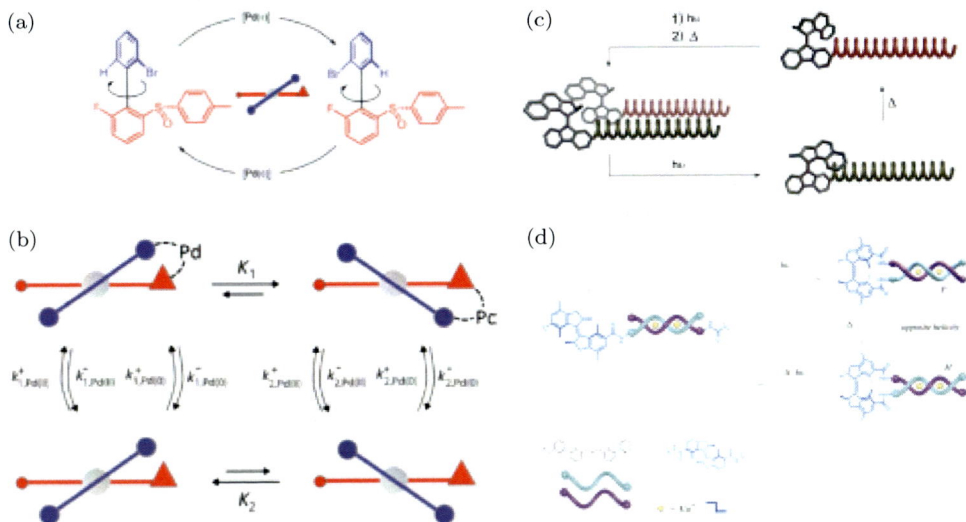

Fig. 1. Control and amplification of rotary motion: (a) structure of redox-driven unidirectional motor; (b) rotary cycle based on palladium metal redox catalysis; (c) responsive polymer driven out of equilibrium with light-powered rotary motor; (d) responsive supramolecular helicates (adapted from Ref. [12]).

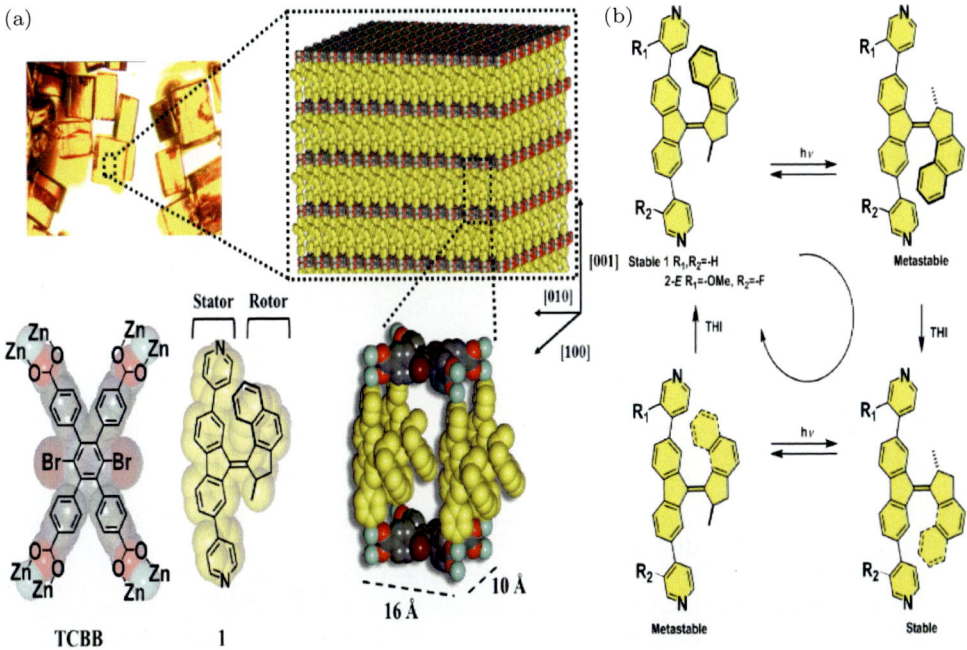

Fig. 2. A unidirectional light-driven rotary motor in confined 3D metal organic framework (adapted from Ref. [18]).

or redox processes, as demonstrated recently in transient induced catalyst formation in a dynamic network [19] and by self-oscillation across length scales in polymer gels to induce peristaltic motion [20], illustrate the fascinating opportunities to couple chemical conversion to control assembly. It should be emphasized that molecular dynamic simulations, as widely applied in membrane assembly and protein channel function among others, offer crucial insights into dissipative self-assembly processes.

Brief Outlook on Future Developments

Bridging the fields of non-equilibrium chemistry and molecular machines, the interplay of computational and experimental approaches will guide the explorer toward complex dynamic molecular systems featuring out-of-equilibrium behaviour. Enhancing the predictive power to better understand (and apply) specific molecular interactions, solvent effects, and dynamics is key to facilitate the choice of components, design multi-molecular systems and coupled catalytic reactions, tune kinetics, and control (dis-)assembly. The prospects toward out-of-equilibrium reaction networks, autonomous systems, and responsive materials are particularly bright, ultimately showing lifelike function. Specifically, for molecular motors and machines, detailed insight into isomerization processes and the excited state energy landscape as well a simulation of molecular machine function will provide invaluable information to design more advanced and alternative systems such as redox- and catalytically driven motors. Addressing the challenges to achieve collective behaviour, synchronization, amplification along length scales, mechanical work, and specific functions mimicking biological machines involved in, e.g., transport, muscle movement, adaptive behaviour, or energy storage, provides fascinating prospects for molecular machines, smart materials, and lifelike systems.

Acknowledgments

Support by the Netherlands Ministry of Education, Culture and Science (Gravitation program 024.001.035) and the European Research Council (Advanced Investigator Grant No. 227897) to B.L.F is gratefully acknowledged.

References

[1] D.S. Goodsell, *Our Molecular Nature: The Body's Motors, Machines and Messages* (Copernicus, New York, 1996).
[2] S. Mann, *Angew. Chem. Int. Ed.* **47**, 5306 (2008); J.M. Berg, J.L. Tymoczko, L. Stryer, *Biochemistry*, 5th edn. (W.H. Freeman, New York, 2002).
[3] N. Kitadai and S. Maruyama, *Geosci. Front.* **9**, 1117 (2018).
[4] W.R. Browne and B.L. Feringa, *Nat. Nanotechnol.* **1**, 25, (2006).
[5] B. Wang and E.V. Anslyn, *Chemosensors: Principles, Strategies, and Applications* (John Wiley & Sons, 2011).
[6] B.A. Grzybowski and W.T.S. Huck, *Nat. Nanotechnol.* **11**, 585 (2016); E. Mattia and S. Otto, *Nat. Nanotechnol.* **10**, 111 (2016); S.A.P. van Rossum, M. Tena-Solsona, J.H. van Esch, R. Eelkema, and J. Boekhoven, *Chem. Soc. Rev.* **46**, 5519 (2017).

[7] J.-P. Sauvage, *Molecular Machines and Motors. Structure and Bonding* (Springer, Berlin, 2001); V. Balzani, A, Credi, and M. Venturi, *Chem. Soc. Rev.* **38**, 1542 (2009); A. Coskun, M. Banaszak, R.D. Astumian, J.F. Stoddart, and B.A. Grzybowski, *Chem. Soc. Rev.* **41**, 19 (2012); B.L. Feringa, *J. Org. Chem.* **72**, 6635 (2007); S. Kassem, T. van Leeuwen, A.S. Lubbe, M.R. Wilson, B.L. Feringa, and D.A. Leigh, *Chem. Soc. Rev.* **46**, 2592–2621 (2017).

[8] F. Wang, B. Willner, and I. Willner, *Top. Curr. Chem.* **354**, 279 (2014); X. Ma, A.C. Hortelao, T. Patino, and S. Sanchez, *ACS Nano* **10**, 9111 (2016).

[9] H.W.H. van Roekel, B.J.H.M. Rosier, L.H.H. Meijer, P.A.J. Hilbers, A.J. Markvoort, W.T.S. Huck, and T.F.A. de Greef, *Chem. Soc. Rev.* **44**, 7465 (2015).

[10] J. Wang, B.L. Feringa, *Science* **331**, 1429 (2011); C. Cheng, P.R. McGonigal, S.T. Schneebeli, H. Li, N.A. Vermeulen, C. Ke, and J.F. Stoddart, *Nat. Nanotechnol.* **10**, 547 (2015).

[11] R. Eelkema, M.M. Pollard, J. Vicario, N. Katsonis, B.S. Ramon, C.W.M. Bastiaansen, D.J. Broer, and B.L. Feringa, *Nature* **440**, 163 (2006).

[12] B.L.Feringa, *Angew. Chem. Int. Ed.* **56**, 11060 (2017).

[13] A. Cnossen, J.C.M. Kistemaker, T. Kojima, and B.L. Feringa, *J. Org. Chem.* **79**, 927 (2014); C.R. Hall, W.R. Browne, B.L. Feringa, and S.R. Meech, *Angew. Chem. Int. Ed.* **57**, 6203 (2018).

[14] A. Nikiforov, J.A. Gamez, W. Thiel, and M. Filatov, *J. Phys. Chem. Lett.* **7**, 105 (2015).

[15] C. Pezzato, C. Cheng, J. F. Stoddart, and R.D. Astumian, *Chem. Soc. Rev.* **46**, 5491 (2017).

[16] S.P. Fletcher, F. Dumur, M.M. Pollard, and B.L. Feringa, *Science* **7**, 890 (2005); B.S.L. Collins, J.C.M. Kistemaker, E. Otten, B.L. Feringa, *Nat. Chem.* **8**, 860 (2016).

[17] D. Zhao, Th. van Leeuwen, J. Cheng, B.L. Feringa, *Nat. Chem.* **9**, 250–256 (2017); J. Chen, F.K.-C. Leung, M.C A. Stuart, T. Kajitani, T. Fukushima, E. van der Giessen, and B.L. Feringa, *Nat. Chem.* **10**, 132 (2018).

[18] W. Danowski, Th. van Leeuwen, S. Abdolahzadeh, D. Roke, W.R. Browne, S.J. Wezenberg, and B.L. Feringa, *Nat. Nanotechnol.* **14**, 488 (2019).

[19] H.-Virgos, A.-N.R. Alba, S. Hamieh, M. Colomb-Delsuc, and S. Otto, *Angew. Chem. Int. Ed.* **53**, 11346 (2014).

[20] R. Yoshida and T. Ueki, *NPG Asia Mater.* **6**, e107 (2014).

THE IMPORTANCE OF CORRECTLY MODELLING CATALYSIS FOR UNDERSTANDING MOLECULAR MACHINES

R. DEAN ASTUMIAN

Department of Physics and Astronomy, University of Maine,
Orono, Maine 04469, USA

Present State of Research on Modeling Catalysis-Driven Molecular Machines

There has been much progress in understanding molecular machines at the single molecule level, where experiments have illuminated the stepping of molecular walkers such as kinesin and myosin [1], details of how the ribosome operates [2, 3], and have provided key insights into the rotation of the flagellar motor and FoF1 ATPase [4]. Computational studies [5] have greatly enhanced the understanding of how structure plays an important role in determining the function of these molecular machines and have shown that many if not all enzymes undergo directional motion during catalytic turnover [6]. One aspect necessary for complete understanding however has lagged behind — the incorporation of chemical catalysis into the developing theoretical descriptions of molecular machines.

Many biomolecular motors — muscle, kinesin, FoF1-ATPase, etc. — are catalysts that facilitate hydrolysis of adenosine triphosphate (ATP) to inorganic phosphate (P_i) and adenosine diphosphate (ADP), ATP $\overset{K_{eq}}{\rightleftharpoons}$ ADP + P_i. Unfortunately, the details of the catalytic mechanism by which chemical fuel is used to drive directed motion and the performance of mechanical or electrical work are often ignored or treated in an at best cavalier fashion by indicating hydrolysis with a lightning bolt and the words "fuel event" [7] or "energy input" [8], and describing the effect of ATP hydrolysis as "providing violent kicks to the motor" [9], all combined with the assurance that the reaction is far from equilibrium. This sloppy treatment is often rewarded by facilitating analysis of "theoretical" models that are consistent with plausible but ultimately incorrect expectation based on macroscopic devices or with experimental results for light-driven or externally driven molecular machines. This apparent consistency leads to the acceptance of a picture in which a conformational change of the protein known as the "power stroke" seems to be important in determining the intrinsic directionality and thermodynamic properties (efficiency, reliability, stall force, etc.) of catalysis-driven molecular machines, including myosin, kinesin, FoF1-ATPase, and the flagellar motor, a conclusion that is further reified by experiments and simulations that focus on individual steps of various molecular machines rather than on complete cycles and which unequivocally show that many biomolecular machines do in fact have "power strokes".

My Work on Thermodynamically Consistent Modeling of Catalysis Coupled to Mechanical Motion of Molecules

Using a thermodynamically consistent description of catalysis, and treating complete cycles of motion and chemistry, I have shown that the idea of a power stroke, while correct for light-driven and externally driven molecular devices, is incorrect as a description of the *mechanism* of any catalysis-driven molecular machine [10].

Kinetic asymmetry

Consider the Michaelis-Menten mechanism

$$S + E \; \underset{k_{-1}}{\overset{k_{+1}}{\rightleftharpoons}} \; E_L \; \underset{k_{-2}}{\overset{k_{+2}}{\rightleftharpoons}} \; E + P$$

Scheme 1.

for catalysis of the reaction $S \overset{K_{eq}}{\rightleftharpoons} P$. The bound state is written as E_L rather than the more traditional E_S in acknowledgement of the fact that up binding, the participants in the chemical reaction lose their individual identities, as emphasized by Terrell Hill [11]. The association constants for S and P are $\frac{k_{+1}}{k_{-1}} = K_S$ and $\frac{k_{-2}}{k_{+2}} = K_P$ and the rate constants obey the relation $\frac{k_{+1}k_{+2}}{k_{-1}k_{-2}} = K_{eq}$, and hence, $\frac{[S]K_S}{[P]K_P} = e^{\Delta\mu}$. These relations hold irrespective of whether the system is or is not at equilibrium. When $\Delta\mu = 0$, the system will relax to equilibrium where $\left.\frac{[E_L]}{[E]}\right|_{eq} = [S]\,K_S = [P]\,K_P$. In general, however, the ratio between the bound (E_L) and free (E) enzyme is

$$\left.\frac{[E_L]}{[E]}\right|_{ss} = \frac{[S]\,k_{+1} + [P]\,k_{-2}}{k_{-1} + k_{+2}} = \frac{[S]\,K_S k_{-1} + [P]\,K_P k_{+2}}{k_{-1} + k_{+2}} = [S]\,K_S \left(\frac{\frac{k_{-1}}{k_{+2}} + e^{-\Delta\mu}}{\frac{k_{-1}}{k_{+2}} + 1} \right) \tag{1}$$

The factor $\left(\frac{\frac{k_{-1}}{k_{+2}} + e^{-\Delta\mu}}{\frac{k_{-1}}{k_{+2}} + 1} \right) \equiv \mathcal{A}$ was introduced by Astumian and Bier [12] in the context of ATP hydrolysis-driven molecular motors and is termed the kinetic asymmetry factor. When $\Delta\mu = 0$ this factor is unity, but away from equilibrium $\mathcal{A}$ plays an important role in determining the steady-state probability for the enzyme to be bound. Note that for the standard irreversible Michaelis–Menten reaction, $S + E \underset{k_{-1}}{\overset{k_{+1}}{\rightleftharpoons}} E_L \overset{K_{+2}}{\to} E + P$ written for initial kinetics (i.e., with $[P] \to 0$), the catalytic velocity is the standard expression $v_{\text{cat}} = \frac{k_{+2} E_{\text{Total}} [S]}{\frac{1}{(K_S \mathcal{A})} + [S]}$ with the Michaelis constant given as $K_M = (K_S \mathcal{A})^{-1}$.

Catalysis-driven conformational changes

Enzymes undergo conformational fluctuations, where different conformational states have different catalytic properties. For example, the enzyme discussed above might have a second conformational state E^* that can also catalyze the reaction $S \overset{K_{eq}}{\rightleftharpoons} P$

$$S + E^* \underset{k^*_{-1}}{\overset{k^*_{+1}}{\rightleftharpoons}} E^*_L \underset{k^*_{-2}}{\overset{k^*_{+2}}{\rightleftharpoons}} E^* + P$$

Scheme 2.

but with different maximum velocity, k^*_{+2}, association constants, K^*_S and K^*_P, and kinetic asymmetry factor, $A^* = \left(\frac{\frac{k^*_{-1}}{k^*_{+2}} + e^{-\Delta \mu}}{\frac{k^*_{-1}}{k^*_{+2}} + 1} \right)$. The ratios of the products of forward and reverse rate constants must still, of course, equal the equilibrium constant $\frac{k^*_{+1} k^*_{+2}}{k^*_{-1} k^*_{-2}} = K_{eq}$. Transitions $E_L \overset{K_b}{\rightleftharpoons} E^*_L$ and $E \overset{K_f}{\rightleftharpoons} E^*$ allow the enzyme to undergo fluctuation between the two conformational states, with a complete cycle described by

$$\cdots \rightleftharpoons (E)_i \underset{\omega_-}{\overset{\omega_+}{\rightleftharpoons}} E_L \overset{K_b}{\rightleftharpoons} E^*_L \underset{\omega^*_-}{\overset{\omega^*_+}{\rightleftharpoons}} E^* \overset{K_f^{-1}}{\rightleftharpoons} (E)_{i+1} \rightleftharpoons \cdots$$

Scheme 3.

The subscript i is a counter for the number of cycles $E \to E_L \to E^*_L \to E^* \to E$ completed in the forward direction. In principle, the system can be arranged such that the conformational changes store work, w (i.e., lift a weight or charge a battery) in the environment for each such completion and of course will dissipate w for each reverse cycle $E \to E^* \to E^*_L \to E_L \to E$. The equilibrium constants for the conformational changes are modified by the load, where $K_b = K^0_b e^{\alpha w}$ and $K_f = K^0_f e^{-(1-\alpha) w}$, with α partitioning the overall load between the two transitions. The kinetic constants for the catalytic transitions, denoted by $\rightleftharpoons$ rather than $\rightleftharpoons$ for the conformational transitions, are $\omega_+ = ([S] k_{+1} + [P] k_{-2})$; $\omega_- = (k_{-1} + k_{+2})$; $\omega^*_- = ([S] k^*_{+1} + [P] k^*_{-2})$; $\omega^*_+ = (k^*_{-1} + k^*_{+2})$. These expressions can be used to derive the directionality, r, given by the ratio of the products of the forward and reverse rate constants

$$r = \frac{\omega_+ k_{+b} \omega_+^* k_{+f}}{\omega_- k_{-b} \omega_-^* k_{-f}} = \underbrace{\frac{K_S K_b^0}{K_S^* K_f^0}}_{=1} \underbrace{\frac{\mathcal{A}}{\mathcal{A}^*} e^{-w}}_{r_0} \tag{2}$$

where the intrinsic directionality r_0 is the ratio of the kinetic asymmetry factors. There is an enormous difference between equilibrium $(\Delta\mu = 0)$ where $\mathcal{A} = \mathcal{A}^* = 1$, and non-equilibrium $(\Delta\mu \neq 0)$, where

$$\text{for } \Delta\mu > 0 \Longrightarrow \begin{cases} r_0 > 1 & \text{if } \dfrac{k_{-1} k_{+2}^*}{k_{+2} k_{-1}^*} > 1 \\[2ex] r_0 = 1 & \text{if } \dfrac{k_{-1} k_{+2}^*}{k_{+2} k_{-1}^*} = 1 \\[2ex] r_0 < 1 & \text{if } \dfrac{k_{-1} k_{+2}^*}{k_{+2} k_{-1}^*} < 1 \end{cases} \tag{3}$$

The ratio of "off" rates, $\frac{k_{-1} k_{+2}^*}{k_{+2} k_{-1}^*}$ that controls the directionality of the catalysis-driven motor is independent of the free-energies of the states and hence of the values of K_b and K_f. Thus, we conclude that the "power strokes" $E_L \overset{K_b}{\rightleftharpoons} E_L^*$ and $E^* \overset{K_f^{-1}}{\rightleftharpoons} E$ play no role in determining the direction of motion of a catalysis-driven molecular machine. This deductive conclusion is based not on a phenomenological model but on rigorous thermodynamics and specifically on microscopic reversibility. Note that the common approximation based on "local detailed balance" $r \approx e^{\Delta\mu - w}$ is valid only in the limit that $\frac{k_{-1}}{k_{+2}} \gg 1$ and that $\frac{k_{-1}^*}{k_{+2}^*} \ll e^{-\Delta\mu}$.

We can perhaps better recognize the import of this result by considering Fig. 1 which is adapted from Ref. [13, Fig. 4(b)].

In Ref. [13], only the bold-faced arrows were shown. These arrows seem to guide the eye to the conclusion that the motor operates by a power-stroke mechanism, but in fact the figure is a *trompe l'ouile* that fools rather than guides the eye. Bier *et al.* [14] derived a simple expression for the ratio between downhill and uphill trajectories on a free-energy landscape, $P(\mathcal{S}_{\text{up}}) = P(\mathcal{S}_{\text{down}}^{\dagger}) e^{-\Delta G}$ where ΔG is the difference in free energy between the start and end points of the trajectory. This result was derived using the Onsager–Machlup theory for trajectory thermodynamics and is based on the principle of microscopic reversibility. When elaborated for the specific case in Fig. 1, we have

$$\frac{\pi(E_L \rightarrow E_L^*)}{\pi(E_L^* \rightarrow E_L)} = \frac{k_{+b}}{k_{-b}} = K_b = e^{G_b(E_L) - G_b(E_L^*) + w/2};$$

$$\frac{\pi(E^* \rightarrow E)}{\pi(E \rightarrow E^*)} = \frac{k_{+f}}{k_{-f}} = K_f = e^{G_f(E^*) - G_f(E) + w/2}, \tag{4}$$

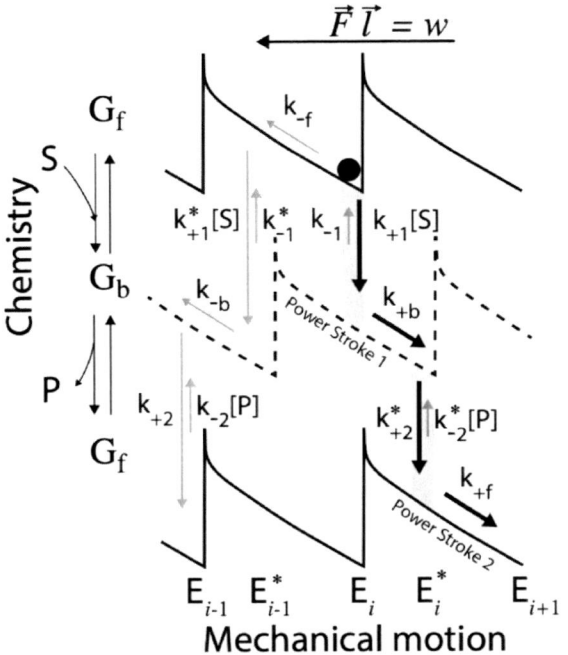

Fig. 1.   Illustration of a "shift ratchet" mechanism for a molecular motor taken from Ref. [11] and based on the flagellar motor where S is proton in the periplasmic space and P is proton in the cytoplasm. The black ball represents the rotor which moves relative to the fixed stator that exerts a force (gradient of the potential) on the rotor. The implication is that chemical energy causes transitions between the "free" (f) and "bound" (b) potentials that are strongly asymmetric and offset by half a period from one another. The sloping potentials then shepherd the rotor to move directionally as indicated by the bold faced arrows. This picture is highly misleading — when rate constant that are consistent with microscopic reversibility are inserted into either reaction reaction-diffusion equation describing the setup in Fig. 1 or kinetic equations such as Scheme 2, we find that the slope of the potential plays no role in determining the direction of motion, which is instead set solely by the kinetic asymmetry factor.

By using the relations $\frac{k_{+1}k_{+2}}{k_{-1}k_{-2}} = \frac{k_{+1}^{*}k_{+2}^{*}}{k_{-1}^{*}k_{-2}^{*}} = K_{eq}$, we arrive at the expression for the ratio of forward to reverse steps, Eq. (2), and the conclusion that despite appearances, the slopes of the potentials labelled power strokes in Fig. 1 have nothing whatsoever to do with determining the directionality of the motor, nor do they play a role in determining the optimal efficiency, the stall force, or the reliability of the motor, all of which can be calculated in terms of r.

Two-state mode

If the conformational transitions $\underbrace{E_L \overset{K_b}{\rightleftharpoons} E_L^{*}}_{b}$ and $\underbrace{E \overset{K_f^{-1}}{\rightleftharpoons} E}_{f}$ are rapid compared to the

binding and dissociation of S and P, we can use the rapid equilibrium approximation to write

$$\cdots \overset{\omega_{+b}}{\underset{\omega_{-f}}{\leftrightarrows}} (f)_{i-1} \overset{\omega_{+f}}{\underset{\omega_{-b}}{\leftrightarrows}} (b)_{i-1} \overset{\omega_{+b}}{\underset{\omega_{-f}}{\leftrightarrows}} (f)_{i} \overset{\omega_{+f}}{\underset{\omega_{-b}}{\leftrightarrows}} (b)_{i} \overset{\omega_{+b}}{\underset{\omega_{-f}}{\leftrightarrows}} (f)_{i+1} \cdots$$

Scheme 4.

where $\omega_{+f} = \frac{1}{1+K_f}\omega_+$; $\omega_{+b} = \frac{K_b}{1+K_b}\omega_+^*$; $\omega_{-f} = \frac{K_f}{1+K_f}\omega_-^*$; $\omega_{-b} = \frac{K_b}{1+K_b}\omega_+$, and we find $r = \frac{\omega_{+b}\omega_{+f}}{\omega_{-f}\omega_{-b}} = \frac{A}{A^*}e^{-w}$, as with the four-state model Scheme 3. Two state models are very common in the theoretical literature describing molecular machines [15–18] because the analysis of the stepping and catalytic conversion in terms of the four rate constants is very simple. Unfortunately, most authors assert that the rate constants obey $\frac{\omega_{+b}\omega_{+f}}{\omega_{-f}\omega_{-b}} = e^{\Delta\mu-w}$ which is incorrect. The error of this expression is partly hidden by the fact that in the limit $\frac{k_{-1}}{k_{+2}} \gg 1$ and that $\frac{k_{-1}^*}{k_{+2}^*} \ll e^{-\Delta\mu}$ the correct expression $\frac{A}{A^*}e^{-w}$ does in fact approach $e^{\Delta\mu-w}$. Note that in the limit $\frac{k_{-1}^*}{k_{+2}^*} \gg 1$ and that $\frac{k_{-1}}{k_{+2}} \ll e^{-\Delta\mu}$, $\frac{A}{A^*}e^{-w} \to e^{-\Delta\mu-w}$.

Why it is important to recognize that describing a molecular machine as operating by a power stroke is wrong

One could argue that the debate about whether a motor does or does not operate by a power-stroke mechanism is simply a matter of semantics. This however is not the case. In order to understand the importance of the distinction, consider the four-state mechanism for a light-driven rotor [19]

$$\cdots \overset{K}{\leftrightarrows} \underbrace{[(M)-7]_i}_{\text{stable}} \overset{\omega_+}{\underset{\omega_-}{\leftrightarrows}} \underbrace{[(P)-7]_i}_{\text{unstable}} \overset{K}{\leftrightarrows} \underbrace{[(M)-7]_i}_{\text{stable}} \overset{\omega_+}{\underset{\omega_-}{\leftrightarrows}} \underbrace{[(P)-7]_i}_{\text{unstable}} \overset{K}{\leftrightarrows} \underbrace{[(M)-7]_i}_{\text{stable}} \cdots$$

Scheme 5.

where P and M denote states of different inherent chirality. The constraints on the rate constant for optically driven transitions are entirely different from those for the thermally activated processes involved in chemical catalysis [20, 21]. According to the Einstein relations for absorption and for stimulated and spontaneous emission of light, the forward and reverse pumped transitions can approach one another ($\omega_+ \approx \omega_-$) in very bright white light which is not the case for thermally activated processes, the ratios of which always have a factor $e^{\Delta G}$. The design principle for the light-driven motor is totally different from that for the catalysis-driven motor. Kinetic gating, whereby state E/E$_L$ is specific for substrate ($\frac{k_{-1}}{k_{+2}} > 1$) and state E*/E$_L^*$ is specific for product ($\frac{k_{-1}^*}{k_{+2}^*} < 1$), is the key design requirement for catalysis-driven directional motion, and the free-energies of the states play NO ROLE WHATSOEVER in determining directionality. In stark contrast, the essential design feature for a light-driven motor is to have states with alternating high and low free energies such that optical pumping with bright light to a high free-energy state (excited state) is followed by thermal relaxation to a lower free-energy state in a process that is well and appropriately described as a power stroke and gives rise to the expression $r = K^2 e^{-w}$ for the light-driven motor — a true power-stroke mechanism.

Outlook to Future Developments of Research on Catalysis-Driven Molecular Machines

The principled analysis of catalysis-driven motion by molecular motors has been greatly facilitated by the recent description of the first catalytically driven synthetic molecular rotor by Wilson *et al.* [22]. The rotor can be designed such that the prediction of the direction based on a power-stroke mechanism is opposite to that based on kinetic asymmetry [23]. Experiment shows that the information ratchet mechanism is correct. Further, all catalytically driven molecular machines operate by an information ratchet mechanism, which is the only thermodynamically consistent possibility.

One may ask why, if the power stroke is irrelevant, do so many biomolecular motors have power strokes. I speculate that this may be revelatory of the evolutionary origins of these machines. Catalytically driven biomolecular machines may have evolved from light-driven molecular machines that do operate by using a power-stroke mechanism. The power strokes are not deleterious to the operation and are possibly important for optimizing the speed of molecular motors, so there has been no evolutionary pressure to eliminate vestigial power strokes from biomolecular machines.

An important point to remember when considering the mechanism of molecular machines is that the conformational transitions are in mechanical equilibrium, i.e., acceleration is negligible. Thermodynamic disequilibrium between substrate and product is manifest only in breaking the global detailed balance that pertains at thermodynamic equilibrium. When substrate is in excess, those trajectories in which substrate binds and product is released are more probable than their microscopic reverse trajectories in which product binds and substrate is released. The essential structurally based symmetry breaking, however, is between forward trajectories in which the molecule steps to the front and substrate is converted to product and backward trajectories, in which the molecule steps to the back and substrate is converted to product [24]. This symmetry breaking depends only on the relative kinetic barriers — kinetic asymmetry — for the forward vs. backward process [25].

Developing a better understanding of how allosteric interactions can be manipulated to control specificities of a synthetic molecular machine depending on the mechanical state is essential for further progress in the design of synthetic catalytically driven molecular machines. This understanding is also key to a better understanding of biomolecular machines. An important element for this program is the detailed study of the energy landscapes for a variety of biomolecular machines [4, 25–27] and how these landscapes can be designed to allow for kinetic gates that shepherd a machine along a preferred pathway given the non-equilibrium concentrations of substrate and product.

References

[1] J.O.L. Andreasson, B. Milic, G.-Y. Chen, N.R. Guydosh, W.O. Hancock, and S.M. Block, *Elife* **4**, 1166 (2015).

[2] J. Frank and R.L. Gonzalez Jr., *Ann. Rev. Biochem.* **79**, 381 (2010).

[3] P.B. Moore, *Ann. Rev. Biophys.* **41**, 1 (2012).

[4] H. Itoh, A. Takahashi, K. Adachi, H. Noji, R. Yasuda, M. Yoshida, and K. Kinosita, *Nature* **427**, 465 (2004).

[5] S. Mukherjee and A. Warshel, *Photosynth. Res.* **370**, 621 (2017).

[6] D.R. Slochower and M.K. Gilson, *Biophys. J.* **114**, 2174 (2018).

[7] W. Hwang and M. Karplus, *Proc. Natl. Acad. Sci. USA* **116**, 19777 (2019).

[8] S.A.P. van Rossum, M. Tena-Solsona, J.H. van Esch, R. Eelkema, and J. Boekhoven, *Chem. Soc. Rev.* **9**, 255 (2017).

[9] J. Liphardt, *Nat. Phys.* **8**, 638 (2012).

[10] R.D. Astumian, *Biophys. J.* **108**, 291 (2015).

[11] T.L. Hill and E. Eisenberg, *Q. Rev. Biophys.* **14**, 463 (1981).

[12] R.D. Astumian and M. Bier, *Biophys. J.* **70**, 637 (1996).

[13] J. Xing, F. Bai, R. Berry, and G. Oster, *Proc. Natl. Acad. Sci. USA* **103**, 1260 (2006).

[14] M. Bier, I. Derenyi, M. Kostur, and R.D. Astumian, *Phys. Rev. E* **59**, 6422 (1999).

[15] J.A. Wagoner and K.A. Dill, *Proc. Natl. Acad. Sci. USA* **116**, 5902 (2019).

[16] J.A. Wagoner and K.A. Dill, *J. Phys. Chem. B* **120**, 6327 (2016).

[17] A.I. Brown and D.A. Sivak, *J. Phys. Chem. B* **122**, 1387 (2018).

[18] A.I. Brown and D.A. Sivak, *Proc. Natl. Acad. Sci. USA* **114**, 11057 (2017).

[19] J. Bauer, L. Hou, J.C.M. Kistemaker, and B.L. Feringa, *J. Org. Chem.* **79**, 4446 (2014).

[20] R.D. Astumian, *Faraday Discuss.* **195**, 583 (2016).

[21] R.D. Astumian, S. Mukherjee, and A. Warshel, *ChemPhysChem* **17**, 1719 (2016).

[22] M.R. Wilson, J. Solà, A. Carlone, S.M. Goldup, N. Lebrasseur, and D.A. Leigh, *Nature* **534**, 235 (2016).

[23] R. D. Astumian, *Nature Comm.* **10**, 3837 (2019).

[24] R.D. Astumian, *Biophys. J.* **98**, 2401 (2010).

[25] S. Mukherjee, R. Alhadeff, and A. Warshel, *Proc. Natl. Acad. Sci. USA* **114**, 2259 (2017).

[26] S. Mukherjee and A. Warshel, *Proc. Natl. Acad. Sci. USA* **112**, 2746 (2015).

[27] R.D. Astumian, *Top. Curr. Chem.* **369**, 285 (2015).

ON THE IMPORTANCE OF NEW COMPUTATIONAL TOOLS FOR THE CONSTRUCTION OF FUNCTIONAL OUT-OF-EQUILIBRIUM REACTION NETWORKS

WILHELM T. S. HUCK

Institute for Molecules and Materials, Radboud University, Heyendaalseweg 135, Nijmegen, 6525 AJ, The Netherlands

Introduction

One of the grand challenges in systems chemistry is the construction of a synthetic living system. Regardless of how such a system would physically look like, it will be powered and controlled by chemical reaction networks. In recent years, a number of bottom-up designed reaction networks capable of a range of behaviours, e.g., information processing, adaptation, and diffusive signaling, have been demonstrated. The design of these networks is inspired by so-called "motifs" found in biochemical reaction networks. In this article, I will argue that in order to transition from the simple networks we have today to much more complex "functional molecular systems", we require new theoretical and computational tools.

My View of the Present State of Research on Designed Functional Reaction Networks

Inspired by how living systems sense their environment, metabolize food, maintain a constant internal environment, or regulate gene expression, systems chemistry aims to design and construct chemical reaction networks that capture some of these capabilities. Although biochemical systems are enormously complex, it has been recognized that relatively simple "motifs" can produce functional outputs such as oscillations or bistable switches. In recent years, the bottom-up construction of chemical reaction networks has made significant progress, with many example networks built around small molecules, enzymes, or DNA now well established [1–4]. In particular, there has been impressive progress in reaction networks based on DNA strand displacement circuits that function as digital and analog computers [5, 6]. Analog computing is a key component in resource-limited living cells, but to capture the richness of functions displayed by living systems, a strong push is need to develop enzymatic reaction networks. To put the desired complexity into perspective, in natural metabolic networks, over 2000 biochemical reactions are connected and coordinated to allow *E. coli* to double its mass every 20 minutes [7]. Recent examples have shown that it is possible to construct and operate complex multi-enzyme cascades and networks that can capture and convert CO_2 into organic molecules,[8] or that show oscillatory function [9]. However, we still lack a general, programmable, and scalable method to design and construct ever more complex functional enzymatic reaction networks.

My Recent Research Contributions

Our group has reported the retrosynthetic design and construction of small enzymatic reaction networks [9–12]. The first step in the design is the choice of a suitable topology, in this case one that produces a clock function, i.e., oscillations. A retrosynthetic approach was employed where the different nodes in the network (enzymes) were connected via small molecules acting as substrates. Crucially, in our method, we ensure that we know all rate constants from individual reactions using purified components, allowing us to construct sets of differential equations that describe the whole of the network. By simulating the network output as a function of starting concentrations, we can pinpoint the phase space in which limit-cycle oscillations will occur. The network is then tested in a flow reactor to maintain out-of-equilibrium conditions, and the output of the reactor was measured using an enzymatic activity assay or determined using chromatographic methods (HPLC), yielding a good fit with simulations.

The strength of our approach lies in the ability to exploit the full power of chemical synthesis to tune the behaviour of the networks by small molecules. Different substituents in the inhibitors enable fine tuning of activation, delay, and inhibition rate constants, leading to subtle changes to the amplitude, periodicity, and parameter space (flow rate, enzyme and pro-inhibitor concentrations) where limit-cycle oscillations can be found [11, 12].

Beyond oscillations, we have developed minimal bistable switches [13] and reaction networks based on simple feed-forward motifs for adaptive response behaviour [14]. Using evolutionary algorithms, we were able to optimize the concentrations of reagents required to balance the inherent trade-off between maximum amplitude and lowest residual fluorescence.

Outlook to Future Developments of Research

A major risk for further progress is that our ability to program "function" is limited. There are ample examples of oscillators and switches, and some examples of adaptation and the construction of logic gates. However, we do not have designs for networks showing homeostasis, metabolism, or many of the other complex traits shown by living organisms.

A first challenge is to address the quality of our models: despite our best efforts, there is always a (considerable) mismatch between models and experiments. The main reason for this is that although the ODEs describe the reactions in the network adequately at the level of single reactions, once multiple reactions occur in the same compartment, cross reactivity, substrate or product inhibition, or substrate competition all impact the ability of the network to accurately describe the dynamics of the network. Although we can account for these effects in simple motifs with limited off-target interactions between reactions, there is presently no general approach to match model and experiment in more complex reaction networks.

The key challenge for a forward-engineering approach of the functional complex molecular system is our inability to rationally design "function", especially if function goes beyond forms of computation. Relying on simple network motifs has not provided insight into how we should design a system that can sustain itself, or maintain the concentrations of all molecules in the network, or adapt their topology in response to changes in the environment. To make this point more clearly, we do not even have a design for creating an oscillator with a pre-defined periodicity! This seemingly simple but extremely challenging problem cannot be solved by rational design, as the periodicity of oscillators results from the interplay of many parameters. Here, we need to take advantage of rapid developments in machine learning and AI, in order to generate new motifs with desired properties. For example, evolutionary algorithms can be used that will build on initial motifs and add feedback loops of different reaction strengths at each "generation", selecting for motifs that generate oscillations with specific frequencies. Thus far, these approaches have only been shown to function *in silico* and future networks based on evolutionary designs should be built and tested.

Acknowledgments

This work is funded by the Dutch Ministry of Education, Culture and Science (Gravity program 024.001.035, Functional Molecular Systems).

References

[1] E. Mattia and S. Otto, *Nat. Nanotech.* **10**, 111 (2015).
[2] B.A. Grzybowski and W.T.S. Huck, *Nat. Nanotech.* **11**, 585 (2016).
[3] G. Ashkenasy, T.M. Hermans, S. Otto, and A.F. Taylor, *Chem. Soc. Rev.* **46**, 2543 (2017).
[4] J. Li, A.A. Green, H. Yan, and C. Fan, *Nat. Chem.* **9**, 1056 (2017).
[5] N. Srinivas, J. Parkin, G. Seelig, E. Winfree, and D. Soloveichik, *Science* **358**, 1401 (2017).
[6] T. Song, S. Garg, R. Makhtar, H. Bui, and J. Reif, *ACS Nano* **5**, 898 (2016), and references therein.
[7] T.J. Erb, *Emerg. Top. Life Sci.* **3**, 579 (2019).
[8] T. Schwander, L. Schada von Borzyskowski, S. Burgener, N.S. Cortina, and T.J. Erb, *Science* **354,** 900 (2016).
[9] S.N. Semenov *et al.*, *Nature Chem.* **7**, 160 (2015).
[10] A.S.Y. Wong, S.G.J. Postma, I.N. Vialshin, S.N. Semenov, and W.T.S. Huck, *J. Am. Chem. Soc.* **137**, 12415 (2015).
[11] A.S.Y. Wong, A.A. Pogodaev, I.N. Vialshin, B. Helwig, W.T.S. Huck, *J. Am. Chem. Soc.* **139**, 8146 (2017).

[12] A.A. Pogodaev, A. S.Y. Wong, and W.T.S. Huck, *J. Am. Chem. Soc.* **139,** 15296 (2017).

[13] B. Helwig, B. van Sluijs, A.A. Pogodaev, S.G.J. Postma, and W.T.S. Huck, *Angew. Chem. Int. Ed.* **57,** 14065 (2018).

[14] S.G.J. Postma, D. te Brinke, I.N. Vialshin, A.S.Y. Wong, and W.T.S. Huck, *Tetrahedron* **73,** 4896 (2017).

MODELLING OF NON-EQUILIBRIUM SYSTEMS: REACTION NETWORKS FOR BIOINSPIRED BEHAVIOUR

ANNETTE F. TAYLOR

Chemical and Biological Engineering, University of Sheffield, Sheffield, S1 2JD, United Kingdom

My View of the Present State of Research on Modelling of Non-equilibrium Systems

Research into the behaviour of non-equilibrium chemical systems has been mainly driven by a desire to understand the principles of emergent behaviour and self-organization in nature [1]. Non-equilibrium states can be modelled by considering a system consisting of a network of chemical reactions confined within a boundary. Typically, in laboratory-based chemical systems with constant supply of mass and/or heat, a single steady state is eventually obtained. The presence of chemical or thermal feedback in a reaction network gives rise to interesting features, such as rhythms and bistable switches, associated with the loss of stability of a steady state. It is not always clear why rhythms occur in living systems, but feedback processes in general are used for self-regulation.

Theory for the selection of non-equilibrium states is based on the structure of the network (topology) and rate of change of species through a pathway (flux) [2–4]. A network for chemical oscillations may involve a positive feedback loop through autocatalysis, where a species catalyzes its own production, combined with delayed negative feedback removing the autocatalyst. Different network topologies for generating oscillations and switches have been compared for their robustness. Since autocatalysis can result in a discontinuous transition between chemical states, reaction networks are sometimes likened to electronic circuits with an "on" (high autocatalyst, high conversion) and "off" (low autocatalyst, low conversion) state and these features have been used to construct logic gates in chemical systems [5]. Autocatalysis has been coupled to phase change processes such as precipitation and chemomechanical swelling [6, 7]. This demonstrates functional possibilities such as information processing, motion, and spatial structuring in multiphase chemical systems.

My Research Contributions to Modelling of Non-equilibrium Systems

Microorganisms such as yeast, bacteria, and slime mold use autocatalytic reactions and oscillations to communicate and synchronize activity. The construction of reaction networks that reproduce these features may shed light on the processes that regulate life but also allow us to develop bio-inspired systems for applications. We have been working on the design of autocatalytic reaction networks for oscillations and on understanding the role of mass transport in the collective behaviour of cells.

Pojman and Epstein wrote in 1997, "We have no doubt that by the time you read these words, many more oscillators will have been discovered" [8]. In fact, there are relatively few new examples and most are based on inorganic redox processes, even though a number of design methodologies have been proposed [9]. We set about designing oscillators that might eventually find applications under mild conditions by using water to generate autocatalysis. Hydrolysis and dehydration processes are acid/base catalyzed and the products can themselves be acids or bases, thus catalyzing their own production. Using simulations to find appropriate conditions, we developed the formaldehyde–sulfite–gluconolactone reaction, involving autocatalytic production of a base coupled with base-catalyzed production of acid [10]. This resulted in the first example of a non-redox oscillator with organic substrates. We also investigated the hydrolysis of urea catalyzed by the enzyme urease [11]. The enzyme reaction has a maximum rate at pH 7 and produces ammonia resulting in autocatalytic production of base. Oscillations were observed that were not obtained in simulations of the reaction, but may involve mass transfer of ammonia to the gas phase.

An alternative method of maintaining a non-equilibrium state involves localization of the reaction, for example, by placing a catalyst-loaded particle in a large bath of substrate. Inspired by the ability of bacteria and other microorganisms to synchronize behaviour in response to increasing cell number or density, we investigated collective behaviour in catalytic microparticles for the Belousov–Zhabotinsky (BZ) reaction [12, 13]. The particles displayed oscillations and produced an autocatalyst which was emitted to the surrounding solution where it degraded. In a stirred system, we found the dynamics depended on the number density of particles. At a low stirring rate, there was a gradual synchronization of oscillators with increasing density: a Kuramoto transition. When a group of particles oscillated together, a pulse of autocatalyst was produced that pulled the other oscillators into the collective rhythm; the strength of the signal increased with the density. At a high stirring rate, none of the particles were initially oscillating as the autocatalyst was quickly stripped from the particle surface. However, above a threshold density, sufficient autocatalyst was produced in the solution to initiate synchronized oscillations across the whole population simultaneously: a dynamical quorum sensing transition. This is a beautiful illustration of collective behaviour that we were able

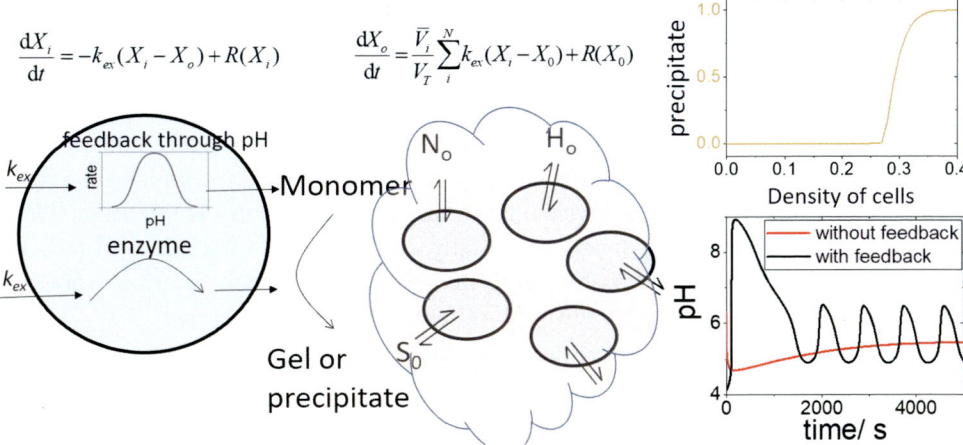

$$\frac{\mathrm{d}X_i}{\mathrm{d}t} = -k_{ex}(X_i - X_o) + R(X_i) \qquad \frac{\mathrm{d}X_o}{\mathrm{d}t} = \frac{\bar{V}_i}{V_T}\sum_i^N k_{ex}(X_i - X_0) + R(X_0)$$

Fig. 1. Model of a single-cell exchanging species with the surrounding solution (k_{ex}) and under-going catalytic reaction ($R(X_i)$), and formation of a gel or precipitate in a group of cells above a critical density. With internal feedback, cells display oscillatory behaviour reaching high pH, whereas without feedback cells reach a steady state at low pH. Adapted from Ref. [15].

to explore in detail using simulations as the chemistry of the BZ reaction is well characterized.

More recently, we have been exploiting enzyme-loaded particles to obtain functional collective behaviour driven by quorum sensing (Fig. 1) [14]. Enzyme-loaded particles in a bath of urea and acid display an ultrasensitive switch, bistable switch, and oscillations in pH. We found that a sharp switch to an oscillatory state or high pH state across the whole population could be obtained with increasing cell numbers. This switch was coupled to base-catalyzed gelation or precipitation of secreted monomers, inspired by the formation of bio-films that protect bacteria from their external environment [15].

Outlook on Future Developments of Research on Modeling of Non-equilibrium Systems

One of the most promising areas of application of non-equilibrium systems is in development of bio-inspired materials. Models that couple reaction–diffusion processes with phase change yield relatively simple structures such as Liesegang patterns, so how can we obtain some of the intricate hierarchical structures observed in, for example, bio-mineralization? Typically, such structures involve the formation and 3D arrangement of nanoparticles, which is yet to be captured in models.

Reaction networks can be designed to program dynamical behaviour in synthetic systems. There is also a fundamental desire to understand and control the output of a network of reactions for synthesis. Traditionally, chemists consider molecular transformations as a series of steps to yield the product of interest.

Living cells are miniature chemical factories able to produce valuable molecules and systems/synthetic biologists are becoming increasingly adept at manipulating pathways toward desired products. So, one research vision in this area is to design bio-catalytic reaction networks for the sustainable synthesis of products/materials, inspired by processes that take place in biological cells, such as lignocellulosic transformations. Examples are emerging, but in general a much better understanding of how to implement a designed reaction network in the laboratory is required. There are often unanticipated side reactions and although there are databases of individual rate constants, these were often determined under specific conditions (optimal pH for enzymes) that are not valid with different complex mixtures.

Models of reaction networks might also be constructed based on non-biological (but realistic) catalytic processes with a view to synthesizing species or reactive compartments through *in silico* evolution. What will such studies reveal? Maybe this will depend on the development of new methods for characterization of non-equilibrium states that emerge from chemical systems to form life.

References

[1]   A.J. Lotka, *J. Phys. Chem.* **14**(3), 271 (1910).
[2]   J.J. Tyson, K.C. Chen, and B. Novak, *Curr. Opin. Cell Biol.* **15**(2), 221 (2003).
[3]   M. Feinberg, in *Applied Mathematical Sciences* Vol. 202, pp. 153–204 (Switzerland, 2019).
[4]   U. Alon, *Nat. Rev. Genet.* **8**(6), 450 (2007).
[5]   O. Steinbock, P. Kettunen, and K. Showalter, *J. Phys. Chem.* **100**(49), 18970 (1996).
[6]   K. Kurin-Csörgel, I.R. Epstein, and M. Orbán, *Nature* **433**(7022), 139 (2005).
[7]   R. Yoshida, T. Takahashi, T. Yamaguchi, and H. Ichijo, *J. Am. Chem. Soc.* **118**(21), 5134 (1996).
[8]   I.R. Epstein and J.A. Pojman, *An Introduction to Nonlinear Chemical Dynamics: Oscillations, Waves, Patterns, and Chaos* (Oxford University Press, 1998).
[9]   G. Rabai, M. Orban, and I.R. Epstein, *Acc. Chem. Res.* **23**(8), 258 (1990).
[10] K. Kovacs, R.E. McIlwaine, S.K. Scott, and A.F. Taylor, *J. Phys. Chem. A* **111**(4), 549 (2007).
[11] G. Hu, J.A. Pojman, S.K. Scott, M.M. Wrobel, and A.F. Taylor, *J. Phys. Chem. B* **114**(44), 14059 (2010).
[12] A.F. Taylor, M.R. Tinsley, F. Wang, Z. Huang, and K. Showalter, *Science* **323**(5914), 614 (2009).
[13] M.R. Tinsley, A.F. Taylor, Z. Huang, and K. Showalter, *Phys. Rev. Lett.* **102**(15), 158301 (2009).
[14] T. Bánsági, Jr. and A.F. Taylor, *J. Roy. Soc. Interface* **15**(140), 20170945 (2018).
[15] T. Bánsági, Jr. and A.F. Taylor, *Life* **9**(3), 63 (2019).

SIMULATION OF MOLECULAR MACHINES

BERND HARTKE

Institute for Physical Chemistry, Kiel University, 24098 Kiel, Germany

My View of the Present State of Research on Simulation of Molecular Machines

There is a considerable amount of simulation work on molecular switches (azobenzenes, diarylethenes, furylfulgides, etc.), covering most of their features in great detail. Unsurprisingly, however, simulation treatments of full-blown molecular machines (in which molecular switches are merely one of many parts) are much rarer, largely due to their size. For molecular switches, much work has focused on light as a stimulus, and far less on other possible stimuli (pH, temperature, etc.). Additionally, and in curious ways detached from the above, there is a considerable community working on Brownian molecular machines, but frequently from a rather abstract, theoretical perspective, focusing on what is possible in principle but rarely delving into the realm of actually realized molecular systems. Although we did some work on light-induced Brownian rotors, I will focus on photochemical molecular switches and their use in simple molecular machines here.

My Recent Research Contributions to Simulations of Molecular Machines

Molecular switches can be used as motors in molecular machines. If the desired machine output is mechanical force, azobenzene has been a frequent choice, since its $Z \leftrightarrow E$ isomerization is conveniently UV/vis-light-driven, reliable in various surroundings, and produces a huge conformational change relative to its small size. Nevertheless, it also has various disadvantages, but these can be overcome by further molecular design: Adding in an additional C–C bridge between the two phenyl rings produces diazocine (Fig. 1(b)), which features improved quantum yields and well-separated excitation wavelengths for the $Z \rightarrow E$ and $E \rightarrow Z$ isomerization directions. Ole Carstensen's in-depth analysis of his diazocine photodynamics simulations in our group [1], however, revealed a significant ground-state barrier in the $Z \rightarrow E$ direction, between the chair and twist forms of the E configuration (analogous to Fig. 1(a)). Reflections from this barrier can even induce temporary returns from S_0 to S_1, significantly lowering the quantum yield. Using further extensive photodynamic simulations, Ronja Höppner in my group therefore examined several variations of this additional bridge, involving longer bridges and hetero atoms

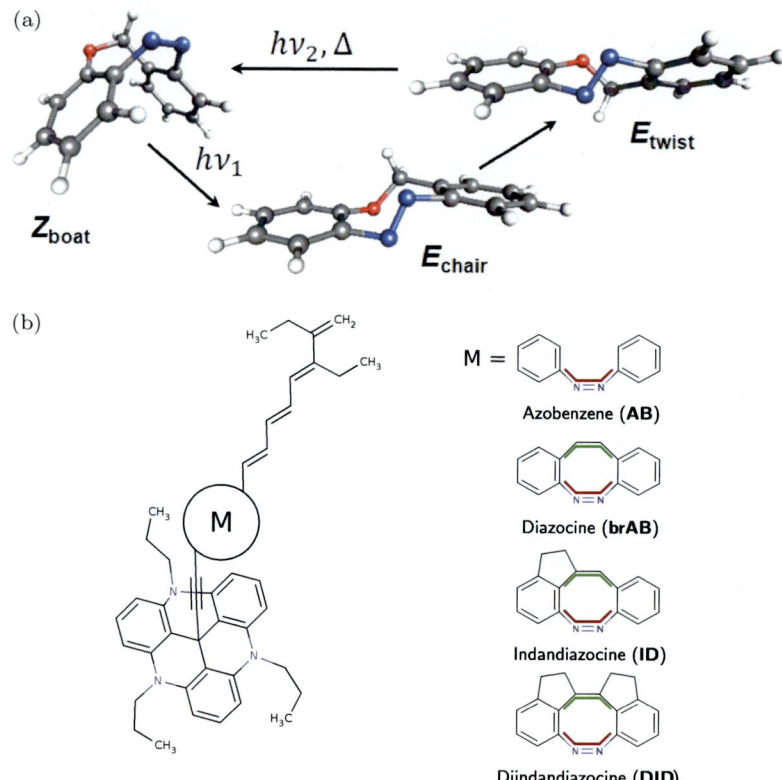

Fig. 1. (a) Oxa-diazocine, an azobenzene derivative optimized in two steps (additional C–C bridge, and replacing one C-atom by an O-atom). (b) left: Artificial cilium, consisting of an attachment unit, a motor unit, and an effector unit; right: Further azobenzene derivatives used as motor units, including (di-)indandiazocines in which the E → Z isomerization can proceed only in one of two directions.

like oxygen. After many failed attempts, one such derivative could be synthesized successfully [2]. Its superior photophysical properties are currently verified by femtosecond spectroscopy [3].

An overarching aim of this collaborative research was to design a molecular cilium [4, 5] (Fig. 1(b)) for directed transport at the nanoscale. It features a tri-azatriangulene (TATA) "platform" (to attach the cilium to any surface, via van der Waals forces and in a perpendicular fashion), a molecular switch as motor, and a suitable effector extension (to increase the motion amplitude and to achieve optimal momentum transfer onto a transport target).

Some parts of the design process are indicated in Fig. 1(b): While in Z → E isomerization normal azobenzenes and diazocines can open only in one direction, they can perform the E → Z motion in two opposite directions, which obviously would be detrimental to directed transport. Adding further hydrocarbon-ring bridges (producing chiral indandiacozine and di-indandiacozine [6]) should block one of

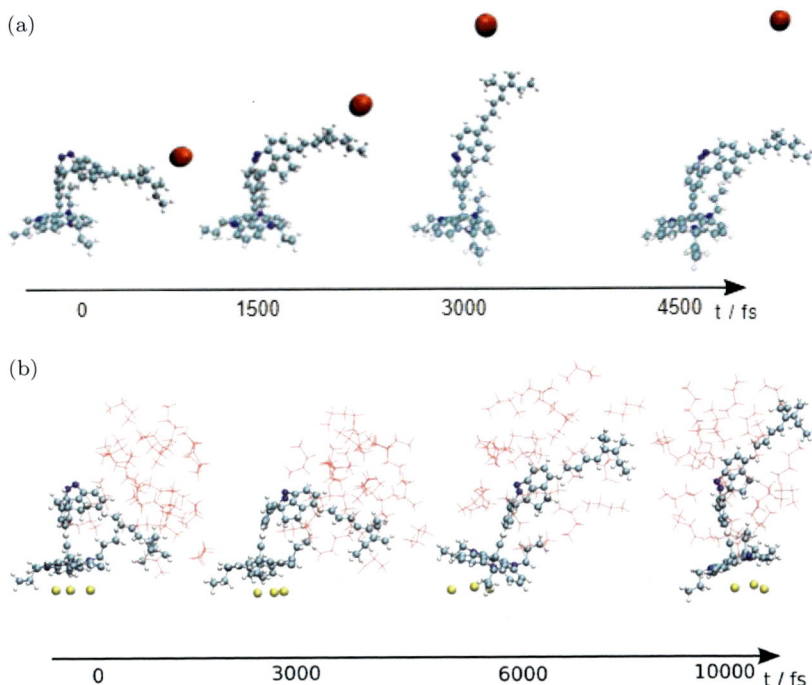

Fig. 2. (a) Simulating the transport of a single helium atom by a photochemically driven molecular cilium of the kind shown in Fig. 1(b). (b) Cilium-driven transport of a cluster of butane molecules, also featuring inadvertent detachment of the cilium from the three-atom surface representation.

these directions due to simple steric reasons. In our simulations [7], we could confirm this unidirectional switching in the full cilium setup, as well as high quantum yields and ultrashort switching times. However, the price to pay is a markedly reduced conformational difference between the Z and E forms.

In contrast to our experimental colleagues, in our QM/MM simulations, we could already demonstrate actual transport by such cilia [8]. As shown in Fig. 2(a), a single, photoinduced cilium hit is sufficient to set a helium atom as target into translational motion.

Interestingly, our original plan of using C_{60} as target failed to work: Despite the alleged "rigid" character of C_{60}, when the cilium effector hits it, there only is rapid internal vibrational energy redistribution (IVR) into its (many) intramolecular degrees of freedom. In other words, C_{60} is heated up but not moved, at least not by a single cilium strike. After such a strike, the cilium remains stuck to the C_{60} surface, resembling Smalley's "sticky finger" problem. However, in simulations of a cilium striking a small cluster of butane molecules [8] (Fig. 2(b)), we do see visible centre-of-mass movement of this butane cluster. This last example, as shown in Fig. 2(b), also demonstrates another problem: The forces (ultimately transferred into the system via the UV/vis excitation) are not only sufficient to set (some) targets into motion but here also to rupture the attachment of the cilium to the surface (in this case modelled by three representative atoms). When we artificially tuned up the van der Waals interaction between surface and cilium, the cilium effector permanently stuck to the surface. With slightly higher forces than those arising from these photoinduced isomerizations, we have also seen covalent bonds breaking in the cilia or in the targets. Of course, one single artificial cilium is insufficient to transport mesoscopically relevant amounts of matter. However, it was demonstrated experimentally [9] that TATA platforms self-assemble into regular monolayers on surfaces. Very recently, we could reproduce this behaviour in our simulations [10], employing global structure optimization by evolutionary algorithms. In contrast to other simulations of molecular assembly on surfaces, we did not need to prescribe adsorption poses or to discretize any other aspect of the problem. Nevertheless, we were able to match the experimentally observed regular superstructures almost exactly and to explain findings that had to be left open by the experiments (TATA orientation, alkyl side-chain conformations, bilayers, etc.).

Future Developments of Research on Simulations of Molecular Machines

Not all of our macroworld-trained mechanical intuition carries over into the nanoworld. One usually overlooked issue is "wear and tear": Macro-machines (or any macro-object) can lose lots of material and suffer crack-like deterioration without overall breakdown, but for a molecular-sized machine there is no gradual deterioration, only a binary one: Either it is fully intact and works perfectly, or it is completely broken. The biological molecular machines in our bodies have the same problem. There, however, it is patched up in a curiously radical and expensive way: Broken (and perhaps also intact) machines are constantly garbage-collected and replaced by newly built ones. As indicated by our results reported above, artificial molecular machines also have a finite lifetime, in particular in the realm of photoexcitation-induced mechanochemistry. Maybe such a constant, large-scale replacement is also the way to go for artificial nanomachines, if they are to reach real applications outside of academic laboratories.

Furthermore, while mechanically acting molecular machines certainly are interesting and inspiring (and inspired by famous examples like ATPase, myosins, and kinesins), they also have inherent limitations, as again exemplified by our findings reported above: The useful force interval between "nothing happens" and "target or machine breaks" is fairly narrow, and even in this narrow range IVR is a nasty distractor. Therefore, it may be helpful to extend the conceptual domain of "molecular machines" to also include other ways of exerting intermolecular influences. Of course, in chemistry, the prime tool of doing that is electrostatics, and this also is what all the other "less mechanical" enzymes employ [11]. In other words, marrying the two emerging fields of molecular machines and of electrostatic catalysis [12] may be an interesting step forward. We are working toward this aim, building upon our earlier work of globally optimal catalytic fields [13], currently extended to include on-the-fly reaction path re-optimizations in a successful application to a prototypical Diels–Alder reaction [14], for which electrostatic catalysis could recently be demonstrated experimentally [15].

Acknowledgments

I gratefully acknowledge generous funding by the Deutsche Forschungsgemeinschaft (DFG) via the collaborative research centre SFB 677 (A1, A5, A7) and via individual grants Ha2498/12-1 and Ha2498/16-1.

References

[1] O. Carstensen, J. Sielk, J. Schönborn, G. Granucci, and B. Hartke, *J. Chem. Phys.* **133**, 124305 (2010).

[2] M. Hammerich, C. Schütt, C. Stähler, P. Lentes, F. Röhricht, R. Höppner, and R. Herges, *J. Am. Chem. Soc.* **138**, 13111 (2016).

[3] D. Bank, F. Renth, M. Hammerich, P. Lentes, R. Herges, and F. Temps, *Ultrafast Phenomena XXII*, accepted (2020).

[4] B. Baisch, D. Raffa, U. Jung, O.M. Magnussen, C. Nicolas, J. Lacour, J. Kubitschke, and R. Herges, *J. Am. Chem. Soc.* **131**, 442 (2009).

[5] M. Hammerich, T. Rusch, N.R. Krekiehn, A. Bloedorn, O.M. Magnussen, and R. Herges, *Chem. Phys. Chem.* **17**, 1870 (2016).

[6] T. Tellkamp, J. Shen, Y. Okamoto, and R. Herges, *Eur. J. Org. Chem.* **2014**, 5456 (2014).

[7] T. Raeker and B. Hartke, *ScienceOpen Research*, 10.14293/S2199-1006.1.SOR-CHEM.ARDTLN.v1 (2015).

[8] T. Raeker, B. Jansen, D. Behrens, and B. Hartke, *J. Comput. Chem.* **39**, 1433 (2018).

[9] S. Kuhn, B. Baisch, U. Jung, T. Johannsen, J. Kubitschke, R. Herges, and O. Magnussen, *Phys. Chem. Chem. Phys.* **12**, 4481 (2010).

[10] A. Freibert, J.M. Dieterich, and B. Hartke, *J. Comput. Chem.* **40**, 1978 (2019).

[11] S.D. Fried and S.G. Boxer, *Annu. Rev. Biochem.* **86**, 387 (2017).

[12] S. Shaik, R. Ramanan, D. Danovich, and D. Mandal, *Chem. Soc. Rev.* **47**, 5125 (2018).

[13] M. Dittner and B. Hartke, *J. Chem. Theory Comput.* **14**, 3547 (2018).

[14] M. Dittner and B. Hartke, *J. Chem. Phys.* **152**, 114106 (2020).

[15] A.C. Aragones, N.L. Haworth, N. Darwish, S. Ciampi, N.J. Bloomfield, G.C. Wallace, I. Diez-Perez, and M.L. Coote, *Nature* **531**, 88 (2016).

NON-EQUILIBRIUM STEADY-STATES IN ARTIFICIAL SUPRAMOLECULAR MATERIALS

THOMAS M. HERMANS

Université de Strasbourg & CNRS, Strasbourg, France

My View of the Present State of Research on Non-equilibrium Systems and Molecular Machines

The most functional, responsive, adaptive, interactive, and in general awe-inspiring non-equilibrium system is life itself. No consensus exists on what life exactly is, but there seem to be common denominators. Living organisms operate in a state of homeostasis, where chemical and physical processes are kept in steady states controlled by complex signaling pathways that sense and correct in continuous feedback loops. Organisms grow by having a metabolism that has a higher buildup (anabolic) than breakdown (catabolic) rate, where overall food is converted into energy, into molecular building blocks like nucleic acids and lipids, and where waste is removed or recycled. The most basic unit — a single cell — can perform all of life's functions as demonstrated by unicellular organisms. More commonly, however, organisms consist of many cells with each having their own function, which become self-organized into larger tissues and organs with the ability to respond to stimuli, adapt to their environment, and reproduce.

Probably the first definition of life as being self-organized was given by Immanuel Kant in 1790 [1]: *This can never be the case with artificial instruments, but only with nature, which supplies all the material for instruments (even for those of art). Only a product of such a kind can be called a natural purpose, and this because it is an organized and self-organizing being.* Kant deduces that life cannot be comprehended, nor be made with artificial components. Modern-day scientists have a more optimistic view I believe, especially in light of scientific breakthroughs of the past century. Take, for example, the top-down approach pioneered by Craig Venter and co-workers. They have created "synthetic life" by synthetizing an artificial genome from scratch, and placing that inside a bacterium[2] or even inside eukaryotic cells [3]. Effectively, they replaced the genetic code of a living organism; the cell was alive, and received a transplant of the information it needs to stay alive. A true bottom-up approach would consist of synthesizing all abiotic (non-living) components of the cell separately, and then putting them together and kick-starting the whole system into being alive. For the moment, however, we have no idea how to do this, but increasingly complex lifelike materials are in the making.

Closest to the top-down approach, we find "engineered living materials" where (optionally) genetically modified cells excrete compounds that form hydrogels or biofilms [4]. Since such systems consist of materials synthesized by themselves, the possibility to have feedback between mechanical/environmental cues and material restructuring exists. For example, *A. aceti* bacteria produce cellulose in oxygen-rich environments and release acetate as waste. When combined with *C. reinhardtii* microalgae that grow in acetate-rich media and produce oxygen as a byproduct, synergy arises and moldable gels are formed [5]. Moving more in the direction of all-synthetic bottom-up systems, one can combine not entire cells, but selected biological components with synthetic materials. For example, FtsZ filaments — the bacterial analogues of microtubules — can be placed inside coacervate droplets, where they grow and lead to droplet splitting similar to cell division [6]. Yet, further away from life as we know it by using familiar building blocks is the use of DNA nanotechnology, where force sensing, transmission, and generation can be achieved using DNA mechanotechnology [7]. DNA nanotubes have been shown to grow and shrink resembling microtubules [8]. In short, borrowing elements from life's toolbox has given rise to an exciting class of partially living or lifelike materials.

The last approach is to take non-natural molecules and devise systems and materials completely from the bottom up. This is one of the key topics in the emerging field of Systems Chemistry [9–11], and is expected to have a large impact comparable to the one synthetic thermoplastics, elastomers, and hydrogels have had in the field of (bio)materials science. Since such materials are abiotic at the start, a key goal is to add some of the key denominators of life (i.e., homeostasis, metabolism, growth, adaptation, responsiveness, reproduction, and evolution) one by one. An early example by van Esch and co-workers showed a metabolic-like reaction cycle that could transiently form an ester bond from an acid moiety, leading first to assembly followed by disassembly of supramolecular fibers [12]. The reaction cycle is powered by a chemical fuel and produces waste, resulting in dissipative self-assembly. Since then, a number of approaches have been explored to transiently form synthetic structures by using (i) bond–making/-breaking reactions, (ii) redox chemistry, or iii) non-covalent binding of adenosine triphosphate, as has recently been reviewed [9, 13].

My Recent Research Contributions to Non-equilibrium Systems and Molecular Machines

The contributions of our research group to the field of Systems Chemistry and more specifically to dissipative self-assembly are focused on homeostasis, that is, to obtain sustained dissipative non-equilibrium steady states. We use small molecules that can be switched between assembled and disassembled states using chemical fuels. By continuously adding fuel and removing waste, one can find steady states, where the distribution of species is dictated by the kinetics and distance from equilibrium, instead of the thermodynamics of the system [14]. That is, the free energies of the

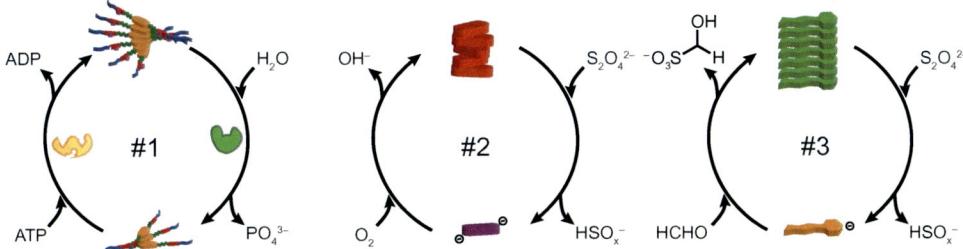

Fig. 1. (Color) Chemically fuelled reaction cycles to control supramolecular assembly and disassembly. Cycle #1: phosphorylation of a serine residue by a kinase enzyme (yellow) using adenosine triphosphate ATP, dephosphorylation using a phosphatase enzyme. Cycle #2: chemical reduction using dithionite $S_2O_4^{2-}$ and oxidation in atmospheric conditions. Cycle #3: reduction using dithionite $S_2O_4^{2-}$ and restoration of the aldehyde moiety using formaldehyde.

assembled versus disassembled states are no longer the most important as they have been in Supramolecular Chemistry traditionally. Instead, the kinetics of assembly and disassembly and those of the involved chemical reactions are dominant.

Specifically, we have developed reaction cycles based on bond making and breaking or by using redox chemistry. In the first case (cf. Fig. 1, #1), we form a new bond by phosphorylation of a serine residue by a kinase enzyme that uses adenosine triphosphate ATP as a fuel [15]. The molecule in question becomes zwitter-ionic due to this phosphorylation, which amplifies its self-assembly by decreasing electrostatic repulsion. The phospho-serine residue can be dephosphorylated using a phosphatase enzyme that releases inorganic phosphate into the aqueous buffer. We used a membrane reactor where ATP fuel and phosphate waste can be exchanged continuously, thus leading to non-equilibrium steady states that depend on the fuel concentration.

In the second case (cf. Fig. 1, #2), we use a perylenediimide derivative that forms highly kinetically trapped aggregates in aqueous buffer. Upon addition of dithionite ($S_2O_4^{2-}$), the perylenediimide core is reduced to the dianion, leading to disassembly because of electrostatic repulsion. Stirring in air leads to oxidation back to the neutral state and re-assembly. When reductant and oxidant are continuously added to a stirred batch reactor, we found spontaneous supramolecular-sized oscillations that we attribute to positive feedback from the nucleation–elongation self-assembly process combined with negative feedback by chemical reduction and disassembly [16]. The same system in non-stirred media can show propagating fronts and centimeter-scale mosaic structures. The latter are far from equilibrium "dissipative structures" in the Prigogine sense [17], where self-organization occurs on length scales that are orders of magnitude larger than the molecules involved. In our case, it is because the self-assembly leads to slight solution density changes that large-scale convective currents result.

In the last system (cf. Fig. 1, #3), we use a functional group transformation that converts the aldehyde moiety of a hydrogelator to a hydroxyl-sulfonate using dithionite [18]. Again, electrostatic repulsion leads to disassembly of the supramolecular structure and loss of the gel state. The aldehyde can be recovered by addition of formaldehyde, which restores the hydrogel. The lifetime of the disassembled sol state can be programmed by tuning the rate of catalytic formaldehyde generation. Surprisingly, the supramolecular structures obtained by heat/cool cycles are very different from those observed after a chemical sol/gel cycle. Moreover, the material properties can be "reset" many times by removing waste products and addition of new fuel.

Outlook on Future Developments of Research on Non-equilibrium Systems and Molecular Machines

It is clear that the field of dissipative non-equilibrium self-assembly is just getting started and that many of the future applications are still unknown. We can build on the advances made in organic chemistry, catalysis, and supramolecular chemistry to design and build bottom-up systems and materials that are increasingly far from equilibrium. To really obtain lifelike materials, we need to develop reactions that are orthogonal and can work under the same reaction conditions. We need to be able to store "pre-fuels" inside the materials that become available to the reaction cycle either by using a catalyst or by a second coupled reaction cycle. Waste produced in the system needs to be recycled either by coupling an auxiliary reaction cycle or by physically removing it by some sort of transport phenomenon or nanopump. It would then be possible to sustain and study steady states inside of supramolecular materials where their non-equilibrium fluctuations should violate the fluctuation–dissipation theorem. Lastly, such lifelike materials should become biocompatible or even bio-orthogonal, so as to be used inside living beings and allow for two-way communication and interaction.

Acknowledgments

TMH acknowledges funding from ERC-Stg "Life-Cycle" (757910) and from the European Union's Horizon 2020 research and innovation programme under the Marie Skłodowska-Curie grant agreement no. 812868 "CReaNet".

References

[1] I. Kant, *The Critique of Judgment* (Simon and Schuster, 2013).
[2] D.G. Gibson, J.I. Glass, C. Lartigue, V.N. Noskov, R.-Y. Chuang *et al.*, *Science* **329**, 52 (2010).
[3] N. Annaluru, H. Muller, L.A. Mitchell, S. Ramalingam, G. Stracquadanio *et al.*, *Science* **344**, 55 (2014).
[4] P.Q. Nguyen, N.-M.D. Courchesne, A. Duraj-Thatte, P. Praveschotinunt, and N.S. Joshi, *Adv. Mater.* **30**, 1704847 (2018).
[5] A.A.K. Das, J. Bovill, M. Ayesh, S.D. Stoyanov, and V.N. Paunov, *J. Mater. Chem. B* **4**, 3685 (2016).

[6]　E. te Brinke, J. Groen, A. Herrmann, H.A. Heus, G. Rivas *et al.*, *Nat. Nanotechnol.* **13**, 849 (2018).

[7]　A.T. Blanchard and K. Salaita, *Science* **365**, 1080 (2019).

[8]　L.N. Green, H.K.K. Subramanian, V. Mardanlou, J. Kim, R.F. Hariadi *et al.*, *Nat. Chem.* **11**, 510 (2019).

[9]　G. Ashkenasy, T.M. Hermans, S. Otto, and A.F. Taylor, *Chem. Soc. Rev.* **46**, 2543 (2017)

[10]　E. Mattia and S. Otto, *Nat. Nanotechnol.* **10**, 111 (2015).

[11]　R. Frederick Ludlow and S. Otto, *Chem. Soc. Rev.* **37**, 101 (2008).

[12]　J. Boekhoven, A.M. Brizard, K.N.K. Kowlgi, G.J.M. Koper, R. Eelkema *et al.*, *Angew. Chem.* **122**, 4935 (2010).

[13]　S. De and R. Klajn, *Adv. Mater.* **30**, 1706750 (2018).

[14]　R.D. Astumian, *Nat. Commun.* **10**, 3837 (2019).

[15]　A. Sorrenti, J. Leira-Iglesias, A. Sato, and T.M. Hermans, *Nat. Commun.* **8**, 15899 (2017).

[16]　J. Leira-Iglesias, A. Tassoni, T. Adachi, M. Stich, and T.M. Hermans, *Nat. Nanotechnol.* **13**, 1021 (2018).

[17]　G. Nicolis and I. Prigogine, *Self-Organization in Nonequilibrium Systems* (Wiley, 1977).

[18]　N. Singh, B. Lainer, G.J.M. Formon, S. De Piccoli, and T.M. Hermans, *J. Am. Chem. Soc.* **142**, 4083 (2020).

AUTONOMOUS CHEMICALLY DRIVEN MATERIALS SYSTEMS INSPIRED FROM LIFE

ANDREAS WALTHER

A³ BMS-Lab,Institute for Macromolecular Chemistry, Stefan-Meier-Str. 31,
University of Freiburg, 79104 Freiburg, Germany
Freiburg Materials Research Center, Stefan-Meier-Str. 21, University of Freiburg,
79104 Freiburg, Germany
Freiburg Center for Interactive Materials and Bioinspired Technologies,
Georges-Köhler-Allee 105, University of Freiburg, 79110 Freiburg, Germany
Freiburg Institute for Advanced Studies, Albertstraße 19, 79104 Freiburg, Germany
Cluster of Excellence livMatS @ FIT – Freiburg Center for Interactive Materials
and Bioinspired Technologies, University of Freiburg,
Georges-Köhler-Allee 105, 79110 Freiburg, Germany

My View of the Present State of Research on Non-equilibrium Systems

An insect lands on an orange, another event of mesmerizing complexity that goes unnoticed. The insect is a dissipative, self-organized molecular system fueled by sugars, and capable of autonomous flight, sensing, and adaptation for hours. Such complex chemical out-of-equilibrium systems are ubiquitous in living organisms. In fact, most life-distinguishing features rely on non-equilibrium dynamics, involving constant energy dissipation, kinetic control, and are orchestrated through feedback loops [1].

We are now at a point in time where the molecular mechanisms for such functions in living organisms are largely understood. They provide an increasing source of inspiration for new generations of molecular materials and systems with unprecedented levels of functionalities — active, adaptive, autonomous, and interactive properties. Temporal control, the ability to accumulate work, for sensing, adaptation, or communication, depends on dissipative chemical systems, and the ultimate goal of lifelike, self-organizing systems is to combine and integrate them using feedback loops.

In contrast, synthetic systems and materials have mostly focused on the generation of equilibrium or metastable (kinetically trapped) structures because this allows one to engineer functional materials with high confidence. The integration of switchable properties has led to responsive materials useful for a diverse set of applications in photonics, bio-medical applications, or for sensors. Fantastic breakthroughs have happened in the last decades.

However, the next disruptive change in the property profiles of soft matter will arise from developing far-from-equilibrium, fueled, and feedback-controlled molecular systems. This will allow for orchestrated, pre-programmed, and autonomous dynamics in the time domain, and provide entirely new capabilities for sensing, adaptation, communication, learning, evolution, and replication. In essence, it will take the present-day "dead" materials to a more interactive, intelligent, and life-like state. Even though this is a great challenge and in parts a distant dream, the understanding of molecular biological mechanisms and the design of synthetic out-of-equilibrium molecular systems are improving swiftly, and exciting times are ahead [1, 2].

My Recent Research Contributions to Non-equilibrium Systems

Our main focus in the past years has been on finding approaches toward autonomous, chemically fueled, and feedback-controlled molecular systems that show transient lifecycles and programmable steady-state dynamics. Those can take different levels of inspiration from the non-equilibrium features of the dissipative self-assembly from microtubules and can either integrate kinetic concerts and feedback mechanisms or implement direct molecular energy dissipation mechanisms. Our focus is to not only provide transient self-assemblies but to also find ways of how we can transduce fundamental advances in autonomous systems design to the materials level.

In recent years, we have introduced mostly two different concepts. On the one hand, we have pioneered the concept of pH-feedback systems (active environment) that allow to orchestrate transient pH curves as pH environments to couple other pH-switchable building blocks in a system approach [3–6]. On the other hand, we have developed an ATP-driven dissipative system (active structures) that uses the balance of ATP-powered ligation using T4 ligase and concurrent cutting using a restriction enzyme to make transient DNA-based systems, in which the steady-state dynamics can be imprinted in the details of the chemical reaction network [7].

The pH-feedback system generally uses a combination of fast or dormant activators in combination with dormant deactivators (such as the urease/urea reaction) to pre-orchestrate transient pH profiles in the acidic and basic regime. This has allowed one to make transient pH profiles that can be programmed in their lifetimes from minutes to days and into which also lag times can be programmed through increasing the complexity or functionality of the underlying feedback-controlled reaction network. We have shown that such pH-feedback curves can be coupled to a diverse set of pH-switchable building blocks ranging from block copolymers, peptides to i-motif DNA sequences and appropriately functionalized Au nanoparticles [3–6]. Although these additional components act as a load on the pH reaction network, the overall behavior of all of these systems regarding programmability is surprisingly robust. One of the key benefits of this approach is the fact that pH-switchable materials are highly developed, which allows one to bridge into materials applications fairly quickly and with high predictivity. This has enabled us to demonstrate the first transient hydrogels, that could be shaped for time-programmed release or which can be used for temporary blocking of fluidic circuits. Additionally, we could showcase transient information storage devices based on block copolymer photonics and long-distance signal propagation on such devices. Due to the robustness of pH-feedback systems, a diverse set of other groups have now started using these approaches.

Our ATP-driven systems take a different approach and also take higher levels of inspiration from the dynamics of the microtubule assembly [7]. They address one of the key aspects of chemically driven self-assemblies that can hardly be addressed with active environments, that is, steady-state dynamics. In our approach, we recently introduced a concept for an uphill-driven chemically fueled dynamic covalent bond that is formed between complementary ends of α,ω-telechelic DNA building blocks. Those feature at their ends a short DNA sequence that a room temperature does not allow for any hybridization. However, once a T4 DNA Ligase and a restriction enzyme are added, these building blocks are dynamically fused together in an ATP-driven ligation step and dynamically cut in a restriction step. Depending on the enzymatic composition and the ATP fuel component, this results in transient DNA nanostructures that are not only programmable in their lifetime, as programmed through the transduction kinetics of the chemical fuel, but also programmable in their steady-state dynamics (exchange frequency) and average degree of ligation as programmed through the amount and ratio of the enzymes. This is a unique feature of the system, as the energy uptake and energy dissipation events directly correspond to structural transitions (built up and destruction), which is hardly possible for supramolecular systems. Notably, next to van Esch's dynamic instabilities in methyl sulphate-driven gelators, this is to the best of my knowledge the only chemically driven system known to date with steady-state structural dynamics. The ability to program the dynamics on a bio-catalytic level is a clear advantage.

In summary, both approaches complement each other and take different levels of inspiration from the dynamics of living systems. Toward a transduction to prototype or real-life applications, it is obvious that pH-feedback systems at present are easier to implement, yet the ATP-driven programmable steady-state systems may in the long term hold fundamental advantageous toward the design of highly adaptive flux-state materials.

Outlook and Discussion Points for Future Developments of Research on Non-equilibrium Systems

The most obvious long-term benefit of developing lifelike soft matter systems is to be able to provide synthetic systems that can interact with living matter in an active and interactive way. This will ultimately pave the way to systems that allow for a bio-synthetic co-evolution and in which synthetic and biological entities in fact adapt jointly toward each other. The prospects are extremely high for active implants, active drug delivery systems, or for tissue engineering constructs. To do so, we still have to go a long way in terms of finding common languages to translate biological signaling and fuel molecules as well as mechanical signals into a language acceptable for adaptive and active synthetic matter and vice versa. This requires in particular that the advances in most complex chemical systems research are brought into a closer interaction with materials to allow for synergetic developments at this interface, and for some faster transduction of fundamental reaction networks to potential applications. One also needs to realize that all autonomous chemical systems and feedback-operated materials are at present bound to water or organic solvents. But, if we look at applications, we can realize that there is a whole materials world without solvents. Hence, we need to identify pathways toward computational interfaces in bulk, such as in solvent-free reaction networks, mechanical metamaterials, or neuromorphic devices, to also empower solid-state materials with higher levels of adaptivity. On route to this, and on a fundamental level, it is also obvious that the design and discovery of increasingly complex chemical reaction network systems will shed more light on how life potentially operates.

Next to these more far reaching outlooks, a discussion needs to be fostered surrounding concepts such as "what is a lifelike materials systems", "non-equilibrium and dissipative systems", or "adaptive vs. responsive vs. interactive materials and systems". This is of profound importance to educate future generations of young scientists who are going to operate in the field and who at present may not be able to see clear terminologies or concepts.

Acknowledgments

A.W. acknowledges funding from the European Research Council in the framework of an ERC Starting Grant (TimeProSAMat; 19 677960), and the Deutsche Forschungsgemeinschaft (DFG, German Research Foundation) under Germany's Excellence Strategy (EXC 2193/1-390951807);

References

[1] R. Merindol and A. Walther, *Chem. Soc. Rev.* **46**, 5588 (2017).
[2] L. Heinen and A. Walther, *Soft Matter* **11**, 7857 (2015).
[3] L. Heinen, T. Heuser, A. Steinschulte, and A. Walther, *Nano Lett.* **17**, 4989 (2017).
[4] T. Heuser, R. Merindol, S. Loescher, A. Klaus, and A. Walther, *Adv. Mater.* **29**, 1521 (2017).
[5] T. Heuser, E. Weyandt, and A. Walther, *Angew. Chem. Int. Ed.* **54**, 13258 (2015).
[6] T. Heuser, A.-K. Steppert, C. Molano-Lopez, B. Zhu, and A. Walther, *Nano Lett.* **15**, 2213 (2015).
[7] L. Heinen and A. Walther, *Sci. Adv.* **5**, eaaw0590 (2019).

SESSION 5: MODELING OF NON-EQUILIBRIUM SYSTEMS AND SIMULATION OF MOLECULAR MACHINES

CHAIR: B. FERINGA

AUDITORS: Y. DE DECKER[1], G. DUPONT[2]

[1] *Nonlinear Physical Chemistry Unit and Center for Nonlinear Phenomena and Complex Systems, Université libre de Bruxelles (ULB), CP231, Bd du Triomphe, 1050 Brussels, Belgium*

[2] *Unit of Theoretical Chronobiology, Université libre de Bruxelles (ULB), CP231, Bd du Triomphe, 1050 Brussels, Belgium*

Discussion among the panel members

Ben Feringa: Okay. As time is limited, we should start immediately. I regret now almost that we did not invite also a philosopher because there are some very fundamental questions. Thank you very much for making this remark that maybe we should discuss if we go to this kind of complexity and far-from-equilibrium behavior. Let's take the time to discuss a little bit. A couple of times we mentioned "predictive simulation" and discussed which circuits to build and how to get systems out of equilibrium? What kind of systems to build? Maybe I should start with Wilhelm maybe he wants to comment on that and add a little bit to what he said already. And I can ask also a few others, because it seems to me that there is something very real. We have to team up with computational scientists and maybe even mathematics and so, and other fields.

Wilhelm Huck: Yes, well, I do think so. I think at the moment we know very much how to make a switch or an oscillator or something that forms a gel and then a liquid again, but I think these are probably not really the functions that we would really be looking for. So, if you look at sort-of living systems, I think the networks do more than just form a clock or a switch: they sense gradients; they change directionality; they maintain internal concentrations like homeostasis; they can up-regulate energy production; they can down-regulate energy production. And I think for many of these things we don't really know how to program this in a small network. And so, I do think we need to learn — probably from computational science or information science — on how you program these kinds of networks in a much better way. That is one thing. And then, if you heard Annette Taylor saying as well, you can also just start at the end of course. You can also try and just have a soup of chemical reactions. But then, I think a soup of chemical reactions at the moment doesn't do anything, it just produces soup. And so how you sort of discover what kind of chemical reactions you need to combine in order for this network to self-organize — to self-assemble — into something that has a sort of well-defined function, is something that we don't know at all. We don't know the driving forces for how networks would become organized. So, I think at the moment we can only go from modular design to something more complex. We certainly can't start at the other end and just hope that we have something emerging from that soup.

Ben Feringa: Andreas, Sure you have some opinions about this as well, because you touched upon that already a little bit. Will you add to that a little bit?

Andreas Walther: I think one of the key features is really that we have to think about what is really the function which we want to generate, and this is a little bit unknown. At the moment I think, you know, people are looking at gels a lot because that's super easy to do. This is one of the key aspects. And then if you really want to improve, let's say, what could be a target... I mean, for me, a target could be if we would have a material which can really have active communication with cells, so that the cell would sort of recognize this as a kind of a living extracellular matrix. This would be very complicated to do, but if we would have, let's say, active information exchange between a cell and a kind of lifelike material, this would definitely be something I would say, "okay, this is really a step forward in this field".

Ben Feringa: Annette, you are probably one of the persons closest to biology. Wilhelm made a fairly interesting comment and also in connection with this discussion. Do we have to look at the biological system or do we just have to see, to look more for artificial systems? I'm just putting it in a perspective of how to build an airplane. We like to look at birds, of course, and I use this in my lectures sometimes. But we don't build a bird according to the materials and the motion of a bird is not the inspiration for the airplane, I think. So, I would like to hear a little bit your thoughts.

Annette Taylor: That's an interesting point. That's, I guess, the biomimetic vs the bio-inspired. Like I think we probably move more and more to the bio-inspired. We take inspiration from biology but we don't necessarily try to mimic exactly the same way that biology does it. We want to take certain functions and certain dynamic states, and we want to be able to imitate those using some underlying principles. So, there are certainly, I think, some core principles that we can learn from biology but then when we reconstruct them in chemical systems I think we're probably going to move more toward the bio-inspired. But it's not necessarily the same, exactly the same structures.

Ben Feringa: Let me ask also a little bit about machines to the gentleman that was talking about motors and switches, and moving objects. Because we heard that computational methods are important, what would be the big challenge now where we really need help? Where we should do much better than we do now? Because, still, our understanding of how motion — how we can get motion — it's still fairly primitive. So, will you?

Dean Astumian: I would say that one of the things that you see is this energy surface, and how it's relatively straightforward to engineer deep wells. But it's very difficult to figure out how to make the chemicals such that they have engineered saddle points on that energy surface. And to add to what was being talked about with regard to function and bio-inspiration: there are two ways of looking at it. You have on the one hand, the structural picture of the F_o-F_1 ATPase. So do you try to mimic that, or do you try and design the two-dimensional energy profile? I would say that it's really the latter rather than the former.

Ben Feringa: OK. Bernd, do you want to add to this?

Bernd Hartke: Yes. I think a key is that we need to find good methods for coarse-graining. So for example, I've not talked at all about the methods we use for computing these photo-switches. We don't use the high-end methods. We have used the semi-empirical multi-reference methods, essentially, which most people in the DFT ab initio business would shy away from. But we have to do that in order to be able to run lots of trajectories for a sufficiently long time and for this we need to really analyze the movement, as I pointed out. We need to have full scale dynamics of these molecular motors and after we have done all of that I have shown, we have to forget it and just pick out a higher level description of the function. So we need to have coarse-graining in space, in time to get to bigger systems like all these nice kinetic networks.

Ben Feringa: We need also to team up much more with the biophysics people that study the mechanics of biological machines, to learn from that. Because still... a lot of chemists I talk also with, we have this idea of mechanical motion, you know, connecting it more to the macro-world. Which is completely wrong, I think, and we could probably learn a lot — but I would also like to hear your opinion.

Bernd Hartke: Yes, we can do that. As was pointed out earlier, we can transform from this quantum chemistry picture into a totally mechanical picture, and this you can also do for electronically excited states. For example, you can model transitions between electronic states with force fields. You can do that. It's not going to be right ab initio, we have to parameterize that, but it can be done. So that you are back from quantum chemistry to something which you think you can understand mechanically.

Ben Feringa: Dean?

Dean Astumian: Let me add one thing with regard, for example, to the F_o-F_1 ATP synthase and the energy diagram. Arieh Warshel has made some interesting predictions regarding what happens if you mutate the protein, and it's relatively straightforward to destroy the function. And you can show how that is. It is changing the energy landscape and how the energy landscape, in a predictable way, no longer preserves that Z shape that is the characteristic of the function. But the question is: can you take, for example, a polymer that gives you no functionality in the 2d or 3d potential energy landscape and then predict how you can achieve function? And so that backward problem seems to be much more complicated than figuring out how to destroy the function given that. And that's what you would like to have if you want to have bio-inspired stuff.

Ben Feringa: I would like to go to Thomas. What strikes me in several of the talks, the talk about networks and dissipative systems and whatever, how you define it. Concentration, confinement in time and space, etc.: we don't often think so much about concentration and so, but these kinds of things play a crucial role and Nature has figured out how to find the right path and to deal with it. Also in what Wilhelm called this "mass of chemical dirt" and so... all these things. Can you comment a little bit on that? Because that's a big challenge, I think. No?

Thomas Hermans: So yes indeed, concentration and confinement definitely play roles. If you look at just self-assembling systems, the propensity of nucleation scales very much with the volume you're doing it in. So, if you go to very small volumes, let's say you take a reaction that is self-assembling. It has a nucleation-elongation event, but if you confine it in very small spaces you know some will nucleate some will not nucleate. So, that's definitely something we have to start thinking about more. I don't think many people realize that the behaviors can be quite different in different-size containers. We see it in our systems. Typically, we first do experiments in big cuvettes like milliliters, and then we go down to microscopy or we go to chips, like our microreactors, the behavior can be quite different. So that's definitely something that's currently not also in the modeling: we just use 0-dimensional kinetic models and it doesn't include any confinement property.

Ben Feringa: So we have to integrate those aspects, otherwise it often doesn't make much sense to design a system without taking that into account?

Thomas Hermans: Exactly. I mean, for the moment what we try to do is we try to limit the influences of those effects as much as possible. So, for example, if we go to a chemically-fueled system we try to either mix it up in the solution state, or if we go to a more "gel" state, we try to make it very, very thin so that the diffusion times are basically fast compared to the reaction times. Because if you start going to macroscale objects then you get reaction-diffusion phenomena on top of your non-linear reaction networks, which makes your life very complicated.

<u>Ben Feringa</u>: What makes it even more complicated, and it was mentioned a couple of times I think, is to go to real devices: solid state materials, polymers, films, surfaces. Who wants to comment on that? You maybe want to start?

<u>Thomas Hermans</u>: I can continue a bit on that. Well, one of the issues is that for the moment we are externally fueling our systems. So, we have a material and we add fuel on top. I think a better approach would be like what Nature does: locally provide the fuel or even locally produce it in a material. So if you think about the reaction cycle and you have a material where clearly you have transport limitations, if you just externally add your fuel it needs time to go inside and by that time, it probably already reacted. So I think what we are working towards now is to have a chemical reaction cycle that produces fuel that drives a chemical reaction cycle that controls self-assembly, which produces waste, and that waste goes into another reaction cycle which recycles the waste and so everything is, let's say, done in a kind of "mini-factory" inside your material.

<u>Ben Feringa</u>: That reminds me very much of my early days of my career at Shell, when we were doing heterogeneous catalysis where things had to go in and out and had to be in confined space in very robust solid materials, at high temperatures and high pressures etc. So your message is we should team a lot more up with the people from the heterogeneous catalysis community? Because there is tremendous knowledge there. We heard already beautiful examples these days and so.

<u>Thomas Hermans</u>: I guess so... I haven't talked much to people in that community, maybe I should.

<u>Ben Feringa</u>: Maybe for another meeting?

<u>Wilhelm Huck</u>: Ben? Maybe I can comment on that. I think catalysis is absolutely crucial. We will not be able to make functional networks that are any bigger if we don't have any catalysis. So, you have to be able to tune the rates of your reactions and you can only do that using catalysis eventually, so that's going to be absolutely crucial.

<u>Andreas Walther</u>: Just to add on to this one. So of course, enzymes have this spectacular feature that you just change the enzyme concentration, you have a good influence on the network and this is a key advantage. But also with enzymes, you can spatially immobilize them and this means you can break the spatial homogeneity of the solution and this will ultimately, of course, lead to higher complexity in what we can do. And I think that this is one of the key advantages. And it also refers a little bit to what you said: can we produce the fuel locally? Yes, sure, we should compartmentalize — maybe the fuel source — and this will bring us to higher levels of complexity.

Ben Feringa: Dean?

Dean Astumian: I was just thinking about the example that you showed.. with regard to Dave Lee's switchable catalysis. I mean, that is at the heart of what allostery is. Depending on the mechanical state of the ring between the two recognition sites, the catalyst is either active or much less active... on or off.

Ben Feringa: Okay, I'm happy to see that so many people will support the idea that when we look at the real — our body — life, that a lot of things, most of it is driven by catalysis, that we have to learn how to integrate there. At least there's one point we agree on that we will have a discussion. Just because we have to finish, because it's time for coffee. I was very much intrigued by this idea of definitions and there will be this issue of advanced materials coming up soon dealing with all these aspects. I know you also contributed but can you, once again, advocate or tell what is the key issue? Are we just spoiled with the terminology and so, or are we working in a very undefined network and chaos?

Andreas Walther: It's not totally chaos of course. But I think there's a lot of people that try to use fancy words for, let's say, very solid and good science. And of course it's fair, I mean everybody is doing this so to some extent, that's OK. But I think in the interest of really developing the field, I think we have to get the terminology straight and, I mean, we wrote this viewpoint as a kind of an opening for discussion. I'm not even saying that everything that is written in this viewpoint is entirely correct, right? Or a different community like synthetic biology might define a few things differently. Soft robotics or robotics community might define interaction and adaptation a little bit differently. But it's important that we start to discuss these things. Yes, we have to bring these communities together and also we need to, basically, learn from the biologists and also distinguish us to some extent from the biologists.

Ben Feringa: It reminds almost of how life probably has started in total chaos but with many trials and errors. It's not fairly defined what the terminology of life would be. We feel maybe a little bit like these early stages where it can go in all different kinds of directions, but I'm sure Bartosz and some others will have some comments on this. I think we have time to think a little bit and think about questions to ask to the panel. We will have a coffee now and we will start again in half an hour.

General discussion

Ben Feringa: We have had already some very stimulating discussions during the coffee break and so let's thank our speakers once again for their contributions. Because they stirred up something, not only in their reaction flasks or in their computer. I would like to invite comments and questions to start the discussion with any of the panel members. Let's give this time the floor to seniority, because Henk was raising his hand so quickly before any of the younger ones, even Bartosz could not beat him. So, Henk?

Henk Lekkerkerker: Thank you Ben. We are here in Brussels and we have the famous Brussels school of thermodynamics. Forty-two years ago Prigogine got the Nobel prize for dissipative structures. I was slightly surprised that I did not hear any reference to his work, his ideas. Of course, Turing was before him, Annette, we have the Belousov-Zhabotinsky reaction, etc, the importance of the auto-catalysis: that, I heard. But the whole framework of Prigogine, is it still alive and inspiring you or is it something of a distant past?

Ben Feringa: Who wants to comment on that?

Thomas Hermans: So of course, personally, I was very much inspired by mainly Prigogine's book from 1978 with Nicolis and I've also been at the Brussels school. I've presented here and interacted with Nicolis, so that's definitely a very big source of inspiration. But, I think that as an experimentalist, if you look at the literature since Prigogine, it's quite confusing. If people talk about entropy production, which is kind of a mystical concept, and then some people say it's minimum, maximum, etc... there's a lot of things. Personally, also in the lab, we have tried to put this on a more practical framework. So we tried to go back to when Carnot was doing thermodynamics: he was just measuring heat dissipation, how much heat is coming out of an engine. And so, for an experimentalist, what helped me a lot was to think about things like entropy production in terms of lost work. If you talk to an engineer and you have a flow through a pipe you have friction, that is lost work and they call that entropy production. So once you think about it like that it is much easier. But to come back to Prigogine, I think that the theory and the nonequilibrium thermodynamics is still quite ahead of what the experimentalists are understanding, because there have been many advances in stochastic nonequilibrium thermodynamics which are beyond my grasp. I think there should be a renewed effort to see how relevant the newer theories are to our current work. But definitely, to give you a short answer, Prigogine is a big source of inspiration and of knowledge.

Hank Lekkerkerker: I'm very happy to hear this and I thank you Thomas.

Ben Feringa: Dean, you want to comment on this?

Dean Astumian: So definitely, Prigogine was a huge inspiration for much of my work. But I would like to emphasize that Prigogine always referred to dissipative structures — structures that form in the presence of dissipation — and unfortunately there has been a shift where now the terminology is sometimes "dissipation-driven" this or that, which seems to me odd. I do not think that Prigogine would have thought that to be sensible. Because dissipation is waste and the point is that you've got dissipation, you cannot get around it, but you want to design the system in a way that it can use the energy flow. And one of the things I note that Prigogine did not focus on were the kinetics, how you could manipulate barriers. And it is very much nonequilibrium "thermal dynamics" and so I would like to emphasize kinetics, because it's a very important part of how you can control the energy flows.

Ben Feringa: We are here in Brussels and we value the enormous contributions of Prigogine. Thanks Henk for asking this question. I would ask in the audience if they want to comment.

Eugene Shakhnovich: Basically, what I wanted to comment on the whole concept, and that is related to what Prigogine was doing. I sort of know what he was doing because I teach part of his work in my nonequilibrium statistical mechanics. But basically, this type of approach, this type of formalism which is based on ordinary differential equations (ODEs) with barriers, with the kinetic rates determined by barriers is kind of maybe correct, but under certain limitations, which I'm not sure that they are valid in the systems under discussion here. For example, if you have mechanical motion you have to consider dissipation and the solvent hydrodynamic interactions, etc., which will bring you to this question of maximization of entropy as a formalism rather than kinetics of reactions as a formalism, and that will give you in several cases very different outcomes. And in this sense, kind of returning to Prigogine, his formulation of one-dimensional projector operators, and maximization of entropy could be much more relevant than equations of chemical kinetics, so, ODEs which are presented here. Specifically, I think that what you have to do in the case of these ratchets and motion, you have to use Kramers' equation rather than equations of chemical kinetics and with stochastic noise, and in this case you can get very different outcomes. For example, the statement that kinetics rules might become true or not true depending on the conditions. So, I would call for considerable caution and re-evaluation of the paradigm here, because I think that in systems with motion and dissipation with the solvent, this is crucial to take into account and stochasticity plays a crucial role. Because then you have coupled fluctuations in the solvent and fluctuations in the system, which will make the transition state theory kind of invalid for these systems. So, that is my comment.

Ben Feringa: First, the floor to Dean, because I'm sure he wants to answer or to comment on this.

Dean Astumian: There are a couple of things. First of, you have motion of course, but it's equilibrium motion. There's no effective friction at the low Reynolds number regime. As you add ATP, for example, does the character of the motion of, say, the F_0-F_1 ATPsynthase change? Of course not. Only the probability for carrying out a trajectory in which ATP goes to become ADP and P_i, relative to the microscopic reverse. Now all of the equations, of course I'm talking about transition state theory, they have been solved also using Kramers, with reaction-diffusion equations and Kramers' equation with passage over parabolic or non-parabolic barriers, and it gives the exact same result — so long as you maintain microscopic reversibility. It's only when you violate microscopic reversibility and start throwing rate constants that don't make sense that you get significant differences between a rate theory and a theory based on Kramers' diffusion reaction.

Ben Feringa: Thank you very much. I would like to go to Lee because he was raising his hand already a few times. Do you want to add something?

Lee Cronin: A comment about Prigogine and also a question. Hopefully as provocative as possible. The comment about Prigogine is that I was really inspired by his work, and also the work of Stuart Kauffman, who basically looked at modern biology and dissipation, and also Manfred Eigen. What Prigogine's work kind of told me is what I've missed, what I've lost. So, the problem of energy dissipation is that we know something interesting has happened because entropy has been dissipated. But it is a bit like the tail wagging the dog. We know something interesting has happened but we don't know why. So this comes to my question. Everybody's been talking about dissipation and energy... I would like to ask all the panelists actually: what role does information that you put into your system in terms of the constraints allow you to genuinely discover new phenomena? And so we take information for granted in biology, and for us and the causal structures which produce experiments. I think we're all missing a fundamental missing problem of our reality which is: how do we use information in a chemical system, not energy? And I think I almost agree with Dean that we have to control the kinetic landscapes to get interesting things to happen. So question to the panelists: can we stop thinking about entropy and start thinking of information.

Ben Feringa: Thank you. A very intriguing question. Now you've asked a question to every panelist so please, let's start. Who wants to start?

Andreas Walther: In general, what we have to say is that we're very poor at information processing in all of the systems we have described so far. So, the systems which can really process information at the moment are DNA computers, I would say. And to some extent your system which you showed. There you can make consensus decisions and you can basically have an adaptive response. So, the question is really if we have information coming into a system, how can we decide how we have these sensory inputs? How can we weight them against each other? For this, we need a chemical reaction network. And then this needs to reshape the energy landscape because this is the only way we can come to an adaptive response depending on information input. That is a super big challenge. I think everyone of us has some ideas on this, but this is not something which can be done very easily.

<u>Ben Feringa</u>: I add one comment to this discussion and questions: I had this discussion with George Whitesides and George cannot be here. He asked the question: what is information in a chemical system or in a molecular system?

<u>Dean Astumian</u>: Landauer had pointed out that information is always physical. And the ring sliding back and forth revealing a catalyst, or not revealing a catalyst, is information — and it is physical.

<u>Ben Ferringa</u>: Okay, point made.

<u>Wilhelm Huck</u>: Maybe the other information you have in your system is your concentration of starting materials. By changing the set of starting concentrations you can tune your information in the system, and of course after the end of your reaction network you have a different set of concentrations of those molecules that are being transformed. So, we know how to process the information. How to program it and how to do something useful with it, I think that is very difficult at the moment. We are thinking about ways that your enzyme would be your hardware, essentially, and your software is this combination of peptide molecules that go into the systems and are being processed. Then if you want to program something you essentially make small changes in that input concentration. Your information is a delta in concentrations, and your output is again somewhere along the curve because what your process is doing is essentially carrying out chemical reactions, so you know where it will end up. What I'm looking for is interesting problems that you would then solve and that, I don't know yet.

<u>Thomas Hermans</u>: Just a very small comment. Of course, people have tried to link information to dissipation. There's the Shannon entropy, but I think what is problematic is that it cannot directly be linked to chemical systems. And the kinetics works so well. To come back to the previous question, people are sticking to ODEs because they work so well and then it's easy to forget about the other.

<u>Lee Cronin</u>: Just one comment about Shannon entropy. I think we should stop thinking about Shannon information. The Shannon entropy was built for user communications, for error correction on phone lines and we misuse it. Shannon information has no meaning in a chemical system. What has a meaning is what Wilhelm was talking about, with the input concentrations and so on. I think lots of people argue about this because they like writing down these numbers and they think they mean something but I've never seen it used for anything.

<u>Annette Taylor</u>: The state of the system is used for information processing. People do construct logic gates with these kinds of chemical systems. You can consider a reacted state as an ON state and an unreacted state as an OFF state and you can do digital processing in that way. Whether you'd want to use that, because biology

works in this fashion, I don't know. I'm not entirely convinced but it's possible to do that kind of information processing.

Ben Feringa: Let me now give the word to Kurt. And then to Bartosz.

Kurt Wüthrich: Relating to the mention of Prigogine, I think that Onsager's flux equations could be a much more direct source for what you have been doing and are doing. But my intervention also targets a different aspect: I am wondering whether the fact that all of you show biological systems is a sign of maturity or immaturity of the field. Should you not aim at going beyond mimicking biological systems? To be specific, I rarely see mechanical bonds in biological systems. When you got the Nobel prize, Ben, everyone was talking about mechanical bonds.

Ben Feringa: I didn't make any mechanical bond.

Kurt Wüthrich: Yes, but others did who shared the prize with you. Maybe mechanical bonds would eventually lead beyond mimicking biological systems?

Wilhelm Huck: Maybe a quick comment. First of all the only really complex molecular system we know is a living system. So in that respect, we do not really have much to go if we do not work with biological systems. I think if you look at the many isolated attempts to go in other directions, the problem is that they all work in different solvents. If we ever want to make something bigger, more complex, we will have to decide on a common solvent for example. There again water seems like a good choice. And if you say what else would you like to make? Yes, you would like to have intelligent materials. But if you ask me what type of intelligent material would you like to make? You would say, well, a material needs to have a function and a property, needs to have certain specifications. And I think that at the moment we have no idea what sort of property/function/specifications we would be looking at. To come back to your first comment, it's probably a reflection of the immaturity of the field that at the moment, we are scrambling to understand living systems at this kind of complex systems level and trying to imitate those, rather than go beyond and invent completely different systems.

Ben Feringa: Just a quick remark.

Andreas Walther: So I think there's plenty of mechanical bonds in Nature, like the von Willebrand factor and all kinds of things which are mechanosensitive bonds. And I think incorporating, for instance, mechano-sensitive bonds into these chemical reaction networks to expose some cryptic sites, or something like this which can then interact with the network, will give rise to very interesting mechano-adaptive materials. And this is something which we are actually trying to do, and this will make actually such kind of materials then more adaptive.

Ben Feringa: Thank you.

Bartosz Grzybowski: I have a simple comment regarding the chemistry of all this entire business of reaction networks. When you start building, let's say, cycles — which is the basic unit of all of this for oscillators and all that stuff — we actually know surprisingly few reactions that decompose molecules. All these cycles are to build, let's say an ester, and then you hydrolyze the ester. When you look at Nature, you build mass and then you cleave the molecule in a non-trivial way. I think there is plenty of chemists who are not taught to be thinking about reactions decomposing big molecules.

Ben Feringa: Thank you, may I immediately react to that as a chairman? One of the big challenges in chemistry will be recycling. We are extremely good at making things from our polymers into God-knows-what. And the challenge for the future, one of the big challenges, is how to recycle things. We need new reactions to decompose into building blocks, so that we can use them. There will be a lot of discoveries and I think that will feed this field as well.

Bartosz Grzybowski: And the second part of the comment was a reaction. I think the approach that the field is taking is mostly modular. Let's say Wilhelm has an oscillator, and Andreas has an oscillator and let's put them together. Let's say that you have three species in the oscillation in the cycle and you have three species, so you have six interactions between them. But you have to take care of nine cross-talk interactions between them that could be conflicting. So it's already very hard to find an oscillator, but putting them together you have to make sure they are compatible chemically and this factor grows linearly with the complexity of the system. So the huge thing will be to avoid cross-reactivity between the components of the systems, and how mother Nature managed to get all of this working by putting it in different compartments is one way of solving the problem. Still, you cannot have a million compartments... So for chemistry, you know how to do it: it will require large data, comparing reactions and making sure everything is compatible in one soup. And I think this is a huge problem.

Ben Feringa: Andreas, do you want to comment on that?

Andreas Walther: II think that if the chemistry is really orthogonal, we could run them in the same pot and basically they would have very little interference. But you are totally right, so as soon as they start to interfere... This probably needs modeling first: we have to understand a little bit about the interference, then some modeling, and then maybe we can come to a reasonable functional output of two interfering networks. It's a significant challenge, I would say.

Bartosz Grzybowski: I think you don't have enough chemistries to actually make them non-interfering. Because if you design two cycles, your cycle will have hydrolysis, and his cycle will have hydrolysis somewhere, and my cycle will have hydrolysis. So they will all use the same major components.

Ben Feringa: Wilhelm?

Wilhelm Huck: Bartosz, I completely agree with you. Hence my sad face during my lecture. This is a major problem and I think there are a couple of ways to solve it. First of all, that's why I mentioned catalysis, because I think one of the big things in Nature of course is that not only do you catalyze reactions, but you can switch catalysis on and off. And so that is one way of ensuring that your similar chemistries become orthogonal. The other one is we have to remember that every cell (if we look at bacterial cells) is very crowded. You are talking about 300g/liter of macromolecules. So you don't need a million compartments, you just need something very viscous. Then, if your molecules are in the right place, your reaction should outcompete diffusion. Or you should sort of think about coupling reaction and diffusion, or competing reaction and diffusion. That would be another way of physically essentially compartmentalizing things even though they are not really compartmentalized.

Ben Feringa: I would like to have you Laura first and then you, and then Joanna.

Laura Gagliardi: I'm going to switch gears. This is a very transdisciplinary meeting. And so I'd like to try to bridge the gap between what we are discussing now and what we heard, for example, yesterday afternoon in which the characterization was done really at a very detailed level. So, when you talk about catalysis for example, do you really know what you are determining at a detailed level? And how can we try to help? Should the theorists, for example, try to give you more detailed information? What type of computational models would you need from us? Because today it's very philosophical, while yesterday and the day before, it was more real for me because my reality is more chemical reality than this, but I was wondering if we could help?

Ben Feringa: A few people that want to comment on this.

Andreas Walther: If we compare, for instance, to the session yesterday — it was super-resolution microscopy. Many of us are working on self-assembling structures where enzymes start to modulate something. But we can only get some sort of ensemble average outcome out of this and we don't really know how enzymes really interact with these things. For example, for the DNA system which I showed, all of these enzymes have a DNA binding domain. Do they only stick there or do they migrate on top of these structures and so on? These would be very interesting things to identify. And this actually needs super-resolution microscopy because that's the only way to access this with sufficient spatiotemporal resolution. So I think that there will be very significant synergies.

<u>Laura Gagliardi:</u> So, is this a grand challenge for the future?

<u>Andreas Walther:</u> Analyzing mechanistic details, how that really works, just to see whether you have dynamic instabilities, for instance, in your synthetic system: this is a grand challenge too, absolutely.

<u>Ben Feringa:</u> If I may add to that. I worked my whole career in homogeneous catalysis, particularly asymmetric catalysis, and I tell you after 30 years that to predict one single rhodium complex or iridium or copper complex that does an asymmetric transformation, and predict which kind of ligand will give you high selectivity for one biphase or the other... it's still hand-wavy and impossible. So yes, we have to much better understand the catalytic mechanisms. And together with the theoreticians, we work on what are the distinctive factors. But this has also to do again with the basics of the dynamics of a catalyst. What is molecular interaction meaning really, in a real live situation? When there are solvent molecules? Or you work in a heterogeneous system, etc? This keeps us busy, the whole catalysis community, and I'm sure Bert Weckhuysen agrees with that because they are struggling also and they try to find out what the catalytic sites exactly actually are, in real life. So thanks for the question. I think we should move on.

<u>Markus Covert:</u> Really interesting, so I appreciate that. First of all, really great session. I was curious about something. Andreas, I think you mentioned specifically going and interacting with synthetic biology or some of these other things and I thought about that. The currency of synthetic biology right now is largely gene expression. But there are other great tools you have. And I think, Wilhelm and Annette, you had ideas in this area that were moving toward connecting these kinds of engineered processes. I think it could be really powerful, just for example, with Bartosz, the idea of having cross-reactivity... You could imagine eliminating by means of selective expression as well as by localization, and so: what is one idea, what do you think the first steps would be that would be exciting? And what do you see as an exciting ten-year achievement in this space as you start to think about bringing these together?

<u>Wilhem Huck:</u> Marcus has an excellent point and maybe it also goes back to what Kurt was referring to earlier. You can see that the solutions to all these problems that we will have when building large networks and orthogonality have been solved by biology, by using essentially very few different chemical reactions but being able to switch them on and off by tuning the rates. And this is why synthetic biology is "easily" scalable. Because you use the same tool over and over again and you can apply it to all these different genes and tune the sort of expression levels. I think that we have to take on board these completely synthetic systems as well. But translating that to a really completely synthetic system is very difficult. We have little inspiration and that is because, as Bartosz said, we don't have a set of chemistries to go backwards and forwards.

Ben Ferringa: We've also been misled a little bit sometimes by biology because it sets the stage along a certain path. So maybe you also want to comment on that because you mentioned it during your lecture, Bernd? To compare artificial systems with biological systems?

Bernd Hartke: I'm totally not a biologist, but my impression is that in lots of biological systems you have features that are probably not optimally designed. Because it evolved, so Nature had to do with what was there, and if we would design optimally the system from scratch it would perhaps look a lot different.

Markus Covert: Can I just add one little point? Because now some of this synthetic biology at least is happening cell-free. Just so we throw this into the discussion. It doesn't have to happen in a cell. You can do quite a bit. The time scales are weird — that is one of the most interesting differences, the time scales are quite different.

Andreas Walther: There are different ways we can interface these two disciplines. The transcription machinery has been used ex vivo already to switch relatively simple materials — and also using some network approaches, so there's progress in this direction. For the DNA computing field, it's still a big challenge. People can calculate relatively complex things already, but the output is typically a fluorescence signal. You need to throw in a huge amount of DNA to basically compute this, and then the question is how to bind this to a material. This is one of the challenges, but this is one of the challenges which we have to overcome, this is very clear. And of course a completely different approach is to really think in the direction of living materials. We can put cells into gels, then of course we're back in gels. But if we can use the intelligence of the cell to produce something even in an adaptive manner and then to trigger a material's response... Of course it is an attractive approach. There are a few people working in this direction and its very worthwhile, because there you have basically the information processing system, alive.

Ben Feringa: Thank very much. Joanna finally!

Joanna Aizenberg: Thank you. I have a couple of comments and also a question related to that. I feel — yes, it's a chemistry conference — but in this particular case, we are a little bit too chemical. And the reason I'm saying that is that in many cases, especially if we talk about materials, at the end we want to use it for a material. Likely it is a solid state and these reactions that we discuss would happen in, let's say, a poro-elastic medium. Maybe the better way to address the problem then is to think about your poro-elastic medium as an example of highly crowded environment, as we are doing in biology. And use this medium to manipulate, control, and provide the way to have signal processing in your system. It's a system! Unlikely you will be able to find one molecule of one reaction to give you everything.

So you have to have sensing elements, you have to have elements that will transduce your signal through the system and you have to have elements that would respond to that. We are discussing the ability to encode information. Encoding information is not necessarily done through molecules that react in your system. Encoding of information could be in your poro-elastic medium. For example, through encoding order parameter that would induce order/disorder transitions in the system. So maybe an interesting way to think about it — especially from this theory point of view— would be not to start with reaction and reaction kinetics but rather with dynamic poro-elasticity where changes are happening in the medium where reactions take place and use these dynamic poro-elasticity with hydrodynamic terms and enhance it with chemical reaction pieces— using ideas of fluxes and things that may happen in this system. So I really wanted to put together this ability to encode information into order and disorder, not only into molecules. And also to think about reactions happening in medium other than a solvent. It's too much focus on a solvent. Our medium is not going to be a solvent if we talk about application and materials that would arise from that.

Ben Feringa: I give the word to Thomas.

Thomas Hermans: You're exactly right, I think. In one of the examples I showed where we have a supermolecular oscillator, it's not just a chemical reaction network of small molecules. Actually, the autocatalytic step in that process is the self-assembly itself. Because the self-assembly is now in the network, it automatically will have, probably, a mechanical feedback. So if you push on something and you change the self-assembly or limit it, it will feedback automatically in the reaction network. I think that's one of the ways to couple things together. Another way, which we found by accident, is that self-assembly can change the density of your medium, and when you have density differences, you can have convective currents, you can have instabilities... all these kinds of things. Also we can observe pattern formation in our cases because the reaction network induces self-assembly, which changes density, which triggers convection, which then changes the transport as well, which influences the reaction cycle. So I think we are trying, or we are starting as a community, to couple not only the reaction network and have a material respond, but also have the feedback loop. Just like what you have been pursuing in your mechanical systems. So I think it is definitely a very interesting point.

Ben Feringa: Anybody else that wants to comment at this point? Otherwise I go to Bert Weckhuysen.

Bert Weckhuysen: I would like to come back to the reference to catalysis. I must say I enjoyed it very much, and that life has the beauty of being able to make this network of chemical reactions. If you look into a plant, if you would go to Antwerp Harbor, you would see all the pipelines and you can somehow get a picture.

I forgot by whom the reference was made but it was something like "that looks like a subway". We can also go to Antwerp Harbor and go to the chemical plants and also see that network. But there the big difference is that we have a series of reactors at every time controlled by T (temperature), P (pressure). And then you do locally in that reactor, you control, actually you maximize your yield. Now here in all the systems I've seen it's room temperature so your T and P, it's atmospheric conditions. So if we want to merge our fields and make a totally new Antwerp Harbor chemical infrastructure, then I want to make one big plant where locally I can control what you all have stated and then make a kind of cell/system where you can do that. And you can tweak... Today I want to make ethylene, propylene, aromatic in that ratio, and I just want my one feed in and my product out that's it.

Wilhelm Huck: Well I think in general you are quite right, that chemical engineers have a network of chemical reactions. They don't have it in one reactor, they have it in a series of connected reactors. And so I think one of the solutions that we can go for is indeed to couple reactions by coupling reactors together. It is not easy to do this, because experimentally it's quite hard to design reactors and how you connect them, and whether you have diffusion between them and all these kinds of things, but it would definitely be a solution. Whether you would like to do this and then go to controlling everything by pressure and temperature and make polyethylene or ethylene gas, I don't know. Because the key thing about living systems is of course to make the right molecule at the right time at the right place, and so not necessarily a lot of something.

Bert Weckhuysen: If we move into circle economy, to circularity, and I think Ben also referred to it, maybe it's a too far stretch to totally redesign our chemical plants. Beyond that let's try to dream and think in a totally different way. And then think about how we could embrace everything what you have stated and try to see how we can push that in a new era of chemical plants. And I'm sure we can inspire each other and try to see how we can move forward and try to make loops and things. There should be a way to get it more flexible, smaller, circular and so on.

Ben Feringa: If I may comment on that also, from a catalysis perspective. What you see of course is that several groups around the world are working on these integrated cascade reactions, where you take one intermediate and then you activate another catalyst for the next step, or integrate enzymatic steps with transition metal-catalysed steps in homogeneous conditions. And there are some beautiful examples where you can do three or four steps in a sequence, also switching catalysts on and off, it was mentioned before, on demand. Now there are the first beautiful examples where you can control stereochemistry and also reactive activity. So you can switch a catalyst on and switch it off, for instance with a light signal or by using a light switch. So, yes I think there is a lot of opportunities there, but maybe somebody else wants to comment on this before I go back to materials.

Julia Yeomans: Ben, at the end of your talk you talked about active droplets, and on the micron lengthscale people have made a lot of machines. They've made active colloids, active droplets, microswimmers. They're driven by everything: by light, by magnetic things. And then there is this huge gap to what I have been hearing about today. Where is the overlap? Should there be an overlap? Or are we sort of trying to get at the same problem from different directions — from top down and bottom up?

Ben Feringa: There is tremendous opportunities there I think to connect to the people who have worked on colloids and these kinds of microdroplets. Because I think there is a lot we can learn from each other if you bring in these dynamic functions: how it interacts with the environment, etc. And we try to do that and I know several groups trying to do that. Not only soft materials, but also nanoparticles and those kinds of materials, integrate them with soft and adaptive functions. Yes, I think there is a whole world out there where we can learn a lot from each other but maybe someone else wants to comment on that?

Andreas Walther: I think you're making a very good point. For instance, if you look at these motors, you can drive them by catalysis. This is one of the most obvious things you can do there. If you can regulate the catalysis using a chemical reaction network, to select how active is the catalyst depending on the inputs, this is something which, of course, could already be made in a much smarter way. This is definitely something where there can be an overlap between these two fields. The chemical reaction network to select and value the information, and then the motor function on the particle, because this is really moving forward. This is definitely where there can be very interesting overlap.

Ben Feringa: If I may just mention two examples. Rafal Klajn at the Weizmann Institute made this beautiful system where he had these adaptive particles that swarm and come together to make a nano reactor, and they are released again. And you have also this swarming behavior with catalytic propulsion that, for instance, Sen at the University of Pennsylvania studies, where these particles come together and they swarm depending on the chemical environment. So, yes, I think there is a lot to be learned to make them more like a system behavior.

Annette Taylor: Another comment. I think we still don't probably mimic with motion. We do not produce synthetic analogues of how self-propulsion occurs at the cell level very well. I think the mechanisms for the self-propelling droplets do not work in biological fluids, do not work under conditions where you have got high ionic concentrations, for example. So I think we are producing mechanisms for motion that are very interesting and we can learn from them but they are not necessarily the same mechanisms that occur in biological systems.

Eva Nogales: I don't know if this is taking things in the opposite direction to where you wanted. But obviously, you want to create complex systems and you want to design them. And you are talking about bio-inspiration because there are obviously interesting, complex things that happen in biology that you would want to be able to design. But I just wonder if in that effort to generate and engineer complexity, you could not at a certain state, from time to time, relinquish a little bit of the control and do what the biological systems do, which are not created by design. That complexity is just created through variation and selection. I'm talking here in a very abstract way, because it would very much depend on the type of system that you do it with, and the pieces you have to play with. Could you not open new venues for your designs by having a variable, something that you could then screen for your activity, that will bring a new influx of ideas that otherwise are not necessarily going to come to you, and then go back to learn something from there. Is there a little bit of space for that kind of evolutionary type of bio-inspiration that can be incorporated in what otherwise would be a true design?

Ben Feringa: Wilhelm?

Wilhelm Huck: That's a great point. The short answer is: no, we can't at the moment. That is the reason why we are now looking more at real networks, because if you see the underlying complexity, it's much larger than the topology that you have. There's a lot of steps in there that are kinetically resolved. A lot of these pathways can easily branch, but they don't branch in reality because something reacts faster than the number of molecules can deal with and therefore it does not branch off into another stream, until you start blocking things. So, yes, you would love to build these kinds of systems that have this sort of unexpected properties but at the moment we can only build really small ones. That's also why I was thinking earlier in terms of theory. We would like to have some theory that guides us in terms of how to make networks that are automatically coming together, that assemble into higher order of complexities without us having to design every step. Because I agree with you, in that design step you lose a lot of new and emergent properties that you would like to see. I completely agree.

Ben Feringa: Thanks Wilhelm. Lee, you will have to wait one second because I give preference to your neighbor because we give preference to people who have not asked questions in this session. So Mark, you first and then Lee. Please go ahead.

Mark Ellisman: I'm not sure this is relevant, but again I'll make an attempt to add something that links activities in biological and non-biological systems, and one thing I did not hear about that I would like your thoughts on. Maybe nobody on the panel knows about something called an "entropic brush"? If you are making latex paints you have a problem keeping them in solution, so what you do is add a whisker that moves around and keeps the colloid in solution. Well, a very creative biologist, Jan Hoh, in the early days of tapping Atomic force Microscopy (AFM) was trying to figure out how neurofilaments — which are an intermediate filament type which don't cross-link with one another — kept at distance from one another, and what he discovered is that the high molecular weight subunit of the neurofilament (it is a 212 kDa protein) makes arms like a bottlebrush and swings around. And he did that by measuring the spring constant with an AFM. So here you have a system in biology, that of course was discovered in biology but the metaphor came from, I guess you would say chemistry of paints. Probably, it is rather universal that you have these different parts of molecules that have very different characteristics. Now, if you fast forward to some of the sessions we have had at this meeting and you think about the implications for water, for Brownian ratchets which we have not talked about, maybe someone on the panel will get activated by that little story and add something that you did not add to your talks.

Ben Feringa: I personally appreciate very much your comments because we work on active coatings and, there, this plays a crucial role of course. When we make these kinds of particles with these kinds of function. But maybe somebody wants to comment on this. Wilhelm?

Wilhelm Huck: I was for more than a decade professor of polymer chemistry and we made a lot of bottlebrush and surface coatings based on polyelectrolyte brushes. Yes, I know these systems; they are also in your knees and the proteoglycans to keep your cartilage from being compressed. I think it's an interesting idea to think about that as a reaction medium — also linking up with what Joanna was saying — as a poro-elastic medium maybe, to use these effects to control where molecules are going. So I think once we take the spatial component into account, then definitely these kinds of environments may be very interesting.

Ben Feringa: Thank you.

Andreas Walter: We have been working on systems a little bit reminiscent of this one, because we want to break the spatial homogeneity of the system. So the DNA systems which I showed you, you can actually grow them as a brush on a surface and what is interesting in this system is that you are in a dynamic steady state. If you add a new building block, it will actually dynamically organize into this grown brush layer. We haven't analyzed how this influences the stability or anything like this. . . But of course by immobilizing this onto these colloids, you will change the reaction network to some extent, because it becomes less and less accessible for the enzyme. So there are interesting effects in these entropic brush layers.

Lee Cronin: I just have a quick question. It was inspired by the question about how we might get complexity into the system or select function, because a lot of us were talking about how we understand things and design them. And I think Wilhelm put it quite nicely when he said we start with biology because we do not have anything else as complex. And we know what biology does. And so my question is to all of you, but particularly the people making adaptive materials: have you considered rather than designing reactions or materials that you think about the problem you want to solve — let's say it's a coating problem — and you code it up and you make a closed loop robot. So you generate a mess of soup, you challenge the soup with your assay. Is it any good? If yes, you then put that composition into a database and you make a sort of artificial closed loop evolutionary system, kind of similar to what Eugene was expressing about some of the protein folding antibiotics stuff. So, you have this kind of high throughput approach. And it seems to me that the adaptive systems are so complex you are never going to engineer everything from the bottom up. So why not engineer a robot with a fitness function to design your complex material?

Thomas Hermans: It is a very good idea. It also alludes to another comment that was made previously. Instead of the robot, you can have a kind of chemistry which is more promiscuous. You could have a range of substrates, which can all react with whatever fuel you're putting in, which can then form structures and then you can do selection by function or structure. But I think a lot of the systems which are currently developed, let's say the activation or deactivation of your building block, are very specific and they do not allow mutations to arise.

Lee Cronin: Just a quick follow up. What I mean is that you are worrying too much about your design. Define what is it you want to achieve with your product, then you challenge the system with that and then you need to close the loop. Biology does the search really well and it persists, so what you have to do is penalize things that do not do what you want. You kill them. You never consider them again and states, processes, objects, concentrations, that give you something drifting in the direction you want, you kind of amplify that and keep them in the system. It is really important to have that...

Thomas Hermans: I think the issue we have is that chemical space is so huge. If I just order 5,000 chemicals from Sigma-Aldrich and chunk them together, how is that practically going to work? We have a problem with defining the initial conditions to be reasonable, to expect at least some outcome.

Thomas Cech: Just following up on that, I agree and I think what biology does is: it has got the ability to amplify a single variant into a large amount of the same variant. And I do not see how that would work in a chemical system. I think that in this mixture that you just pointed out, even if you had some winners in there, that other molecules in the vast mix would poison the reaction. And that is why I think it's not easy to have a selection-amplification system in chemistry in the way that has been done in both in vitro, in Systematic Evolution of Ligands by EXponential enrichment (SELEX), in RNA functional systems, evolutionary systems, and also in actual life.

Dean Astumian: One of the problems that Lee's approach would have is that you actually would have to do the experiments. You have to assay whether or not the molecule that comes out does in fact what you want it to do. Whereas if you can understand, for example, the free energy profile or landscape in terms of the function that it would produce, you are doing everything computationally. And then you can select few possible variants and actually go and do the experiments. It is much quicker to do an experiment with a computer where you do not actually have to manipulate tests tubes.

Ben Feringa: Thanks Dean.

Raimund Ober: A much more general question from a non-chemist. Since this is a computational conference, we have heard about different types of modeling approaches: some ODE-based ones, some other ones. So I wanted to ask you: if someone were young and starting to get into your field, where do you think the gaps are? Where do you think things should go from a modeling point of view, which types of methods of modeling do you think require significant work and attention? And then, one additional point which was already brought up before: do you actually have the experimental capability to discriminate between models? How much can experiments really inform us as to which models really work and which do not work?

Ben Feringa: Thank you very much and now you challenge all of us. Who is the first one who wants to comment?

Annette Taylor: I think one issue is: we often try to distill out chemical networks down to three variables, because it is a lot easier to track steady states with low numbers of variables than it is with a lot of variables. You end up with a lot of very messy phase diagrams when you have a lot of variables that are contributing to the behavior that you are interested in. There is a computational challenge there, which is trying to track steady states and trying to pull out very generic dynamical features when you have a lot of variables contributing to the behavior. And it is interesting to me still, that we can have in a reaction vessel, like the Belousov-Zhabotinsky reaction for example, hundreds of reactions taking place and yet we can still distill that into a two variable system, and we can quite nicely reproduce a lot of that behavior in a generic sense. But if we want to find behavior when looking in a large region of parameter space, obviously then it becomes a much more challenging problem. So, searching parameter space as well when we have a lot of variables contributing to the behavior is also a challenge.

Ben Feringa: Thank you very much. Wilhelm?

Wilhelm Huck: Combining your question with the comment that Tom Cech made earlier, I think the big problem we have in these chemical systems is that we cannot select one mutant that has a certain property and then generate a whole set of new mutants that have similar related properties, but sort of evolve further. Computationally, by combination with Artificial Intelligence or other things, what if you could actually generate this new library of reactions that you could all use to replace the reaction that you just had, but have slightly different reaction kinetics? At the moment if we have one reaction that might do the job, but it just needs to be a thousand times faster, then there's nothing we can do. If you have a Diels-Alder reaction, how do I then replace that Diels-Alder reaction by something else that has the right rate constants and the right products, or by a whole range of them so I can start a new selection. That's maybe where computational tools can come in to really simulate those new libraries that you could eventually test, because I cannot test them all, so I need to eventually only use the ones that look promising.

Ben Feringa: Thank you very much. I think we have to come slowly to a close but, as the chairman, I want to take the opportunity to ask a final question to Joanna. She was discussing about soft materials and as you are more into materials, I was wondering what we hardly discussed or touched about, is that there is this huge community of semiconductor physics, redox materials, etc. which changed our world. Now semiconductor physics materials and this kind of things are not particularly invented by living systems. But should we not team up, also, with those communities to see and to take advantage of redox processes, electrochemistry, and all this kind of things? We see them of course in some enzymes and in the photosynthesis machinery, but are there not tremendous opportunities to integrate them with soft materials, the things you are talking about, and to give us completely different opportunities?

Joanna Aizenberg: Thank you Ben, for this question. Coming back to the idea of optimization in biology, I really do not think there is any optimization in biology. Biology is trapped, at best, in suboptimal solutions because it is seriously limited in materials. It is seriously limited by the type of conditions biology lives in. And indeed, we've done a lot of interesting things in semiconductors; we've done interesting things in metals, in steel. Just so many things that we know how to do very well that have nothing to do with biology or with dynamics. I truly believe that if we now think about it as a hybrid system (and not the way we consider hybrid in purely chemical terms but hybrid by taking a system where you have multiple elements that are optimized to do what they know how to do), for example, sensing elements could be made not necessarily from trying to come up with a sensing molecule, we can use a sensing element. That could be a nanoparticle, that could be a semiconductor particle or something else that is already done by the electrical engineering community. But then, if you need to transduce information you can use optics. There are so many things. I honestly believe that chemistry is wonderful, and I am a chemist, but we should combine chemistry with advances in mechanical engineering, in electrical engineering, in materials science in general. We should not be shy of taking the advantages and combining them with what we can do best, and what we can do is actually improve what their approach is by providing better molecules or better materials. The idea of systems chemistry — systems biology was mentioned here —, thinking about systems chemistry, or even systems materials, especially if we talk about responsive dynamic adaptive materials, we need to have sensors, transducers, and action items and to be able then to create new robots or whatever it is where function is important. So, thank you again for this question and I hope that chemists would be more and more open to working with these communities, not only with modeling communities but actually with engineers who are coming from other fields.

Ben Feringa: Thank you very much for this prospective view to the future. Yes, chemistry has a lot to discover and to invent. It is a bright future, I think for all of us, if we are daring enough to think a little bit broader and to team up with our colleagues from other disciplines. I greatly appreciate the discussions, the contributions of all of you. I thank the speakers of course, but all of you to bring in your views and perspectives which will move our fields forward to the future. Thank you so much.

Session 6

Computers in Interactive Structural Biology
Leading to Modeling of an Intact Biological Cell

INTRODUCTION TO MODELLING OF AN INTACT BIOLOGICAL CELL: STRUCTURAL BIOLOGY, FLUORESCENCE MICROSCOPY AND COMPUTATIONAL METHODS

THOMAS R. CECH

BioFrontiers Institute and Howard Hughes Medical Institute, University of Colorado, Boulder CO 80303, USA

Seeing is Believing

Biologists are no longer content with observing organisms in action. They dig down to understand the structure and function of individual cells within the organism and of the biomolecules that enable life. In the last two decades, unparalleled advances in three areas have revolutionized cell and molecular biology. First, genomics and the other "Omics" (including transcriptomics and proteomics) have advanced to the point where a single student can catalog the status of many thousands of genes, RNA molecules, or proteins in a biological sample in a week. In many cases, this can be done at the single cell level. Second, the field of genetics has been transformed by genome editing, allowing a researcher to alter the letters of the genetic alphabet with surgical precision to test hypotheses. In the near future, this genome editing will allow us to improve human health, with ethical implications that should be discussed widely and deeply. Third, imaging at multiple scales — from the atoms within biological macromolecules to the molecules within intact cells, tissues, and organisms — has surpassed traditional limits. All three of these new fields produce such vast datasets that they rely on computer science for data acquisition and analysis.

The theme of this session concerns the third of these topics: biological imaging at all scales and the computational techniques that enable it. Certainly, "seeing is believing" is not a new theme in cell biology. Biological insights have long been driven by imaging, from Anton van Leeuwenhoek, the pioneer in light microscopy who first observed "animalcules" (ca. 1674–1683), to Santiago Ramon y Cajal, who inferred the function of neurons from his stunning microscopic images of brain tissue (Nobel Prize in 1906). Yet, the advances of the past 20 years represent a quantum leap in our ability to image biological molecules, cells, and tissues.

Single-particle Cryo-electron Microscopy

The field of Structural Biology has been built on NMR spectroscopy and X-ray crystallography, which continue to be extraordinarily useful. As time went on, however, molecular biologists began to realize that much of what drives the biology of complex organisms involves enormous complexes of multiple proteins, often bound to DNA or RNA. Typically, these complexes are large and also heterogeneous either compositionally (not all complexes contain the same subunits) or conformationally (flexible portions allow multiple shapes), making them very challenging subjects for NMR and X-ray crystallography. Cryo-electron microscopy (Cryo-EM) solves these challenges because it analyzes single particles, so if the sample contains multiple types of particles, then it may be possible to obtain the structure of each type in a single experiment. Furthermore, large particles are advantageous for cryo-EM. Pioneering work leading to the current revolution in cryo-EM was recognized by the Nobel Prize in Chemistry to Jacques Dubochet, Joachim Frank, and Richard Henderson in 2017.

The cryo-EM advances have required new instrumentation, such as the direct electron detectors, and also computational advances that allow automated picking of particles, sorting them into classes, calculating class averages, generating a 3-D model, and fitting the electron density with a molecular model. Our speaker Dr. Eva Nogales has reviewed these advances [1]. Dr. Nogales is best known for solving structures of fundamental biological interest, such as human RNA polymerase II caught in the act of opening the DNA double helix to initiate transcription [2].

Critical information that cannot be gleaned from single-particle cryo-EM is the location and orientation of different protein complexes within a living cell. Our speaker Winfried Denk has addressed the problem of identifying macromolecules in crowded environments [3]. A powerful solution to the problem relies on the projected 2D electronic "signature" that is unique to a protein complex, allowing it to be identified in a cryo-EM image of a cell [3].

Live-cell Imaging by Fluorescence Microscopy

As powerful as cryo-EM techniques are, they yield static images. To capture the motion of molecules and organelles within a living cell or an organism, fluorescence microscopy is the method of choice. Our speaker Dr. Jennifer Lippincott-Schwartz has developed techniques for localizing and watching the dynamics of proteins that reside in or transit between cellular membranes. Recently, she and her colleagues have tackled the problem that fluorescence imaging is typically limited by the number of different labels that can be distinguished. They therefore developed a multispectral image acquisition method for live cells that uses confocal and lattice light sheet microscopes and a new five-step imaging informatics pipeline [4]. They have been able to determine organelle numbers, volumes, positions, and speeds, as well as dynamic inter-organelle contacts, in living cells [4]. Live-cell imaging of organelles has also shown that the endoplasmic reticulum contacts mitochondria and constricts them, leading to mitochondrial fission [5].

Biological chemists and biomedical scientists often want to know where individual molecules reside in cells and their dynamic motions, as well as how they are affected by environmental changes (heat, stress) and pharmaceutical drugs. Fluorescence is sensitive enough to detect single molecules, and specific labelling can be achieved either by genetically encoding a fluorescent protein module [6] or adding cell-permeable small-molecule fluorophores that bind to a specific protein [7]. My own research group has benefitted from Luke Lavis's fluorophores [7], which allowed us to measure the dynamics by which the telomerase enzyme is recruited to chromosome ends in human cells [8].

Like many of our speakers, Dr. Xavier Darzacq has contributed both to technology development and by using the technology to discover new biology. On the computational front, single-particle tracking is essential for obtaining quantitative data from live-cell imaging experiments (Fig. 1), and Darzacq's SPOT-ON software improves the analysis by accounting for certain biases, including molecules moving out of focus [9]. My own research group has utilized SPOT-ON to analyze the dynamics of a chromatin-modifying complex in human cell nuclei [10]. In terms of biological insights, Darzacq and Tjian have applied imaging to show how

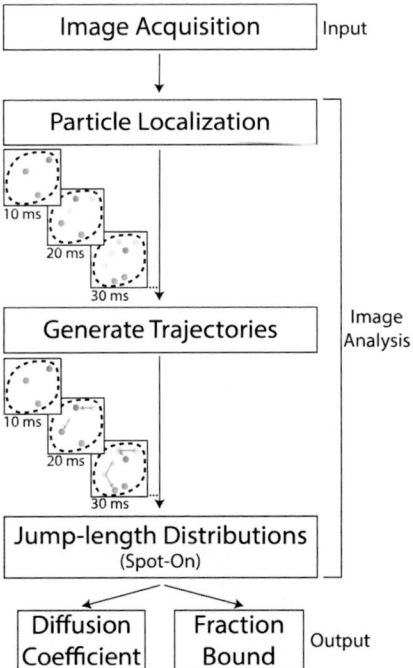

Fig. 1.   Computational pipeline for determining biophysical properties of macromolecules from single-particle live-cell imaging.

interactions of low-complexity protein domains help regulate gene transcription in cells [11]; their model is distinct from the currently very popular idea that transcription hubs represent liquid–liquid phase transitions within the cell.

Computational Approaches to Quantitative Analysis of Cellular Images

The central role of computers in quantitative image analysis is further emphasized by the final two speakers of our session. Among many other accomplishments, both Dr. Markus Covert and Dr. Mark Ellisman have tackled the problem of image segmentation: how can a computer determine where one cell stops and the next one begins in an image of densely packed cells, and how can a computer identify the various compartments — nucleus, nucleolus, mitochondria, etc. — within a single cell?

Covert has applied machine learning, specifically deep convolutional neural networks, to segmentation of live-cell fluorescence microscopy images [12]. He begins with a manually annotated image where each pixel has been identified as a cell boundary, cell interior, or background (non-cell). This provides a training dataset that contains representative images of each class. Machine learning then finds a classifier that distinguishes the three classes in the training dataset, and the classifier is then applied to new images.

Ellisman is fascinated by the "tyranny of scale". We need enormous datasets to capture the spatial and temporal aspects of living cells and the macromolecules that reside within them. However, the tsunami of data requires computational methods to identify components, their interactions, and their dynamics, all of which encounter great challenges. Toward that end, Ellisman uses deep learning for image segmentation and analysis of 2D and 3D imaging datasets from light, X-ray, and electron microscopy [13].

Outlook on the Future: Modelling of All Components of Intact Biological Cells

A future goal might be to localize all of the macromolecules in a living mammalian cell, identifying each protein, gene, and RNA molecule with precision and measuring how they move and interact in real time. Such an accomplishment would bridge biology with chemistry (the biomolecular structures) and physics (their diffusion coefficients, trajectories, and binding events). It would require vast computational resources to track everything and store the data. Alternatively, an integrated experimental–computational model might be more easily obtained. Indeed, Covert's group has reported a whole-cell computational model of a bacterial cell that includes all of its molecular components and their interactions [14].

I remember that twenty years ago many of us thought that finding every macromolecule within a living cell was theoretically possible, but we could see no pathway by which it might be attained. Now, with the advances in cryo-EM, advanced microscopy including super-resolution microscopy, fluorescent probes, and innovative computational methods, the problem is still daunting, but solutions are forthcoming.

Acknowledgments

I thank my colleagues Joe Dragavon, Jens Schmidt, C. J. Lim, and Dan Youmans for helping me understand imaging, Dan Youmans for preparing Fig. 1, and the students in my Quantitative Optical Imaging course for asking stimulating questions. I am an investigator of the Howard Hughes Medical Institute.

References

[1] E. Nogales, *Nat. Meth.* **13**, 24 (2016).
[2] Y. He, C. Yan, J. Fang, C. Inouye, R. Tjian *et al.*, *Nature* **533**, 359 (2016).
[3] J.P. Rickgauer, N. Grigorieff, and W. Denk, *eLife* **6**, e25648 (2017).
[4] A.M. Valm, S. Cohen, W.R. Legant, J. Melunis, U. Hershberg *et al.*, *Nature* **546**, 162 (2017).
[5] J.R. Friedman, L.L. Lackner, M. West, J.R. DiBenedetto, J. Nunnari *et al.*, *Science* **334**, 358 (2011).
[6] N.C. Shaner, R.E. Campbell, P.A. Steinbach, B.N. Giepmans, A.E. Palmer *et al.*, *Nat. Biotechnol.* **22**, 1567 (2004).
[7] J.B. Grimm, B.P. English, J. Chen, J.P. Slaughter, Z. Zhang *et al.*, *Nat. Meth.* **12**, 244 (2015).
[8] J.C. Schmidt, A.J. Zaug, and T.R. Cech, *Cell* **166**, 1188 (2016).
[9] A.S. Hansen, M. Woringer, J.B. Grimm, L.D. Lavis, R. Tjian *et al.*, *eLife* **7**, e33125 (2018).
[10] D.T. Youmans, J.C. Schmidt, and T.R. Cech, *Genes Dev.* **32**, 794 (2018).
[11] S. Chong, C. Dugast-Darzacq, Z. Liu, P. Dong, G.M. Dailey *et al.*, *Science* **361**, eaar2555 (2018).
[12] D.A. Van Valen, T. Kudo, K.M. Lane, D.N. Macklin, N.T. Quach *et al.*, *PLoS Comput. Biol.* **12**, e1005177 (2016).
[13] M.G. Haberl, C. Churas, L. Tindall, D. Boassa, S. Phan *et al.*, *Nat. Meth.* **15**, 677 (2018).
[14] J.R. Karr, J.C. Sanghvi, D.N. Maklin, M.V. Gutschow, J.M. Jacobs *et al.*, *Theory* **150**, 389 (2012).

CRYO-EM VISUALIZATION OF MACROMOLECULAR STRUCTURE AND DYNAMICS

EVA NOGALES

Molecular and Cell Biology Department, University of California Berkeley, Berkeley, CA 94720, USA

Structure Determination of Biological Macromolecules and the Role of Cryo-EM

Visualization of biological macromolecules, such as proteins, nucleic acids, and complexes of both, is considered an important step toward the mechanistic understanding of how such macromolecules carry out their biological function. The final goal of generating an atomic model has the potential, not only of providing fundamental knowledge about the biological process of interest but also of guiding therapeutic strategies that modify, increase, or eliminate a specific molecular function to alter the biology of the system, such as improving plant biomass for food production, promoting bioremediation, or fighting human disease, to cite a few practical applications. For many years, X-ray crystallography has been the major structural biology technique, leading to the steady growth of entries in the Protein Data Bank. Unfortunately, certain biological samples of critical biological and medical relevance had traditionally been poorly represented in the PDB due to difficulties in their crystallization. These included integral membrane proteins, large macromolecular complexes, biological polymers, and more generally samples that could not be produced in large amounts for crystallization trials, or that due to intrinsic flexibility either did not crystallize or gave rise to poorly diffracting crystals. NMR is another major technique to obtain structure and dynamic information on proteins and nucleic acids, but its use in ab initio structure determination is limited by the need of large amounts of labeled sample and is restricted to molecules of relatively small size. Cryo-electron microscopy (cryo-EM) has been an alternative method for macromolecular visualization that could overcome the bottleneck of crystallization, but for years was limited in its applicability and resolution. Recent technical developments have revolutionized the cryo-EM field, making it a major structural biology method with very broad applicability, that can reach atomic resolution, and with the capability to describe compositional and conformational heterogeneity.

Briefly, cryo-EM is a modality of transmission electron microscopy in which the biological macromolecules are observed in a frozen-hydrated state obtained by very fast freezing and vitrification of the water solution containing the sample on an EM grid. In order to avoid radiation damage, images are taken using very few electrons and are consequently very noisy. Extensive image analysis is carried out to classify, align, average, and ultimately reconstruct the 3D structure of the molecular sample. In recent years, the use of "direct electron detectors" has resulted in images with a dramatic improvement in contrast and resolution with respect to the previously used film or scintillator-coupled CCDs, with the consequent gain in the resolution and throughput of structures. Furthermore, recent developments in software tools, specially the Bayesian implementation of maximum likelihood principles during image processing, have allowed the researcher to take full advantage of the image quality obtained with the new detectors. The information obtained often includes a description of conformational and compositional mixtures in the sample that otherwise would have hampered structural analysis and now add functional meaning to the studies. The recent technical developments in detector technology and in powerful software tools of easy usage [1, 2] mean that high-resolution (atomic or near-atomic) and high-throughput cryo-EM has come within reach of many structural biologists. As a result, cryo-EM is being adopted by an ever-growing number of researchers, and the PDB is expanding quickly to include samples that previously had been refractory to structure determination.

Cryo-EM Visualization of the Human Transcriptional Machinery on DNA: Structure and Dynamics

An example that illustrates the power of cryo-EM in the study of challenging systems that cannot be accessed structurally by any other means is that of the large molecular machinery involved in eukaryotic transcription. Transcription of protein-coding genes starts with the recognition of core promoter regions on DNA by TFIID and the consequent assembly of a sizable pre-initiation complex (PIC) that includes the large RNA polymerase II (Pol II). TFIID is itself a very large protein complex (>1 MDa) comprising the TATA-binding protein (TBP) and 13 or 14 different TBP-associated factors (TAFs). TFIID binds to all protein gene promoters (the site where genes start), where it loads TBP onto the DNA to initiate PIC assembly [3]. TFIID is also involved in regulation of protein gene expression via its interactions with gene-specific activators and repressors and histone modifications associated with active regions of the genome [4, 5]. Thus, TFIID must act as a molecular hub that integrates different regulatory cues into a certain level of transcriptional output, possibly through changes in its capacity to bind DNA and/or recruit Pol II. Structure determination of a full TFIID has been challenging because, due to its size, it is hard to produce recombinantly and endogenous sources are very limiting. TFIID is also fragile and highly flexible [6]. Cryo-EM has played a critical role in the study of this complex by overcoming the requirements of crystallization, and its use has led to both a description of TFIID architecture and of the conformational changes that are critical for its function.

Human TFIID has a horseshoe shape made of three main lobes, named A, B, and C. The B and C lobes form a fairly stable core, but lobe A moves across the complex

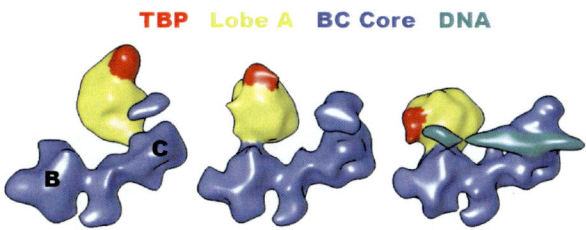

Fig. 1. Flexibility of human TFIID. Lobe A, and TBP with it, moves with respect to a more rigid BC core (modified from Ref. [8]).

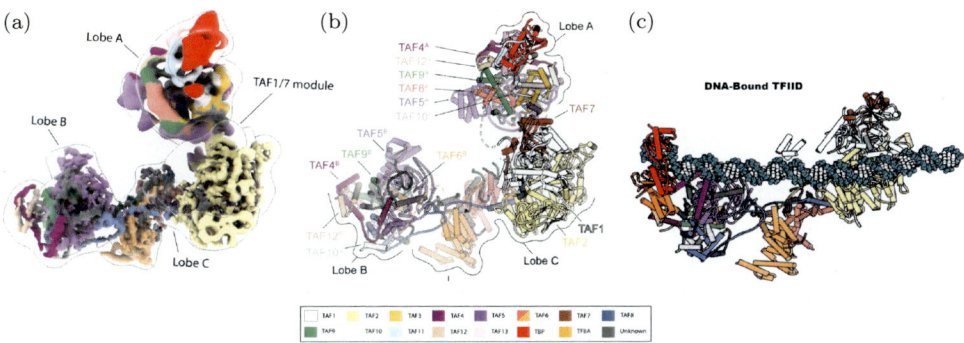

Fig. 2. Structure of human TFIID. (a) Cryo-EM map of TFIID, with BC core at 4.5-Å and lobe A at ∼9-Å resolution. (b) Atomic model of apo TFIID. (c) Atomic model of DNA-bound TFIID. Maps and models colored by subunit as indicated below (modified from Ref. [8]).

[6, 7], from interacting with lobe C to interacting with lobe B, a displacement of ∼150 Å (Fig. 1).

Obtaining a complete architectural model for human TFIID and its interaction with DNA required huge cryo-EM datasets that were analyzed using complex and iteratively optimized image processing schemes. One of the outcomes was a density map of most of the BC core at about 4.5-Å resolution, and of lobe A at about 9 Å [8] (Fig. 2(a)). The cryo-EM structures of both free and DNA-bound TFIID were complemented with crystallographic structures, homology models of proteins or domains, and chemical crosslinking-mass spectrometry (CX-MS) data, to generate atomic models of TFIID in multiple functional states (Figs. 2(b) and 2(c)).

The transition from the apo states of human TFIID to the form stably bound to the core promoter involves the release of TBP from lobe A, likely in stages that overcome the inhibition of different functional surfaces on TBP (Fig. 3). The interactions of the TAND1 and TAND2 regions of TAF1 with TBP [9] are likely to be released (i) as lobe C engages downstream DNA and positions the upstream promoter region such that it can be "scanned" by TBP, and (ii) as TFIIA joins the complex via interaction with lobe B and further stabilizes the location of TBP for DNA interaction. The final engagement of TBP with the promoter, with a

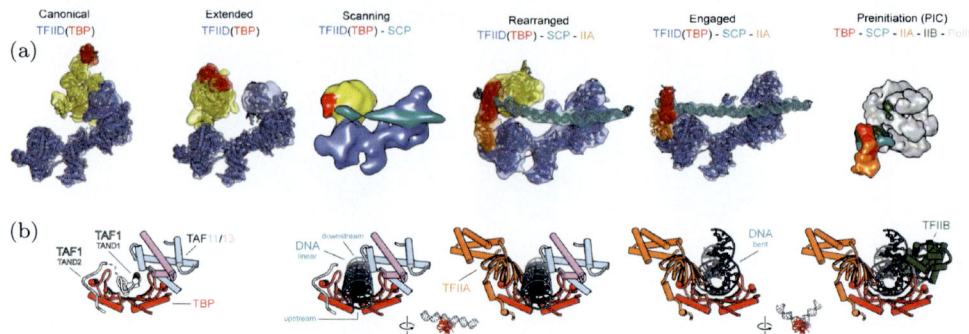

Fig. 3.   Functional states of human TFIID visualized by cryo-EM. (a) Structural transitions in TFIID in the presence of TFIIA (IIA) and super core promoter DNA (SCP). The last panel corresponds to an early state in PIC assembly observed in a study in the absence of TAFs. (b) Proposed states of TBP interactions with TAFs, TFIIA, and DNA for the corresponding cryo-EM states above (modified from Ref. [8]).

concomitant bending of the DNA, is sterically incompatible with the TBP-TAF11 interaction that connects this protein to the rest of lobe A, and therefore leads to TBP detachment from Lobe A and the opening of the binding site for TFIIB, which in turn will engage Pol II, leading to the progression of PIC assembly [8] (Fig. 3).

Remaining Challenges and Opportunities

Some critical bottlenecks still preclude rapid and systematic atomic characterization of biological samples by cryo-EM. It is quite common that biochemically sound samples break apart, aggregate, or simply disappear when they are placed on an EM grid, blotted to a thin layer, and quickly frozen. The physical origin of these problems appears to stem from the way biomolecules interact with the air–water interface, which can partially or totally denature proteins that rely on hydrophobic forces to fold and to interact. While there are a number of strategies that can be tried to overcome these problems, the solution, if found, seems to be different for different samples. Thus, innovative strategies, hopefully more generally applicable, need to be proposed and tested.

There is another major issue that can make obtaining atomic resolution of highly coveted macromolecules hard to reach. While available software packages can detect and characterize conformational mixtures in the sample, if that intrinsic flexibility is of a continuous nature, it will challenge present software tools. Overcoming this problem today requires some compromises, some ingenuity, and generally very large datasets. This is an area where computational innovation is badly needed.

Acknowledgments

This work was funded by NIGMS grant R35 GM127018. E.N. is a Howard Hughes Medical Institute Investigator.

References

[1] E. Nogales and S.H.W. Scheres, *Mol. Cell* **58**, 677 (2015).
[2] G. McMullan, A.R. Faruqi, and R. Henderson, *Meth. Enzym.* **579**, 1 (2016).
[3] B.F. Pugh and R. Tjian, *Genes Dev.* **5**, 1935 (1991).
[4] S.R. Albright and R. Tjian, *Gene* **242**, 1 (2000).
[5] R.H. Jacobson, A.G. Ladurner, D.S. King, and R. Tjian, *Science* **288**, 1422 (2000).
[6] M.A. Cianfrocco, G.A. Kassavetis, P. Grob, J. Fang, T. Juven-Gershon *et al.*, *Cell* **152**, 120 (2013).
[7] R.K. Louder, Y. He, J.R. López-Blanco, J. Fang, P. Chacón *et al.*, *Nature* **531**, 604 (2016).
[8] A.B. Patel, R.K. Louder, B.J. Greber, S. Grünberg, J. Luo *et al.*, *Science* **362**, eaau8872 (2018).
[9] M. Anandapadamanaban, C. Andresen, S. Helander, Y. Ohyama, M.I. Siponen *et al.*, *Nat. Struct. Mol. Biol.* **20**, 1008 (2013).

BUILDING WHOLE-CELL COMPUTATIONAL MODELS TO PREDICT CELLULAR PHENOTYPES AND ACCELERATE DISCOVERY

MARKUS W. COVERT and ERAN AGMON

Department of Bioengineering, Stanford University,
Stanford, CA 94305, USA

The Present State of Research Regarding the Modelling of Biological Cells

Large-scale computational models have transformed nearly every scientific discipline, including aerospace engineering with simulations of flows over airfoils and through engines [1], astrophysics with simulations of globular clusters and black holes [2], climate models for studying long-term climate effects [3], and simulations of macromolecules for uncovering molecular mechanisms [4]. By analogy, such models in cell biology have the potential to transform basic science, with significant implications for medicine. This potential is only beginning to be realized.

The development of a "whole-cell" computational model — one that incorporates everything that we know about the genes and molecules in a given cell, together with their annotated functions and mechanisms of action, into a unified mathematical framework which can be used to simulate cellular behaviour — has been considered a "grand challenge" for the past several decades, a vision that was perhaps initiated by none other than Francis Crick and Sydney Brenner [5], but shared by many others [6–8].

Most modelling efforts in computational and systems biology have focused on highly reduced representations of cellular function — only incorporating the interactions of few sub-cellular mechanisms, or supra-cellular behaviour with cells abstracted away from their molecular underpinnings. These efforts have been highly successful at elucidating the molecular mechanisms of distinct cell processes. The challenge of whole-cell modelling lies in the integration of these diverse representations, such that many unified mechanisms can together transform the global cell state.

The large-scale integration effort required for whole-cell modelling would depend on mathematical approaches that can both represent diverse biological processes mechanistically as well as accommodate many millions of data points, which exhibit an extraordinary amount of heterogeneity. Accordingly, extensive research has been directed toward building mathematical models that can represent the complexity of an entire cell. This has been approached from multiple perspectives. One approach is focused on assembling physical models with many individual molecule structures [9, 10], which has higher spatial resolution but is less comprehensive with respect to functionality; another set of efforts focuses on models that are more comprehensive in functionality but (at least for now) less spatially resolved. Such network-level models have included models based on ordinary differential equations [11–13], linear optimization strategies [14, 15], and stochastic physics [16], stochastic physics with spatial compartments [17–20], together with hybrid models which combine multiple of these strategies [21–23].

Our Recent Research Contributions to the Modelling of Biological Cells

In 2012, our lab reported construction of a whole-cell model for *Mycoplasma genitalium*, the simplest culturable bacterium [24]. To build this model, we developed a new and highly integrative modelling framework which enabled the entire cell to be simulated as a set of multiple sub-models, each one representing a particular biological functionality (e.g., metabolism, transcription, DNA supercoiling, or protein decay) with the most appropriate mathematical representation. These sub-models were integrated into a unified simulation as they ran independently from one another during short time steps, and communicated with each other between those steps. In total, roughly 1,900 experimentally determined parameters were incorporated into the model, withdrawn from 900 publications, each of which was manually curated as part of an extensive literature search. The finished model produced detailed simulations of a single *M. gentialium* cell progressing through a single division cycle, generating detailed time courses of such diverse properties as the transcription and translation state of all the genes, metabolic fluxes running through all of the enzymes, the location of all the DNA-binding proteins on the chromosome, and the like.

We found these simulations to be extraordinarily rich in biological insight, particularly with regard to complex, multi-factorial cellular behaviours. Most often, these insights regarded complex, multi-factorial cellular behaviours that could not be tested using current experimental methods. For example, the simulation output suggested that collisions on the DNA occur far more frequently than had previously been appreciated, predicted the existence of an emergent control of cell-cycle duration that was independent of genetic regulation, and provided a quantitative assessment of energy production and usage throughout the cell's life cycle.

This richness also enabled us to make several predictions about *M. genitalium* that were verified experimentally. The most notable of these concerned the prediction of gene disruption phenotypes, which further led to the model simulations predicting a number of molecular properties for enzymes that had not previously been characterized in *M. genitalium* [25]. This first demonstration of "model-driven discovery" in cell biology using a comprehensive model pointed to the possibility of using model and experiment together in a synergistic way, to systematize, acceler-ate, and even eventually automate the discovery process.

Several other products had to be developed to complement the model, including the following: an organism knowledge base which contained not only the param-eters and mathematical equations used but also the mechanistic knowledge which was curated to create the *Mycoplasma* model, and was machine-readable with open access [26]; interactive data visualization tools to enable user engagement with the massive datasets generated by each simulation [27]; and a simulation database to further facilitate data access and analysis [28]. The model has been of interest to the synthetic biology community, where it is envisioned as a design tool for genetic engineering [29]. The large number of parameters which must be identified, and in some cases fit, to enable viable simulations has also drawn the attention of parameter sensitivity experts [30], which catalyzed the development of new param-eter estimation approaches [31] and an international competition, the DREAM8 challenge [32].

Outlook on Future Research Developments Regarding Cellular Modelling

The ultimate goal of whole-cell modelling as a field is to perfect the experiment–model cycle, with models constructed to explain experimental results and predict new results, and experiments to check model validity and gather new data. Whole-cell models aim to encompass an increasing number of cell types, with richer internal complexity and a capacity for multi-cell interactions. This trajectory began with minimal cells, and now looks toward larger bacterial cells, eukaryotic cells, and ultimately multi-cellular assemblies such as tumors and tissues.

Toward the goal of increased model complexity, our lab has been developing a whole-cell model of *Escherichia coli*. This popular model organism poses a formidable challenge due to its size (10 times more genes and 100 times more molec-ular species than *M. gentialium*), its capacity to grow in a wide variety of environ-ments, and its extensive self-regulation. Our model consists of several modules that represent processes such as metabolism, transcription, translation, and complexa-tion. Each module operates on some of the 16,000 molecular species in the cellular state (i.e., the translation module operates on RNA, ribosomes, amino acids and pro-teins). It will be important for future efforts to integrate representations of spatial organization [18, 33], cell shape [34], motility [35], and compartmentalization [36].

To support cell–environment and cell–cell interactions, whole-cell models will require a deeper integration with models of cellular micro-environments. Cellular environments consist of molecules including nutrients, toxins, signals, and additional cells of the same and different species. The combined effect of these factors signif-icantly impacts the cell's growth, division, and its overall phenotype throughout both a single-cell cycle and evolutionary timescales, and thus accounting for these influences is essential. Agent-based models have been used to simulate multi-cell interactions with simplified cell models to explore biological tissue patterning [37]. Cellular Potts models specialize in modelling cell shape and tissue formation [38]. Accordingly, our lab has been developing a hybrid whole-cell/agent-based model for simulating cell–environment interactions with cell models that retain full kinetic details.

Finally, whole-cell modelling will become further integrated with experiment and analysis pipelines [39]. The previous two decades have witnessed the rapid cataloging of gene expression profiles, proteomics, metabolomics, lipidomics, and epigenomics, among others. Recent techniques in artificial intelligence can automate some aspects in the whole-cell modelling pipeline [40], but the construction and integration of models of molecular mechanisms remain a primarily human activity. Every cellular strain is unique, with immense amounts of complexity and no clearly automatable path toward model construction. Thus, large scientific collaborations, with many experts contributing their specific talents, software, analysis tools, and data, will be required to fully realize the vision of whole-cell modelling — with all of its transformative potential.

Acknowledgments

Many thanks for Paul G. Allen's generous support through an Allen Distinguished Investigator award and an Allen Discovery Center award.

References

[1] A. Jameson, *Encyclopedia of Computational Mechanics* (2004).
[2] S.F.P. Zwart, H. Baumgardt, P. Hut, J. Makino, S.L. McMillan, *Nature* **428**, 724, (2004).
[3] G. Flato, J. Marotzke, B. Abiodun, P. Braconnot, S.C. Chou *et al.*, in *Climate Change 2013: The Physical Science Basis.* Contribution of Working Group I to the Fifth Assessment Report of the Intergovernmental Panel on Climate Change (2014).
[4] M. Levitt and A. Warshel, *Nature* **253**, 694, (1975).
[5] F.H.C. Crick, *Perspec. Biol. Med.* **17**, 67 (1973).
[6] H.J. Morowitz, *Isr. J. Med. Sci.* **20**, 750 (1984).
[7] M. Tomita, *Trends Biotechnol.* **19**, 205 (2001).
[8] M.L. Shuler and M.M. Domach, in *Foundations of Biochemical Engineering*, H.W. Blanch, E.T. Papoutsakis, and G. Stephanopoulos (eds.) (1983), p. 93.
[9] Yu, T. Mori, T. Ando, R. Harada, I. Jung, Y. Sugita, and M. Feig, *eLife* **5**, e19274 (2016).
[10] M. Feig and Y. Sugita, *Ann. Rev. Cell Develop. Biol.* **35**, 191 (2019).
[11] M.M. Domach, S.K. Leung, R.E. Cahn, G.G. Cocks, and M.L. Shuler, *Biotechnol. Bioeng.* **26**, 203 (1984).
[12] M.L. Shuler, P. Foley, and J. Atlas, *Meth. Molecular Biol.* **888**, 573 (2012).
[13] M. Tomita, K. Hashimoto, K. Takahashi, T.S. Shimizu, and Y. Matsuzaski, *Bioinformatics* **15**, 72 (1999).
[14] J.D. Orth, I. Thiele, and B.Ø. Palsson, *Nat. Biotechnol.* **28**, 245 (2010).
[15] A.M. Feist, C.S. Henry, J.L. Reed, M. Krummenacker, A.R. Joyce *et al.*, *Molecular Syst. Biol.* **3**, 121 (2007).

[16] A. Arkin, J. Ross, and H.H. McAdams, *Genetics* **149**, 4 (1998).
[17] E. Roberts, A. Magis, J.O. Ortiz, W. Baumeister, and Z. Luthey-Schulten, *PLoS Comput. Biol.* **7**, e1002010 (2011).
[18] E. Roberts, J.E. Stone, and Z. Luthey-Schulten, *J. Comput. Chem.* **34**, 245 (2013).
[19] T.M. Earnest, R. Watanabe, J.E. Stone, J. Mahamid, W. Baumeister, E. Villa, Z. Luthey-Schulten, *J. Phys. Chem. B* **121**, 3871 (2017).
[20] T.M. Earnest, J.A. Cole, and Z. Luthey-Schulten, *Rep. Prog. Phys.* **81**, 052601 (2018).
[21] M.W. Covert, E.M. Knight, J.L. Reed, M.J. Herrgard, and B.Ø. Palsson, *Nature* **429**, 92 (2004).
[22] M.W. Covert, N. Xiao, T.J. Chen, J.R. Karr, *Bioinformatics* **24**, 2044 (2008).
[23] P. Labhsetwar, J.A. Cole, E. Roberts, N.D. Price, and Z.A. Luthey-Schulten, *PNAS* **110**, 14006 (2013).
[24] J.R. Karr, J.C. Sanghvi, D.N. Macklin, M.V. Gutschow, J.M. Jacobs *et al.*, *Cell* **150**, 389 (2012).
[25] J.C. Sanghvi, S. Regot, S. Carrasco, J.R. Karr, M.V. Gutschow *et al.*, *Nat. Meth.* **10**, 1192 (2013).
[26] J.R. Karr, J.C. Sanghvi, D.N. Macklin, A. Arora, and M.W. Covert, *Nucleic Acids Res.* **41**, D787 (2013).
[27] R. Lee, J.R. Karr, and M.W. Covert, *BMC Bioinform.* **14**, 253 (2013).
[28] J.R. Karr, N.C. Phillips, and M.W. Covert, *Database (Oxford)* **2014**, bau095 (2014).
[29] O. Purcell, B. Jain, J.R. Karr, M.W. Covert, and T.K. Lu, *Chaos* **23**, 025112 (2013).
[30] A.C. Babtie and M.P.H. Stumpf, *J. R. Soc. Interface* **14**, 20170237 (2017).
[31] J.C. Mason, M.W. Covert, *J. Theoret. Biol.* **461**, 145 (2019).
[32] J.R. Karr, A.H. Williams, J.D. Zucker, A. Raue, B. Steiert *et al.*, *PLOS Comput. Biol.* **11**, e1004096 (2015).
[33] K.C. Huang, Y. Meir, and N.S. Wingreen, *PNAS* **100**, 22 (2003).
[34] K. Keren, Z. Pincus, G.M. Allen, E.L. Barnhart, G. Marriott, A. Mogilner, and J.A. Theriot, *Nature* **453**, 475 (2008).
[35] N. Vladimirov, L. Løvdok, D. Lebiedz, and V. Sourjik, *PLoS Comput. Biol.* **4**, e1000242 (2008).
[36] A. Regev, E.M. Panina, W. Silverman, L. Cardelli, and E. Shapiro, *Theoret. Comput. Sci.* **325**, 141 (2004).
[37] B.C. Thorne, A.M. Bailey, and S.M. Peirce, *Brief. Bioinform.* **8**, 245 (2007).
[38] J.A. Glazier and F. Graner, *Phys. Rev. E* **47**, 2128 (1993).
[39] D.N. Macklin, N.A. Ruggero, and M.W. Covert, *Curr. Opin. Biotechnol.* **28**, 111 (2014).
[40] J. Ma, M.K. Yu, S. Fong, K. Ono, E. Sage, B. Demchak *et al.*, *Nat. Meth.* **15**, 290 (2018).

TOWARD HIGH RESOLUTION INTRACELLULAR MAPS IN SPACE AND TIME

MATTHIAS G. HABERL*, RAFAEL ARROJO E. DRIGO*,†,

MARTIN W. HETZER† and MARK H. ELLISMAN*,†

* University of California San Diego,
National Center for Microscopy and Imaging Research,
9500 Gilman Drive, San Diego, California, USA
† Salk Institute for Biological Studies, Molecular and Cell Biology Laboratory,
10010 N Torrey Pines Rd, La Jolla, California, USA

Introduction

We address the tyranny of scale in biology by advancing strategies to visualize and analyze the structural organization of cells, sub-cellular compartments, and protein complexes, quantifying how these components change over their lifetime. This represents one of the major challenges of modern biology. Current technical advances in biomedical imaging and computational analysis provide us with multi-scale microscopy tools that allow the examination of large-scale multi-dimensional morphological organization principles in space and time. However, a number of grand challenges remain to be addressed. At the forefront of those challenges are the resolution gaps between different multi-modal microscopy techniques. Furthermore, applying imaging and analysis tools at scale poses previously unimaginable demands for the user, hardware and software level. Here, we outline emerging avenues of research questions that are enabled through recent technical advances. Below, we highlight some of the challenges within this domain that will require concerted community effort to realize the full potential of emerging imaging capabilities.

Revealing Intracellular Organization Principles

New automated volume electron microscopy technologies abound and enable the investigation of intracellular organization at scale and beyond individual cells. Here, we outline a set of goals which need to be addressed to fulfill the promise of identifying intracellular organization patterns and their physiological relevance beyond individual cell types. Sharing imaging data not only as raw data but also at different stages of analysis, from segmentation to 3D models, is critical [1, 2]. To enhance the lifespan and propel reuse of individual datasets, to advance achievement of broader long-term goals, more substantial documentation of image data is critical, since, e.g., future notions of cell types and classes may shift and previously neglected information may become the center of a new definition or distinction. We expect that large coordinated multi-national efforts will emerge to provide hubs of research data collections, in order to facilitate meta-analysis, such as the identification of cellular organization patterns and their influence on functional capabilities of the cell and its subcellular domains, across cell types.

New Concepts Broadening the Utility of Scientific Software

Over the last decades, the development of image analysis software entered into a new era, as the capabilities of data processing and rendering increased drastically (from gigabyte- to terabyte-sized image volumes) further supported by a new generation of computer scientists that gained interest in fueling the capabilities of the biomedical domain. However, while the computational capabilities can, for example, be drastically increased by employing parallelization on high-end GPUs, designing computer-guided workflows fully utilizing available resources to process large data is often a complex task and usually out of the realm of typical imaging laboratories. At the forefront of continuously developed image analysis tools enabling broad and free access to complex image-processing tasks are the ever-increasing capabilities of ImageJ [3] and newly developed python packages [4].

Yet, the sheer size of individual datasets can quickly exceed what can be processed on regular analysis workstations or within reasonable timescales. Therefore, future efforts will have to go beyond facilitating the distribution of processing packages, or software (such as containerized installations) toward distributing easily scalable workflows. This requires that software be distributed in a state that simultaneously provides access to high-performance compute nodes or supercomputer clusters and also configures to use the specific capabilities specific to many of these advanced computational platforms.

To this end, we recently pioneered the use and distribution of a deep learning-based tool for the segmentation of large-scale image volumes on high-performance cloud compute nodes [5]. While cloud computation is typically a paid-per-hour utility, it has the unique advantage of enabling immediate access to high-end performance nodes for groups around the world, unlike supercomputer centers which have diverging standards across countries, require specific scheduling systems and technical expertise on the end-user side, and are more difficult to customize and debug. New solutions to provide scientific cloud resources or strategic developments with commercial cloud providers are critically needed, in order to facilitate sharing cutting-edge image analysis tools and workflows in full operation with the biomedical community and to reduce numerous parallel implementations of similar tasks.

Following Cellular Events in Temporal Scale: A Historical Perspective

Harold Urey's discovery of deuterium (^{2}H) in 1932 [6] paved the way for a new field in biochemistry, allowing scientists to dissect temporal aspects of cell metabolism, including how quickly proteins are recycled *in vivo*. The prevailing dogma at the time dictated that organisms and their internal components (e.g., cells and tissues) were static and suffered from "wear and tear" as the organism aged. Working with David Rittenberg and Rudolph Schoenheimer, Urey discovered that rats fed with ^{15}N-labelled amino acids incorporated ^{15}N into protein-based material in several organs at different rates [7]. These experiments revealed that proteins indeed turn over, in a process they called "metabolic regeneration". Seventeen years later, James Arnold discovered that the slow-decay rates of the radioactive carbon isotope ^{14}C could be used for determining the turnover rates (or age) of longer-living biological structures by quantifying the amount of ^{14}C incorporated after the dawn of above-ground nuclear tests [8, 9]. Together, these landmark discoveries show that the homeostasis of various biological structures occurs at different scales of space (i.e., protein-to-whole organism) and time (minutes to years).

Mapping the Age of Protein Complexes and Cells in Space and Time

Following the steps of Urey and Arnold, recent studies based on ^{14}C or ^{15}N measurements in living organisms have revealed that the body contains cells and protein complexes with remarkable longevity, some of which can last an entire lifetime without turning over [10–13]. These findings highlight that whole-organism homeostasis is a multi-scale process that combines "metabolic regeneration" and "wear and tear" of specific components, from protein complexes to entire tissues. However, these discoveries have been made in bulks of tissue, and therefore lack the spatial resolution needed to reveal the exact location of long-lived cells (LLCs) and long-lived proteins (LLPs) *in situ*.

To address this technological gap, we have developed a multi-scale imaging platform, called MIMS-EM, which integrates electron microscopy (EM) with multi-isotope mass spectroscopy (MIMS). MIMS uses a caesium ion (Cs$^+$) beam to ionize surface ions of various isotopes — such as ^{15}N — and that can be quantified at nanometer resolution *in situ* [14]. When applied to tissues labelled with stable isotopes, MIMS-EM creates maps of tissue, cell, and protein super-complex architecture combined with age information for virtually any cell or protein super-complex. Using MIMS-EM, we discovered that the longevity of cells in somatic organs is tissue- and cell-type specific and vastly different. While specific cell types can be remarkably long-lived and as old as neurons, others turnover significantly [15]. Similarly, elements within long-lived structures such as the basal body of the primary cilium and the nuclear pore complex (NPCs) display heterogeneous turnover rates within a single complex and between complexes of the same type [15, 16]. This combination of young and old components at different scales is a basic organizational principle we have termed "age mosaicism".

Future Implications of Age Mosaicism

We propose that this age mosaicism is a cellular organization system analogous to the concept of stigmergie, first described in the context of insects (termites) and their nests, where the organized behaviour of a swarm is indirectly coordinated by long-lasting signals deposited in the environment by members of the colony [17]. In the stigmergie of age mosaicism, the longest-living components in protein complexes, cells, and tissues would form a multi-scale cellular positioning system to guide the structure-function of more frequently replaced cells and subcellular molecular constituents. Further, this principle would also be important for tissue development, where progenitor cells follow chemical gradients (like insects) made of longer-lasting morphogens and deposited by migrating cells to achieve, in a coordinated fashion, different cell fates — as predicted by Alan M. Turing in 1952 [18, 19].

Concluding Remarks

While the data volumes per experiment continually increase, many groups are switching from local to cloud storage with the goal of mitigating costs. A positive effect is that data sharing has technically become easier, yet, to benefit from publicly accessible data more efforts have to be devoted to improving standards in data annotation. However, we should cautiously monitor the developments in offerings and costs for durable, long-term storage with years to decades worth of experimental data at stake, which may not align with the priorities of commercial cloud providers, but will be critical for driving scientific progress. Tightening budgets and lack of centralized support for long-term storage of experimental data put us at risk of losing highly valuable experimental data. With those goals in mind and knowing how influential the early discoveries and publications were in guiding us toward age mosaicism and the hypothesis of cellular stigmergie, we expect that plenty of new directions will emerge from the wealth of information contained in advanced microscopy techniques.

Acknowledgments

Funding sources: Research described here was supported by multiple NIH grants 3P41GM103412-30S2, 5R01GM082949-12 and 3R01GM082949-11S1 to Mark H Ellisman supporting the National Center for Microscopy and Imaging Research (NCMIR) and the Cell Image Library (CIL), respectively. This work was further funded by NIH Transformative Research Award (R01 NS096786), the Keck Foundation, and the NOMIS Foundation to Martin W Hetzer (M.W.H). Rafael Arrojo e Drigo (R.AeD.) was supported by an American Diabetes Association postdoctoral fellowship (1-18-PMF-007). M.W.H and R.AeD. received funding from National Institutes of Health and Human Islet Research Network (1U01DK120447-01). Matthias G Haberl was supported by a postdoctoral fellowship from an interdisciplinary seed program at UCSD to build multi-scale 3D maps of whole cells, called the Visible Molecular Cell Consortium (VMCC). This work benefitted from the use of credits from the National Institutes of Health (NIH) Cloud Credits Pilot, a component of the NIH Big Data to Knowledge (BD2K) program.

References

[1] M.E. Martone, A. Gupta, M. Wong, X. Qian, G. Sosinsky *et al.*, *J. Struct. Biol.* **138**, 145–155 (2002).

[2] D.N. Orloff, J.H. Iwasa, M.E. Martone, M.H. Ellisman, C.M. Kane, *Nucleic Acids Research.* **41**, D1241–50 (2013).

[3] C.T. Rueden, J. Schindelin, M.C. Hiner, B.E. DeZonia, A.E. Walter *et al.*, *BMC Bioinformatics.* **18**, 529–26 (2017).

[4] S. van der Walt, J.L. Schönberger, J. Nunez-Iglesias, F. Boulogne, J.D. Warner *et al.*, *PeerJ.* **2**, e453 (2014).

[5] M.G. Haberl, C. Churas, L. Tindall, D. Boassa, S. Phan *et al.*, *Nat. Meth.* **15**, 677–680 (2018).

[6] H.C. Urey, F.G. Brickwedde, and G.M. Murphy, *Phys. Rev.* **39**, 164–165 (1932).

[7] R. Schoenheimer, *Yale J. Biol. Med.* **14**, 677–677 (1942).

[8] W.F. Libby, E.C. Anderson, and J.R. Arnold, *Science* **109**, 227–228 (1949).

[9] J.R. Arnold and W.F. Libby, *Science* **110**, 678–680 (1949).

[10] J.N. Savas, B.H. Toyama, T. Xu, J.R. Yates, and M.W. Hetzer, *Science* **335**, 942–942 (2012).

[11] B.H. Toyama, J.N. Savas, S.K. Park, M.S. Harris, N.T. Ingolia *et al.*, *Cell* **154**, 971–982 (2013).

[12] A. Ori, B.H. Toyama, M.S. Harris, T. Bock, M. Iskar *et al.*, *Cell Syst.* **1**, 224–237 (2015).

[13] K.L. Spalding, R.D. Bhardwaj, B.A. Buchholz, H. Druid, and J. Frisén, *Cell.* **122**, 133–143 (2005).

[14] C. Lechene, F. Hillion, G. McMahon, D. Benson, A.M. Kleinfeld *et al.*, *J. Biol.* **5**, 20–30 (2006).

[15] R. Arrojo, E. Drigo, V. Lev-Ram, S. Tyagi, R. Ramachandra, T. Deerinck *et al.*, *Cell Metabolism.* **30**, 343–351.e3 (2019).

[16] B.H. Toyama, R. Arrojo, E. Drigo, V. Lev-Ram, R. Ramachandra, T.J. Deerinck, *et al.*, *J. Cell Biol.* **218**, 433–444 (2019).

[17] P.P. Grassé, *Insectes Soc.* **6**(1), 41–81 (1959).

[18] A.M. Turing, *Phil. Trans. R Soc. Lond. B* **237**, 37–42 (1952).

[19] A.M. Turing, *Bltn. Mathcal. Biol.* **52**, 153–197 (1990).

SESSION 6: COMPUTERS IN INTERACTIVE STRUCTURAL BIOLOGY LEADING TO MODELLING OF AN INTACT BIOLOGICAL CELL

CHAIR: T.R. CECH
AUDITORS: M. PREVOST[1], H. REMAUT[2]

[1] *Structure et Fonction des Membranes Biologiques (SFMB), Université libre de Bruxelles (ULB), Bd du Triomphe, 1050 Brussels, Belgium*
[2] *Structural & Molecular Microbiology, VIB/Vrije Universiteit Brussel, Pleinlaan 2, 1050 Brussels, Belgium*

General discussion

Bernd Rieger: I was wondering about some things that you did not mention. The first is the use of the phase plate in cryo-EM, does this really help? Maybe you can address that? And the other thing you did not really talk about is Correlative Light and Electron Microscopy [CLEM]. Is this something that would make a difference in the future, given it offers the possibility for the combination across different scales?

Thomas Cech: Eva, would you like to answer this?

Eva Nogales: I didn't have time to go into the details but in order to generate contrast at the low spatial frequencies that are used for the alignment of the particles and the build-up of the signal at high resolution, we have to defocus the image. It is kind of counterintuitive but that is the way contrast is generated in the transmission electron microscope [TEM]. Now, what has been happening in the last few years, is the development of phase plates. They act just like the half lambda wave plates in the optical microscope to basically change the phase of the unscattered vs scattered electrons by 90 degrees. At that point you generate low frequency contrast at low resolution so you don't have to defocus. It basically means more contrast in your images and the idea is that would be very useful in cases when you are limited by signal, which is true for small particles. But this is especially true for cryo-electron tomography, where the maximal dose you can give the biological sample has to be split over all the images that you take of the one object during the tilt series collection. What has happened is that there is now an implementation that you can buy called the Volta Phase Plate [VPP]. I think almost everybody buys them, but only 2 or 3 labs have used it successfully and published data using it. It has worked for us very well, but I have to tell you that in many cases it does not work. You often have to throw away half your data and the correction of CTF

[Contrast Transfer Function] still doesn't work very well. This is because, the VPP works in a very weird way that has to do with generating an electrostatic field that changes during the process of collecting the data. There is light at the end of the tunnel however, as my colleague Holger Müller in the physics department at UC Berkeley is now developing a laser-based phase plate that will be analytically well understood, tuneable and completely stable. When this is used, it will become very relevant to study, not only small particles but also the study of things that are relatively small and move within a larger complex, and that would also be essential for electron tomography. I think there still is a long way to go but it has the potential of being a much better, technically robust implementation of the phase plate.

Thomas Cech: Thank you but there is a second half of the question which is: how do we bridge the scales from electron to light microscopy? Mark, would you like to take that on?

Mark Ellisman: I will certainly take that on. I was very lucky in the last 30 years to have my lab and Roger Tsien's lab work together and we worked quite a bit on probes for correlative light and EM, both genetic and non-genetic probes. So almost everything that I showed you was part of a correlated project and I would say that personally I react negatively to the acronym "CLEM". Just the sound and the fact that it defines a limit in what we think about as correlatable images. So I prefer, as our chairman said, to speak of this as multiscale multimodal microscopy. In the middle of all of this now, we generally use X-rays at some point but the most evolved forms of this are doing Light microscopy in 4D, so 3D and time. So using let's say lattice light sheet microscopy or something like this that is light efficient and has good time resolution, and then try and stop the action by high pressure freezing, freeze substitution. And then use some of the methods that I did not have time to show you in using something that uses the initial fluorescent signal to something that gives you an electron scattering signal. Some of the most evolved ones we are using now is some of the last things Roger Tsien was making, which are lanthanide-containing compounds. There is no need to use osmium or anything, we now have lanthanum caged and we can easily get a signal in cryo-EM from the lanthanum against the ice. So I think this is going to evolve very quickly, but again I would be very careful to adopt this very easy 4-letter acronym because it is really multi-modal microscopy and data fusion that I think is what we need to be thinking about.

Winfried Denk: We have always, whatever the 4 letter acronym here, thought about the correlation microscopy. We have found that it usually requires compromises in the preparation technique. One has to be really sure that it is worth doing that. Because the first impulse is to say that "More information, how can that be bad?". But there is a price that we have to pay for getting more information and you have to really understand if it is required for your question. Because in many cases, if you modify one of the two techniques differently, you can look in super resolution microscopy and not need the EM or you can try to identify whatever you want to look at in the EM. There is of course a great incentive for the microscope companies to sell you these things [CLEM set-ups]. It all sounds good but think hard before you actually go into this area because I think that it has its costs.

Kurt Wüthrich: I want you to clarify some issues on the interpretation and presentation of cryo-EM data. Cryo-EM has become so dominant in structural biology, that we in NMR have to fight for life. One thing is, as you showed explicitly in your second slide, that in all preparations for cryo-EM the molecules are under the regime of Brownian motion for a short time. So the computational analysis of the data in the first place has to sort this out. If you had only one structure, one rigid structure, you would find that all species superimpose, right? The second thing is that you then showed a movie where one domain in a large molecular structure moved a lot. But this is not what you see. This movie is made from many frozen structures and you just assume that they go over into each other. But you don't have activation energies or rates for this. That is of course what we are measuring by NMR.

Eva Nogales: Kurt makes a very good point. When we image these molecules, they are all doing their own thing. Upon freezing, each one of them will be in a different state that is statistically corresponding to a sampling, an equilibrium of just one molecule going through these different states. The way we put them together is just based on physical constraints, because the idea is that the lobe is going to go from here to here and not going to dematerialise and then appear magically in another place. But what we have completely lost is rates. Rates are completely missing for us. For that, we need techniques that really have that kind of information as a read-out. There is no information on the rates. When I showed the little movie where we were seeing the thing moving, that is a superimposition of several class averages in order to see the range of the motion. Rates are totally invisible for cryo-EM. It is true that if something could be triggered to actually synchronise your process, and someone would have done these experiments and tell me at which point we would have different species, we could then try to capture them, triggering their reaction with the freezing. But again, we are not the ones that are providing any kinetic information.

Stefan Hell: We hear a lot about electron microscopy and light microscopy of course, but we just mentioned once the use of X-rays. Is there any potential in going down to those short wavelengths and do anything with it?

Mark Ellisman: Mark Ellisman here, San Diego. We would have been called UC La Jolla but they though it was too pretentious, so we stuck with UC San Diego.

In any case, the laboratory scale machines are actually pretty decent for modest resolution of about $0.4\,\mu$m isotropic. We use those as the intermediary technique. In trans-cranial imaging using two-photon microscopy for example, we can localise the area of interest in the sample and carry it to the next high resolution EM. There are many people who are pushing synchrotron technologies these days, particularly in the connectome area, the area that Winfried Denk helped catalyse, to try and look at the wiring level details. I know less about how far one can do this in context of cells. The work that folks like Carolyn Larabell are doing in the water window (X-ray wavelengths where water is transparent) with these X-ray optics are not giving us the resolution we want. Nevertheless they are giving us information that is of native frozen samples, so you have a density based on something other than the way we look at it. I think it is very promising, I just worry it is something that will be yet another synchrotron-related discipline. My preference however is democratisation, where hardware giving you information can be distributed to lots and lots of laboratories. Even though I think it is important to promote this, I worry about fuelling, at least in the US, more department of energy beamlines. But that is a political statement.

Winfried Denk: The basic problem with X-rays is that if you really want to go to high resolution and in particular when you want to dispense with a heavy metal, the cross sections are very unfavourable. To get really to molecular resolution, as with single molecule, is unlikely I would say. Thinking from the connectomics point of view, the very useful thing is that EM and X-ray have similar contrast. So it can be very useful if you have large sample which you plan of using in your machine for a year-long acquisition for example. To first see whether there are problems with your sample, which you do not want to discover after half a year. X-ray spectroscopy is thus in some sense a helper technique from the connectomics perspective. There is a small chance I think that the synchrotron-based techniques will get to high enough resolution but at this point I don't think that they will win the race. Nevertheless it is too early to really handicap them in a serious way.

Lee Cronin: This is a question for Markus Covert about the cell molecule modelling stuff which I think is really exciting. I am quite confused about it but I will try to ask a non-confused question. It has nothing to do with you. There is this kind of connection between people looking at images, architecture at the molecular level and sample averaging of the molecules on one hand, and then phenomena of what goes on in the cell on the other hand. I am referring to you recent Cell paper, but just as a more general question: Do you think there is room for having a hierarchy in your model where you can put in an architectural design component and say: "Right, I am going to test postulates about conformations and rates and things and then run your model"? That is one question. The second question I have is: Can you then extend your model to a single cell multi-organelle creature like slime mould, which is kind of easier to grow and image and play around with?

Markus Covert: Thank you for the question. I appreciate the comment. So, for the first question: In retrospect when I look back on this session, I should have announced or at least mentioned that there is something exciting going on in San Diego, La Jolla. Arthur Olson who is there does these beautiful kind of CellPack simulations where they take all the structures and they put it in the context of the cell. You might have seen the David Goodsell water colours, right? They do this in 3D architecture. They have done this for Mycoplasma, which is the first organism that we went after. Over the last couple of years we had some funding together in which, and I think this is the first necessary step, they have taken all of our simulations together and they are making one for many time points rather than just one static one. They are going to stich those together so pretty soon there will be actually a pretty fascinating connection between these two types of modelling. When they were talking to me about it, I think that the general idea was that we have no idea what to expect when we put these together so let's just start by making one for each time point and then see how we can interpolate from that. I am super excited about it and I don't even really know what it is going to produce but I believe it will be very interesting.

Lee Cronin: If you can almost look at it like virtualisation, and be making virtual machines you could keep those models and that would be very interesting.

Markus Covert: The second part of your question would basically be: How to drill down into a simulation when you are looking down on an individual protein or complex. So I totally think that these kinds of things will be fascinating and I was especially watching yesterday's session thinking about these exact things. How can I take use of some of the amazing chemistry and chemical modelling that has been done? I have a neighbour Zev Bryant, who does molecular motors, some of you might know him. Usually when we speak together to students Zev will say; "Markus has just explained to you how he can understand a whole cell, I will now proceed to explain why we can't understand even a single protein in that cell." And there is truth to that. That is a very interesting way to kind of think about those problems. In my case, one thing that is fascinating and that I think I can lay out as a challenge: We can predict mutations and replications and stuff like that. People have asked: "Can you predict an evolutionary outcome?" I was very excited about that and went to literature but I was shocked by the lack of information or examples and maybe will lay this down as a challenge. There is so little for me to go on, once I know that there is a mutation. So I know there is a mutation but usually I don't know what that means for the structure and I almost never know what that means for the function, in terms of constants or whatever. That link would be incredibly powerful and would give a lot to what you suggest.

Todd Martínez: I just want to react to something Eva Nogales has said, not having the information about rates. There are some people that would claim that if you have enough snapshots then, by virtue of a statistical hypothesis, you would actually know what the free energy barrier was and therefore you would know the rate. What do you think about this?

Eva Nogales: If you would double both the on and off rate (e.g. of a protein-protein interaction), we would not see a difference in the distribution of the particles. The rates, honestly, I don't think we can deduce them unless there is something very obvious that I am missing.

Thomas Cech: What about an equilibrium constant then?

Eva Nogales: Yes, we could get an equilibrium constant. There is a little caveat to that. I don't know if I want to open this can of worms, but I mentioned the fact that we are worried about the air-water interface. We cope with it the best way we can by using surfactants and surfaces that absorb the molecule to stop it from diffusing and hitting the air-water interface in the time scale just before freezing. But the truth is, when you look at papers' material and methods, there are cases where as much as 90% of the particles in the dataset are thrown away. We call them "bad" particles. It basically means that they are not consistent with each other and the idea is that they have been partly destroyed. So if every state is destroyed proportionally the same, then we are still getting the equilibrium right. But if there are protein states that are particularly sensitive to the air-water interface, because they expose a hydrophobic surface or whatever, then we are not seeing that equilibrium. So, I would normally not claim "this state is more abundant than the other", because I am not using 100% of the particles and I have to be cautious. I honestly think rates are invisible to us. I can tell you more about why I am pretty sure about that.

Thomas Cech: Would you care to comment about this question, Xavier?

Xavier Darzacq: Yes, I think in comparison with electron microscopy experiments we have a complementary approach where we can get some rates. However, the gap between the two techniques today is to be able to jump in between the high resolution visualisation of particles and the ability to effectively be able to see two different parts of a molecule, or two different parts of a complex using our imaging approach in a reliable manner. It is a gap that is being filled slowly, but there are still a few technical problems such as seeing different colours in the same experiment in a statistically relevant manner. But we can imagine the near future, where we start to measure distances in between parts of a complex in living systems and hopefully feed it back into these cryo-EM measurements and hopefully bring the dynamics along.

Thomas Cech: Winfried would also like to comment.

<u>Winfried Denk:</u> I am a little confused about this issue, why you cannot measure rates in the electron microscope. Of course if everything is in equilibrium, that would be true, but we know that a lot of the processes are driven by cycles of biochemical energy dissipation. At that point, from the occupancy of the rates you kind of know what the transition state to the next state is. Because if you have a barrier that is the rate limiting step, everything should pile up in front of that. I think it is not quite as hopeless as it might sound if you confine yourself to true equilibrium situations.

<u>Kurt Wüthrich:</u> Well, what you are suggesting is doing, for example, a temperature or pressure jump experiment using the EM as a detector.

<u>Winfried Denk:</u> Even without doing a jump experiment. The crucial problem here is the extraction of molecules from the cell. If you don't extract things from the cell and just let the cell run, then you take a molecular snapshot of the proteins in the cell. Every molecule in there is somehow frozen in their tracks. So if there is a stage in a reaction cycle that is very slow or that is blocked then everything would be piled up in the state before that transition. Moreover, if you would have a balance between different rates, they would pile up respectively. The ratio between different weighting room occupancies would give you the relative transition rates.

<u>Thomas Cech:</u> I think it is correct what you say but maybe qualitative rather than quantitative.

<u>Winfried Denk:</u> Why would it not be quantitative?

<u>Thomas Cech:</u> Because I don't think you can observe enough molecules in each state. Also, you don't know what the free energy landscape looks like. You are assuming a single transition state.

<u>Winfried Denk:</u> It is not a trivial thing, but what I wanted to contradict is the idea that it would be impossible to get rates if you take static images.

<u>Sinan Keten:</u> You mentioned that there are many spatiotemporal gaps. One of them is the coupled internal conformational changes that exist in proteins, which perhaps we can get some sense from statistical data in cryo-EM, and on the other hand you have the rigid body-like motions that you can track. But in reality, there is coupling between these things, for allosteric behaviour, for ligand binding and other things, the internal motions are coupled with the larger scale diffusive motions. And this is a general question to the panel. Do you think this can be addressed by more statistical analysis? If we would have an all-atom model of course it would be addressed, but perhaps reduced-order models, coarse grained models or NMR data can supplement some of the internal dynamics data and couple it with the larger

scale protein-protein interactions and larger scale assembly movement. So what is your perspective on how to bridge that spatiotemporal gap?

Xavier Darzacq: What I find fascinating in the kind of data that as a group we are producing is the fact that for the molecules that are diffusing in space, we know that they are exploring different conformations. What we see is that in the target-exchange process, they are not just going to be diffusing and interacting with their final partner but are interacting with a myriad of other partners through extremely weak interactions. These very weak interactions, today, we don't know exactly what they are. We know that there are these low complexity domains of proteins and we know they are under our detection limit. They are far under the millisecond immobilisation time in between two partners. By definition we know that they exist and therefore they are going to constrain the conformational space that these molecules are going to adopt during the target search. As of today, I have no way with the techniques that I am using to go into these conformations. Eva has ways to go into these conformations but she doesn't have the fine dynamics. So I wonder if modelling could be the way to go until we can make the measurements to really dive into this. And maybe I can add to this the fact that we are completely disregarding the organised water at these interfaces. Again, I have no idea how to measure that.

Eugene Shakhnovich: My question is to Xavier. You show this kind of mutual condensation of Bicoid with this large protein Zelda or whatever is the name. This is a very intriguing possibility and I think it is very appealing. The question though is: it can condense in the right place or in the wrong place? You would expect that there should be some specificity or some guidance for these condensing factors. Do you have any clue as to whether these kinds of disordered proteins interact somehow with DNA or are themselves oriented by their diffusion? How do they do this and how do they help Bicoid be properly rather than improperly condensed?

Xavier Darzacq: Clearly when you see an accumulation of a factor in a place in the nucleus, it could either be sequestered away from an active site or it could be fuelling locally activity. I didn't have time to show this data, but in the case of Bicoid for example we see these local accumulations in the posterior region. This led to the hypothesis that maybe this accumulation is what is driving expression of a few genes and then we could show by chromatin-IP (chromatin immunoprecipitation, ChIP) for example that cells in the posterior are driving a Bicoid-mediated transcription at a few genes. This is something that can be tested in particular cases. I think it is a mode of regulation that in this case will be used for activation, but it can also be used to sequester factors away from the site of activity, absolutely.

Mischa Bonn: My favourite topic came up a few times, which is interfacial water. I have a specific question for Eva Nogales; you used the phrase "disruptions of proteins at the water-air interface". I would like to understand better what you mean by "disruption". I would like to tell you that we have ways of looking at ensemble-averaged protein structures at interfaces and compare those with protein structures in solution, so that might be a way of quantifying how much the presence of the interface affects the secondary structure.

Eva Nogales: So, regarding the effect of the air-water interface. We can see, for example, complexes falling apart. I had some beautiful examples of an oligomeri-sation depended on a hydrophobic type of interface, and these molecules were sequestered away as monomers at the air-water interface. One of our favourite complexes is the Polycomb complex that is used for gene silencing. Polycomb has a certain domain that relies on a hydrophobic patch to fold and to recruit some of the other components, and when hitting the air-water interface it disappears. It is not that the molecule is gone, or has been chopped, but that part has now unfolded. In some cases we also see aggregation because the molecules have time to unfold and aggregate in the process. Even in cases where the molecule is pretty robust, if it has an exposed hydrophobic patch that makes it orient in one way, then we are in trouble because we need all different orientations for solving the structure. These are the ranges of disruptions that I am telling you about. As I said, we can play with surfactants, but they have to be compatible with the stability of the molecule and not all of them work for all of the proteins.

Thomas Cech: So as Mischa ensured, we don't want to study how the protein inter-acts with this air-water interface, we want to avoid it, right?

Eva Nogales: But it is an interesting process. Maybe if we understood it better, we could control it better and stop it from happening.

Mischa Bonn: Exactly, that was my point. At breakfast, you also mentioned that your initial electron microscopy frames are sort of blurred, for reasons you didn't quite understand. I thought about that a little bit and I was wondering whether that wasn't simply stress release from the fact that you flash froze your system.

Eva Nogales: This has to do with the fact that there is what we call a beam-induced motion. When these highly energetic electrons go through this thin layer of water, they cause some kind of motion that some people think is like a drumming motion of the water, and as a result it blurs the images. The motion seems to happen in two regimes. A very early regime for which the motion is really fast, and then, a second regime when the motion is slower. One of the things that the new direct electron detectors have is a very fast read-out, which allows us to split the total dose which we put through the molecule, typically spread out over a few seconds, into frames. If you have enough signal, you can then align one part of the frame to the equivalent in the other frames and correct for that slower phase, but not for the fast phase. So, this motion has nothing to do with the freezing, it has to do with the effect of electrons going through this thin layer of water. It seems right now to be kind of unavoidable. There are ways of reducing it, which have to do with the substrate that you use for your ice and making sure that the substrate and the ice respond very similar to the freezing. Maybe that is what you are thinking about? If you use gold grids, which sound fabulous and are very shiny, the response to the freezing is much more similar to the freezing of the ice than if you use copper support, and you have less of this buckling motion as they are being hit. You can reduce the motion, but not eliminate it.

Yoav Schechtman: My question is about the cryo-EM single particle analysis. When we look at these fields of view with many particles, they obviously contain a lot of information and the challenge is how to extract it efficiently. A very attractive, but maybe naïve thought would be to throw this into a neural net machine learning approach and have it extract several 3D structures. How far away are we from having that result?

Eva Nogales: So, last night at 1 am, when I found out that Alán Aspuru-Guzik was not going to be here, I actually gave him my presentation and I told him that, especially when we have a very complex mixture of states, it is now very time-consuming to even determine in which way the states are different. With the present software capabilities we try to separate and discriminate each state and align one module with respect to another. It ends up being this very convoluted pipeline, but that pipeline was one that we chose out of trying fifty different ones for which every step requires a number of decisions. I told him: "Can we get some help?". This is now something we are going to be exploring. I think we need to be able to speak each other's language. He told me this is the kind of thing that he believes, the kind of principles that were applied to the analysis of chemical reactions and taking the right pathways could be used in this context. I hope that, and I love listening to all of you of course, this is what I can get from coming to this meeting, it would already be very useful. So it is the kind of thing we are thinking about, and this should be a great influx of new thoughts and new methodologies going into cryo-EM.

Thomas Hermans: Stefan Hell of course very nicely outlined that according to him the resolution limits of optical microscopy are within reach. So, I am wondering if there is anything to be gained for example if you look at these light field cameras which detect not only intensity but also the direction of, in this case the photon. Do you think there is still room for improvement of the detectors by not only detecting intensity but also the direction of the scattered electron? In this way maybe going to even to lower dose, which would perhaps enable liquid EM without doing cryo-EM]?

Eva Nogales: There is still room for improvement on the detectors, especially as we want them to be even faster and improve the MTF [modulation transfer function] even more. In conjunction with the phase plate, that early stage where we cannot deal with the blurring due to motion could be actually used, because we would have a faster reading rate during those periods with higher contrasts that would help us for the alignment of the frames. This is because those early frames, where the sample is less damaged, are about time from zero to few milliseconds. In addition, we want larger fields of view, because using these microscopes is very costly and if we double the number of pixels, we quadruple the number of particles that we get in one image. So there are a number of things to optimise. You were talking about something else I think, where you want to detect something to go beyond what we are doing right now, which is counting electrons as they arrive on the pixel of the detector. I am not sure that in the context of imaging and not diffraction mode, as we are not imaging the diffraction plane, whether we are gaining a lot. But maybe I am not understanding the physics.

Thomas Hermans: In a light field camera you actually record instantaneously all the planes and it seems that you, in EM, need to be out of focus to get contrast but you lose resolution as you are out of focus. In a light field camera you actually can, after recording the image, because you detect all the directions of the photons in that case, you can get all the focal planes in one shot. I am just wondering if that is applicable to EM as well.

Eva Nogales: Now I understand what you mean. The defocus has nothing to do with the recording plane, but there is one case in which electrons are in different places, and it has actually to do with their energy. In the case of chromatic aberration, you end up with electrons that came from the same point in the image actually arriving at different heights in the detector. But this has to do with the fact that they have different energy and now the lenses are focussing out of plane. It is important to realise that our images are projections because the focal length is really huge, which means the full thickness of our sample is in focus. That is different from defocusing the image to generate contrast. If you are thinking about confocal microscopy and going through layers, that is not what is happening as we are dealing with a projection, having all layers squashed in one image.

Thomas Cech: Mark has an answer.

Mark Ellisman: Not really an answer, I just want to take the heat off Eva a little bit, by maybe taking it in a slightly different direction as I heard a different aspect of your question. You said something about wet-EM for one, right?

Thomas Cech: Wet-EM? Like in, doing it in liquid?

Mark Ellisman: Yes, indeed. There is activity that is mainly in material science field for pulsed-EM, and then there are some historical activities to get rid of chromatic aberration, by adding a second aperture and a second back-focal plane, it is almost like an energy filter. These are electro-optical designs, that haven't made it into the mainstream but are part of the history of how people have explored microscope designs. The pulsed-EM is very interesting and people are trying to do this with wet chambers. What you do is, you activate your electron source as a photo-cathode. You send a pulse of photons in order to get a bolus of electrons that come down the column in one shot. You can simultaneously deliver, through the way you set up your laser, something that activates the sample. More precisely, by using different wavelengths but time-linked to the pulse. Then, what some of these investigators, Nigel Browning and others are doing, is that they take advantage of some of these direct detectors, for example the high speed ones that we originally made and they use electrostatics at the level of the projector to within that pulse time put images tiled across the detector. This means that you can see the response of the specimen in a kind of cine across tiles. The problem is, it is noisy, you are in water, and the optics are not great. There is hope, but very little funding relative to the structural biology efforts, which are so successful on platforms that are more mainstream. But this should be explored more, I think.

Kurt Wüthrich: Cryo-EM prides itself of showing multiple conformations within one experimental data set, which is in stark contrast with almost all crystallographic data. You mentioned during the discussion that sometimes you only use 10% of the species in your sample. Which tools do you use in order to decide what is meaningful? For example, destroying the complexes in the liquid-air interface could create new, artifactual structures that you see. Who decides how many meaningful structures you see within one data set?

Eva Nogales: The cases where you end up throwing away 90% of the sample, are typical examples such as small integral membrane proteins that are in detergent or just poorly behave. They are subjected to extra strain. The way it is decided is completely objective, it has to do with dealing with self-consistency. You could have only 5% of your molecules that belong to a certain state, but they could be fine because they are self-consistent. When there is no self-consistency, the idea is that you have the protein unfolded to a degree that there is no possible consistency in between them. A different thing is that you have self-consistency but in different states that correspond to either different compositions, e.g. with a ligand bound or not bound, or with conformational states, where you see that one module has moved from one position to another. There are many ways in which you can see how relevant this is. You can repeat your experiments or you can add a factor and see how the equilibrium is shifted in a way that makes physical sense. So, I knew that I would open a can of worms when I mention there is data that is thrown away. That is data that is not self-consistent in any way, meaning you cannot group to give rise to a structure that has physical meaning, and it just gives you blurred snowflakes. Everything else that is self-consistent will have either one or several structures that have physical meaning and that you can tweak in a way that is predictable and makes biological sense by adding say, DNA or a drug or something like that.

Kurt Wüthrich: Is it the human mind or the computational powers that make the decision?

Eva Nogales: It is the computational powers. We are not picking particles and doing anything, this is done in a way in which there is no real human intervention. There is a human intervention in the sense that you have to inform the program. Right now, with the pipelines that we have and maybe that can be improved, you have to indicate to the program, e.g. "I think that there may be 5 different states". I don't want to fractionate my data to more than 5 states, because I won't have enough signal. Then it will spit out something, where if 3 classes are the same, you know you went too far. You forced the system to classify more than needed. Whilst, if you say five and there is still blurriness you know that you have not classified enough. That is the only human intervention. Otherwise it is just self-consistency of the data.

Winfried Denk: First of all, once we have a structure, or a structure you think is the correct structure, it is not so difficult to test whether it is real. You can basically take a vitrified cell and we can mill it down to 100 nanometres or so with a FIB and then basically try to find those structures in the cell. If we have two different structures and we don't know which one is the correct one, we will easily see the one which has the bigger cross correlation. We have now a way to close that loop. The whole discussion to know which of the two is the correct structure is almost mute, because we can experimentally test it.

Bernd Rieger: I was wondering now if you look a few years down the road, then maybe these new detectors will always be run in single electron counting mode. That means you will not get a pixelated image but these detectors will also return the position of the electron that hit the detector, so in principle you will get a localisation, which is not exactly the same as what we have currently in localisation microscopies. So the output of the detector in the end would be a list of coordinates where the electrons hit the detector that was computed based on the response of the electron in the material. So I was wondering if there are ways you can think about how imaging processing would change, or if there are things that we could do? This is what we are thinking on the light microscopy side, if we could use image processing on points or localization and not pixel-based images. So that could for example reduce the amount of data that you have to store, or the amount what you have to process. Do you any have ideas on that?

Mark Ellisman: The same guy that gave you the detectors for X-ray crystallography, Nguyen-Huu Xuong, was the guy who has had the idea to work on these direct electron detectors. When he and our combined students did this, we realised immediately that if we could do this in single electron mode, you could also compute subpixel, since there is always a little bit of spill-over to the adjacent pixel. So then, it is in his thesis and it is in the patents, you can get four times the resolution if you are in electron counting mode. But the detectors are so good now and the speed is coming up, there are already 8k detectors, that if your application does not require that kind of single electron counting they are still the detector of choice. These detectors are now commonly used for virtually all electron microscopy, because they have the equivalent of a reasonable bin depth, you read them out rapidly and you get high dynamic range. So they are quite radical and are broadly applicable. I gave you one example, where people were trying to use it in this kind of dynamic pulsed EM. There is an awful lot of processing opportunities and analytical opportunities on the data. They are using them in scanning transmission EM [STEM] now and in 4D-STEM, where you turn them into dark field detectors. If you follow some of the applications, particularly in material science for these detectors, you will see that the major effort is on the computation and analysis side of the data that is streaming out of them.

Mischa Bonn: To come back to my favourite topic, interfacial water. I am sorry, but you guys did mention it. I would like to ask Mark. You mentioned water at the interface of the dendrites, and that this is possibly linked to the function. We have techniques that allow us to look label-free at water with high interfacial sensitivity and selectivity, but not lateral. Do you think it would be possible to come up with a model system that is, let's say at the water-air interface, laterally expanded and has the same properties and maybe partly the same function that would allow us to study the water in more detail?

Mark Ellisman: I can't think about the air-water interface, except for the horrible forces that it puts on things that find their way there. But my comment, what I was trying to do is, in some ways a bit catalytic and off-the-charts in terms of predictions, because I thought that would be a good thing to do in this environment. It comes from watching the evolution of our understanding of how crowded is the cytoplasm or nucleoplasm and this controversy which is at least fifty years running about organized water, water in the fourth phase so to speak, and what that does when you actually make that part of what cells use in order to manage some dynamism. In other words, water is not always as available in all locations, not just in hydrophobic domains, but you can change diffusion rates for some things based on having water organized along surfaces. You think the river is big but the river is actually narrow [metaphor]. That is why we chose that dendritic spine because there is so much actin, and actin is acidic and there is very little myosin, so there the negativity is exposed. That is an environment where you have these multi-millisecond activity of the spine. You have cations enter and cations released from a stuffed sac, the endoplasmic reticulum. So you quickly change what would be the masking if you like, of those negative charges and potentially someone should try and model it as it is impossible in first principles. You almost have like an avalanche diode of increase in the activity of water in that space. Maybe I am just having too many piped dreams or something, but it seems to me that that would be something that biology would utilise, because all of a sudden you have all of this energy released by the activity of this water to do some kind of work almost for free.

Bernd Hartke: I am just curious whether one of the premises of this session is really true, in that you know all the basic information for the cell, like you have the complete genome, and you have the complete transcriptome and so on. Are there things that you don't know? For example what about glycohydrates, sugars on the cell surface that people tell me are very important but poorly understood? Or what about junk DNA, does it really have no role, or does it have a role?

Thomas Cech: Yes, those are among a long list of things that we don't understand. For example, these glycoproteins are mostly on extracellular proteins and so those are exported, so maybe we can excuse ourselves for our limited knowledge about glycosylation. But certainly the small molecules, that's why I particularly said that the goal of this session was to locate every atom in every macromolecule, because if you want to know where all of the ATP is in 3 dimensions or 4 dimensions, that is much more challenging.

Bert Weckhuysen: I would like to come back to a comment which Stefan Hell made on X-rays and I would like to address it. I think here, life sciences can learn from material science. Batteries, catalysts are studied heavily with X-rays and I would like to make the statement that lab-based instruments and synchrotrons are moving into real nanoscopy, that you can do it nanometer scale. So, current systems achieve 10 nanometers, if you want to do 3D, 20–30 nanometers, and if people can build the synchrotrons they want, then we will also move to 1 nanometer, something like that. That is a comment I would like to make and as a question: we have seen multimodal microscopy. There was a quest for seeing atoms everywhere in 3D and all that long sentence which you made. What I missed is vibrational methods, like infrared and Raman spectroscopy, which also are trying now to go into diffraction-limited approaches.

Thomas Cech: Mark, as you didn't have much chance to speak, to you want to handle that one?

Mark Ellisman: If you look at the laboratory-scale platforms, the handiest ones are without X-ray optics, they are projection systems. The ones that I think you are referring to have phase-optics and one of the problems in applying those on the things that most of us are doing is that the depth of field is very short. So I think they are extremely important platforms but I haven't seen them outside of materials and biological labs yet. I think there is an opportunity, but I haven't seen their useful application at this point.

Thomas Cech: So at this point the organisers would like to have an opportunity to close the meeting, so I would like to bring this discussion session to a close and thank all of you for your very fine questions.

INDEX